D 26

INDIVIDUAL-BASED MODELS AND APPROACHES IN ECOLOGY

INDIVIDUAL-BASED MODELS AND APPROACHES IN ECOLOGY

Populations, Communities and Ecosystems

Donald L. DeAngelis and
Louis J. Gross, Editors

Chapman & Hall
New York   London

Proceedings of a Symposium/Workshop: Knoxville, Tennessee, May 16–19, 1990.

First published in 1992 by
Chapman and Hall
an imprint of
Routledge, Chapman & Hall, Inc.
29 West 35th Street
New York, NY 10001-2291

Published in Great Britain by
Chapman and Hall
2-6 Boundary Row
London SE1 8HN

Library of Congress Cataloging in Publication Data

Individual-based models and approaches in ecology: populations, communities, and ecosystems/editors, Donald L. DeAngelis and Louis J. Gross; assisted by L. W. Barnthouse . . . [et al.].
    p. cm.
  "Proceedings of a symposium/workshop, Knoxville, Tennessee, May 16–19, 1990"—T.p. verso.
    ISBN 0-412-03161-2 (HBK)—ISBN 0-412-03171-X (PBK)
    1. Population biology—Mathematical models—Congresses. 2. Biotic communities—Mathematical models—Congresses. 3. Ecology—Mathematical models—Congresses. 4. Population biology—Research—Methodology—Congresses. 5. Biotic communities—Research—Methodology—Congresses. 6. Ecology—Research—Methodology—Congresses. I. DeAngelis, D. L. (Donald Lee), 1944–II. Gross, Louis J. III. Barnthouse, L. W. (Lawrence W.) QH352.I53 1992
574.5'248—dc20
                                            91-19060
                                              CIP

British Library Cataloguing in Publication Data

DeAngelis, Donald D., *1944–*
  Individual-based approaches in ecology. -I. Title II. Gross, Louis J., *1952–* 574.5

ISBN 0412031612 hb
ISBN 041203171X pbk

Contributors

D. P. Aikman
Horticulture Research International
Littlehampton West Sussex
United Kingdom

James J. Anderson
Center for Quantitative Science and
  Fisheries Research Institute
University of Washington
Seattle, WA

Larry Barnthouse
Environmental Sciences Division
Oak Ridge National Laboratory
Oak Ridge, TN

L. R. Benjamin
Horticulture Research International
Wellesbourne
Warwick
United Kingdom

Louis W. Botsford
Department of Wildlife and Fish
  Biology
University of California at Davis
Davis, CA

Hal Caswell
Biology Department
Woods Hole Oceanographic Institution
Woods Hole, MA

J. S. Clark
Department of Botany
University of Georgia
Athens, GA

Larry B. Crowder
Department of Zoology
North Carolina State University
Raleigh, NC

J. M. Cushing
Department of Mathematics
University of Arizona
Tucson, AR

Donald L. DeAngelis
Environmental Science Division
Oak Ridge National Laboratory
Oak Ridge, TN

A. M. de Roos
Department of Pure and Applied
  Ecology
University of Amsterdam
Amsterdam, The Netherlands

E. D. C. Ford
Center for Quantitative Science in
  Forestry, Fisheries and Wildlife
University of Washington
Seattle, WA

Louis J. Gross
Department of Mathematics
University of Tennessee
Knoxville, TN

W. S. C. Gurney
Department of Statistics and Modelling
  Science
University of Strathcylde
Glasgow, Scotland

J. W. Haefner
Department of Biology
Utah State University
Logan, UT

Thomas Hallam
Department of Mathematics
University of Tennessee
Knoxville, TN

Michael Huston
Environmental Sciences Division
Oak Ridge National Laboratory
Oak Ridge, TN

A. Meredith John
Office of Population Research
Princeton University
Princeton, NJ

S. A. L. M. Kooijman
Department of Biology
Free University
Amsterdam
The Netherlands

Ray Lassiter
Environmental Research Laboratory
U.S. Environmental Protection Agency
Athens, GA

Jia Li
Department of Mathematics
University of Alabama
Huntsville, Alabama

Adam Łomnicki
Inst. of Environmental Biology
Jagiellonian University
Krakow, Poland

Neil MacKay
Center for Limnology
University of Wisconsin
Madison, WI

Elizabeth Marschall
Department of Zoology
North Carolina State University
Raleigh, NC

McCauley, E.
Department of Biological Sciences
University of California
Santa Barbara, California

William McKinney
Department of Mathematics
North Carolina State University
Raleigh, North Carolina

J. A. J. Metz
Inst. of Theoretical Biology
University of Leiden
Leiden
The Netherlands

Thomas Miller
Department of Biology
McGill University
Montreal, Quebec
Canada

William Murdoch
Department of Biological Sciences
University of California
Santa Barbara, CA

Nisbet, R. M.
Department of Statistics and Modelling
  Science
University of Strathclyde
Glasgow, Scotland

John Palmer
Southern California Edison Company
Environmental Research
Rosemead, CA

M. Phipps
Department of Geography
University of Ottawa
Ottawa Ontario, Canada

W. M. Post
Environmental Sciences Division
Oak Ridge National Laboratory
Oak Ridge, TN

Les Real
Dept. of Biology
University of North Carolina
Chapel Hill, NC

James A. Rice
Dept. of Zoology
North Carolina State University
Raleigh, NC

Bernadette Roche
Dept. of Biology
University of North Carolina
Raleigh, NC

K. A. Rose
Environmental Sciences Division
Oak Ridge National Laboratory
Oak Ridge, TN

Edward J. Rykiel
Biosystems Group
Industrial Engineering Department
Texas A & M University
College Station, TX

Kristin Sorrensen
Center for Quantitative Science in
    Forestry, Fisheries, and Wildlife
University of Washington
Seattle, WA

R. Andrew Sutherland
Horticulture Research International
Wellesbourne, Warwick
United Kingdom

R. A. J. Taylor
Dept. of Entomology
The Ohio State University
Ohio Agricultural Research and
    Development Center
Wooster, OH

Earl Werner
Department of Biology
University of Michigan
Ann Arbor, MI

Webb Van Winkle
Environmental Sciences Division
Oak Ridge National Laboratory
Oak Ridge, TN

Table of Contents

Foreword

These proceedings derive from a symposium/workshop that focused on the current and anticipated directions of theoretical and applied research into populations and larger-scale ecological systems based upon characteristics of individuals. This emphasis on the individual-based approach has been stimulated by the disappointing performance of more aggregated models. Successful application of the individual-based approach is very dependent upon biological data on individual organisms and upon powerful computational capabilities. Recent advances in both of these areas have made the detailed analysis of these complex systems more feasible. Applications of this individual-based approach to environmental problems such as the effects of toxic chemicals on ecosystems, global scale problems, and resource management, as well as the theoretical approaches and the numerical techniques required for investigation of these types of problems, were debated.

The Electric Power Research Institute is in the midst of a multi-year program on Compensatory Mechanisms in Fish Populations (COMPMECH). The long-term goal is to improve predictions of fish population response to disturbance. One fundamental assumption in the design of COMPMECH, and the basis for enthusiasm about co-sponsoring the workshop and these proceedings, is the premise that to better understand the dynamics of a population, we need to place a greater emphasis on understanding what is happening at the level of the individual fish. As we saw during the workshop and as is evident in these proceedings, the individual-based approach offers a similar promise in a number of fields besides fisheries.

Jack S. Mattice
Electric Power Research Institute
Palo Alto, California

Robert G. Otto
R. G. Otto & Associates
Arlington, Virginia

Webb Van Winkle
Environmental Sciences Division
Oak Ridge National Laboratory
Oak Ridge, Tennessee

Preface

A major problem in the application of modeling and theory to field research and experimentation in ecology is that mathematical modeling in ecology requires simplifying assumptions, most of which are not compatible with the reality of ecological systems. One of the most important of these assumptions is that individual members of populations can be aggregated into a single state variable representing population size. Many classical models in ecology, such as the logistic equation and the Lotka-Volterra equations, assume that all individuals in a population are identical and can be lumped together.

This assumption that many individuals may be aggregated into a single state variable transgresses two basic tenets of biology. The first tenet is that biological individuals are unique, differing from each other physiologically and behaviorally, depending on their specific genetic and environmental influences. The state variable approach may be viewed from either of two perspectives. In one view it assumes that all individuals represented by a population number are, for all practical purposes, identical. A second view is that there is no loss of information by considering just population averages. This view acknowledges that individual differences among organisms exist, but that they are not of concern to the issues under consideration, so that models which average over the full distribution of types in the population are sufficient.

The second tenet transgressed is that of "locality," which means that a given organism is primarily affected only by other organisms in its spatial-temporal neighborhood. State variable models without spatial dependence implicitly assume that every member of the population has an equal influence on every other member of the population.

One response to this acknowledged inadequacy of models in which the population is represented by a single variable for size has been the intensified interest in modeling approaches that focus on age or size class structure and dynamics in populations, as evidenced by books such as that of Caswell (1989), Metz and Diekmann (1986), and Ebenman and Persson (1988). A second response has been to avoid aggregation altogether and develop models based on processes at the level of individual organisms. Łomnicki (1988) explored this approach at an analytic level and Huston et al. (1988) reviewed a large number of papers in which computer simulations were used to follow large numbers of individuals simultaneously to describe populations. These two alternative approaches are both referred to as "individual-based modeling" approaches, but have been distinguished by Metz and Diekmann (1986) as, respectively, individual state (or i-state) distribution and i-state configuration models.

Leaders in the development of both types of approach were invited to a Symposium/Workshop on "Populations, Communities, and Ecosystems: An Individual-Based Perspective," held in Knoxville on May 16–19, 1990. The theme of the meeting was that a key to better understanding of population, community, and ecosystem phenomena lies in describing mechanisms at levels below that of the whole population; for example, mechanisms at the level of the size class, age class, or individual organism.

The symposium had five general objectives. The first was to review the state-of-the-art in application of individual-based modeling approaches to theoretical and applied problems in ecology. Ecologists in many separate subdisciplines are independently pursuing individual-based approaches, often using similar methods, but with little awareness of parallel efforts being conducted elsewhere. Many field and empirical ecologists are unaware of recent developments in modeling. Communication is necessary to make the new modeling approaches accessible to the entire field of ecology and to infuse biological realism into these new models.

The second objective was to present case studies demonstrating the utility of studying higher order ecological processes by basing theoretical developments at the physiological or individual level. The temporal and spatial scales at which processes at the individual, population, community, and ecosystem levels operate need to be explored and compared. The organizers anticipated that detailed case studies focusing on individual-based concepts could offer a means to compare and extrapolate concepts useful at one level of the organizational hierarchy to other levels.

A third goal was to analyze recent developments of techniques for integrating information from the individual level to investigate processes at the population, community, and ecosystem level. Traditional transition matrix and partial differential equation approaches continue to dominate the mathematical modeling of the i-state distribution approach, but new elaborations of such models and new numerical methods of for solving the equations continue to be developed. Monte Carlo computer modelng is the primary methodology used in the i-state configuration approach, but both new software (e.g., object-oriented modeling) and new hardware are appearing at a rapid rate. Any review of the field of individual-based modeling must pay attention to these technical developments.

A fourth goal was to foster increased interaction among theoretical and empirical scientists of different disciplines who are actively working on extrapolating information from individuals to higher levels of organization. Through formal and informal discussions, the participants were encouraged to explore commonalities in problems and approaches.

The final objective was to identify critical needs for future research. A prime objective of the meeting was to focus attention on the key aspects of the individual-based approach that are crucial for future research of broad scale problems.

The first of these objectives was addressed through a series of overview papers, which were assessments of the current state of individual-based modeling by recognized leaders in the field. Shorter technical papers and poster sessions provided progress toward Objectives 2 and 3. Besides the presentations, a number of small, concurrent group discussions were held to try to foster interactions (Objective 4) and identify critical needs for future research (Objective 5).

The volume is structured into five parts. The first part, the Introduction, contains three general overview papers. The next three sections contain papers that focus, respectively, on Modeling Techniques, Animal Population and Community Models, and Plant Population and Community Models. In each of these three sections, the first two papers are primarily review papers. The placement of Modeling Techniques as the second section is in accord with the emphasis the meeting placed on familiarizing ecologists with the growing array of modeling tools that are available. Sections III and IV reflect the fact that much ecological research is still divided according to the plant and animal kingdoms. This is changing, however, and as more studies of species interactions are pursued, such as Real et al.'s study of pollination ecology, the necessity of considering simultaneously the population structures of organisms of all types will increase. The final section is a summary of the issues discussed in the working groups. Some of the main problems currently affecting the use of individual-based models, as well as future directions, are described.

It will be clear to the reader that this volume includes only a limited view of the work on individual-based models. Notably, no detailed papers dealing with genetics or epidemiology are included, though several contributions briefly discuss these topics. In addition, there are relatively few contributions concerned with specific problems of resource management. It is our hope that this volume will be a major influence in fostering the cross-fertilization that may lead to significant work on the genetics/ecology interface, as well as providing the impetus for modelers to consider individual-based approaches when faced with a management problem.

Finally, it is a pleasure to thank the many scientists and others who helped with the symposium/workshop and with the production of this volume. Those who helped with local arrangements, field trips, and audio-visual equipment during the meeting included G. Canziani, J. Jaworska, D. McDermott, L. R. Pounds, L. Provencher, J. Silva, and A. S. Trebitz. The reporters were M. Bevelhimer, R. Lassiter, S. A. Levin, L. Real, E. J. Rykiel, L. R. Taylor, and E. E. Werner. We thank those who anonymously reviewed the papers included in this volume. The editors greatly appreciate their assistance, and the resulting volume owes much to their efforts. We also thank the contributors for their (general) adherence to our deadlines, allowing this volume to be brought to fruition in a timely manner.

D. L. DeAngelis
L. J. Gross
Knoxville, March 1991

References

Caswell, H. 1989. *Matrix Population Models: Construction, Analysis, and Interpretation.* Sinauer Associates, Sunderland, MA.

Ebenman, B., and L. Persson (eds.). 1988. *Size-structured Populations: ecology and evolution.* Springer-Verlag, Berlin.

Huston, M. A., D. L. DeAngelis, and W. M. Post. 1988. *New computer models unify ecological theory.* BioSci. 38:682–691.

Lomnicki, A. 1988. *Population Ecology of Individuals.* Princeton University Press, Princeton, NJ.

Metz, J. A. J. and O. Diekmann (eds.). 1986. *The dynamics of physiologically structured populations.* Lecture Notes in Biomathematics 68. Springer-Verlag, Berlin.

Acknowledgments

This symposium/workshop was jointly sponsored by the Environmental Sciences Division of the Oak Ridge National Laboratory, the Science Alliance Center of Excellence of The University of Tennessee, and the Electric Power Research Institute. Environmental Sciences Division Publication No. 3832.

Part I

Introduction

1

Population Ecology from the Individual Perspective

Adam Łomnicki

ABSTRACT. The individual-based approach in ecology can be regarded as an application of reductionist methodology, i.e. deriving the properties of a system from the properties and interrelations among elements of the system. The reductionist method has generally proved to be very successful in science, and there are good reasons to believe that the same is true in ecology. The use of an individual-based approach is also justified in view of the recent progress in evolutionary and behavioral ecology.

Although difficult, the progress achieved by the individual-based approach allows us to see more clearly the mechanism of population stability as determined by scramble and contest competition. What is more important, it allows us to determine how population stability is related to unequal resource partitioning and resource monopolization. There is a chance that several important ecological problems can be solved by the application of an individual-based approach, especially in the area of population ecology.

Introduction

During the last five years, three reviews (Metz and Diekmann 1986, Łomnicki 1988, Huston et al. 1988) have been published on the individual-based approach in ecology. The idea seems to have been developed independently in several places. Therefore, this conference, which gathers the authors of these reviews as well as many others interested in this approach, is very timely. We have a real chance to integrate our efforts and to accelerate progress in the field.

I feel honored to give the first lecture at this symposium, but I do not promise to present a well-balanced view of all investigations based on the individual approach. Rather, I am inclined to show why this approach has been neglected for a long time, and why it is not yet widely accepted now.

I shall also attempt to present some difficulties we encounter when applying this approach and to outline a perspective for further research within this area of ecology.

Reductionism and Holism in Ecology

The essence of the individual-based approach is the derivation of the properties of ecological systems from the properties of the individuals constituting these systems. The idea that the properties of any system are derived from the properties of its parts and the relations between them is not new. On the contrary this is an old and fundamental principle of empirical science known as reductionism. It is, therefore, worthwhile to consider why reductionism, widely accepted in other fields, is questioned by many ecologists. Recently some outstanding biologists (for example, Mayr 1982) rejected reductionism as a concept which does not take into account the consequences of evolutionary theory, ignores interrelations among organisms, and claims that molecular biology, biochemistry and biophysics are the most important fields of biology.

These views seem to be based on misunderstandings (Caplan 1988). The success of any field of science is due to application of the generally accepted scientific method. If there is a system about which we know very little, so that we are unable to predict and control its behavior, we usually use the same approach, irrespective of whether it is an individual animal, a cell, or a molecule. We try to identify its elements and to find out the properties of these elements and the interrelations among them. On this basis we try to predict how the entire system behaves. This is the essence of reductionism. The critics of reductionism who claim that, according to the reductionist approach, the properties of the system are based on the simple summation of the properties of its elements, are obviously wrong. Physicists, chemists and molecular biologists, who apply reductionist methods, do not ignore interrelations among the elements of their systems.

If the reductionist approach is so widely accepted in other branches of empirical sciences, why does it come so late to ecology? I think that there are some reasons for this, which I am going to discuss further. It should be noted, however, that ecology is not totally deprived of reductionist methods. Some old models of theoretical ecology, like predator-prey equations by Volterra and Lotka (May 1973), represent not phenomenological descriptions of the predator-prey system, but attempts to derive the properties of ecological systems from the properties of their elements. In the case of the predator-prey system, the properties of individuals are included in the model in a very simplified form. Nevertheless, these properties are essential for an individual prey or predator, and the character of the model makes possible future inclusions of more realistic details. I believe that this model represents a true reductionist approach.

Large ecological units like forests, lakes or seas are very complex systems. It seems virtually impossible to identify all the elements of such a system and the interrelations among them. For a biologist at the beginning of this century, who was not familiar with mathematics and did not have computers at his disposal, there were two choices: either to ignore most of the elements and study only one or a few of them which seemed to be the most important, or to describe the structure of such a system, e.g., lists of species, and to compare this structure with that of other similar systems. It is well known that a knowledge of the external morphology of a single organism, even without investigation of its anatomy and physiology, is very useful. Therefore, application of a similar approach to large ecological systems seems to be justified.

No wonder, then, that ecology has started with an epiphenomenological description of plant and animal communities. Ecologists have also found a theoretical reason for this approach: a hierarchical structure of nature from the molecules through cells, tissues, organs, organisms, populations, communities, ecosystems, up to the biosphere. Since it is useful and justified to study the external morphology and behavior of a single organism, without studying the details of its internal anatomy and physiology, then consequently, it seems useful and justified to study a forest community as a list of species and their abundancies without knowing interrelations among these species. Applying the same holistic approach one can study a single population as a unit characterized by its density or age structure, without considering differences and interrelations among individuals.

There are two arguments against the holistic approach described above. The first is a pragmatic one. A population or an ecosystem is usually a much more vaguely defined unit than an individual plant or animal. In the field, we see individual plants and animals. However, we have to make some assumptions and to accept some theoretical concepts in order to determine the boundaries and characteristics of a population or a community. It is quite the opposite situation when a single individual is concerned: it is a well defined unit, although the identification of its elements requires a special study of anatomy and physiology. Therefore, it seems justified to study behavior and external features of individual plants or animals ignoring their elements, but it does not seem justified to ignore the knowledge we have already about individuals which are the elements of populations and communities. This could be justified only under special circumstances when, for instance, the individuals are not easily distinguished (e.g., among clonal organisms), or if our knowledge of the entire community or ecosystem (e.g., soil ecosystem) is more extensive than our knowledge of organisms making up such an ecosystem.

The second argument against the holistic approach in ecology is theoretical. Among all the levels of the hierarchy from molecules to the biosphere, an individual has something special and unique to it. This uniqueness comes from the fact that only individuals can be considered units of selection. Below the level of the individual, down to cells, we deal with genetically identical units; above this level we deal with a mixture of different genotypes, so that among these higher units natural selection either does not operate at all or it is much less effective than the selection among individuals. Therefore, the characteristics of the levels in the hierarchy above individuals are much more vaguely defined.

To these two arguments I can add a historical one. During the first half of the 20th Century, there was an extensive development of community ecology. There were diverse approaches to community investigations, for example community succession in time, the study of plant associations according to the Braun-Blanquet (1932) school of phytosociology, and others. The common concept applied at that time, either explicitly or implicitly, was community or population as a "superorganism." Recently, the term "superorganism" has almost completely disappeared from the ecological literature and all the concepts developed on its basis now seem dead. The recent development of plant phytosociology is symptomatic; its investigations are either reductionist analyses of ecological factors determining species composition of plant associations or they go into the domain of plant micro-geography.

On the other hand, some very early mathematical models that were used to predict the behavior of ecological systems from the properties of their elements (e.g., the age-specific model of unlimited population growth by Lotka or the predator-prey models mentioned above) are still very much alive and still form some of the basis of ecological theory.

Ecology seems to be lagging behind other biological sciences in capitalizing on the strength of the reductionist approach. The recent reviews, recent laboratory and field studies, and this symposium attest to the shift that is now occurring toward the reductionist approach.

Evolutionary Ecology and Population Ecology

The concept of "superorganism" lies far away from evolutionary theory. However, ecologists who developed this concept did not see it that way. Many of them believed that not only individuals but also populations and communities as well as entire species are the units of natural selection, so that there could be adaptations advantageous for species, populations and communities.

This approach has changed during the last 30 years. The concept of group selection by Wynne-Edwards (1962) has initiated discussion about the possible levels of selection and the understanding of the evolutionary basis for animal behavior. Kin-selection theory (Hamilton 1964) was the starting point of the development of sociobiology. New sociobiological concepts have promoted many field studies in behavioral ecology. This progress in behavioral ecology and sociobiology has given population ecology quite a different perspective.

Before this sociobiological revolution, many ecologists believed that the real progress in population and community ecology could be made by studying populations where large numbers of individuals can be easily caught. There was a belief that natality, mortality and age distribution are the phenomena which would explain all ecological processes. The easiest way to investigate these phenomena is to catch and to examine the largest possible number of individuals, so that good statistical analysis of demographic processes can be made. Insects seemed to be the ideal object of population studies, while among mammals, small rodents seemed to fit the requirements. When studying ecology in the late fifties, I was taught that the best way to learn about the regulation of animal numbers in nature was to study small mammal population cycles.

Now, after more than 30 years, we see that something quite unpredictable has happened. Large collections of insects made by ecologists in many different communities turned out to be of little help for understanding ecological processes. Lists of species, with the number of individuals for each species, can be analyzed by elaborate taxonomic methods in order to distinguish and to classify animal communities, or can be used to test niche theory. However, the studies based on such lists have not brought about any breakthrough in ecology.

Small rodents have also proven to be most difficult populations to study. After more than 50 years of investigations, the mechanism underlying small rodent cycles still eludes our understanding. Many hypotheses concerning these cycles have been put forward and enormous research efforts made, but the mechanism remains an enigma.

There has been remarkable progress, however, in the study of birds and large mammals. Birds and large mammals are not small like insects or nocturnal like small rodents, so that their behaviour can be more easily monitored. This progress shows how really important it is in ecology to have detailed knowledge of animal behavior and of interrelations among population members. This is the kind of knowledge collected, for example, by Clutton-Brock et al. (1982) in their study of red deer in the Isle of Rhum. It is remarkable that population dynamics of the deer was not the main concern of the authors; they were much more interested in sociobiological differences between two sexes. Nevertheless, it was the detailed knowledge of the behavior and general biology of these animals that allowed the authors to understand their population dynamics. Similarly, we know more about the population dynamics of birds, so easily seen and studied, compared with the nocturnal small mammals living underground during daytime.

The examples presented above show that population studies, based on sociobiology and on reliable knowledge of the biology of the species studied, tell us more about population dynamics than the classic ecological methods which consider individuals as identical molecules subjected to mortality and natality. These examples clearly prove how fruitful the individual-based approach to ecological problems can be. What we are witnessing now is the beginning of the application of sociobiological methods to ecological problems. Many more attempts have to be made in this field in the near future.

There is a widespread illusion among ecologists that mathematical models can always serve as the explanation of ecological mechanisms and processes. This may not be so: a mathematical model of population growth may be nothing more than a phenomenological description of the net result of many processes which are poorly understood or not understood at all. On the contrary, some detailed studies of two individuals competing for limited resources, made applying the concept of the evolutionarily stable strategy (ESS), may show the essence of the competition and help one to predict population dynamics. For example, Korona (1989a, 1989b) has studied under which circumstances the females of flour beetles, *Tribolium confusum*, compete for limited resources in an oviposition site or withdraw in search for other sites. By applying an ESS model, he has shown that it depends both on differences in body size and population density in relation to oviposition sites. Such studies, based on ESS models applied to conflict between two individuals, cannot replace models of population dynamics. Nevertheless they provide a solid theoretical basis for a new generation of population dynamics models.

Difficulties of Individual-based Approach

I am afraid that there were some serious reasons why population ecologists have ignored the individual-based approach for such a long time. If one intends to address more detailed and precise questions applying this approach, one indeed encounters both experimental and mathematical difficulties.

The crucial issue of the individual-based approach is to estimate resource partitioning within a population. This can be easily done for a single isolated individual as well as for an entire group of plants or animals kept in the laboratory. Therefore, there are plenty of reliable data about average resource intakes measured either for separate individuals or for entire groups. However, it is very difficult to find out the resource intake of a particular individual living within a group of other individuals. It is virtually impossible in a group of plants, and it requires quite a lot of ingenuity, time and effort to do this for most animals.

The best and easiest way to estimate individual resource intake, especially among plants, is to monitor individual growth. This is, however, a very crude and biased method. The maintenance cost is not linearly proportional to food intake and some other complications can also occur. It is the same for animals. For example, the individuals of Everglade pygmy sunfish which defend territories and get more food show much slower increases in body size due to higher maintenance cost (Rubenstein 1981).

Therefore, in order to estimate the process of intra-specific competition, either one should study detailed resource partitioning within very small groups, say two individuals, or one should get an approximate estimate of the net result of resource partitioning within the population, reflected in body weight. Either way, to elucidate the process of intra-specific competition is still much better than a general description in which individuals are assumed to be identical with respect to their food intake.

To estimate the importance of intra-population variation in fitness, one should be able to identify individuals throughout their life. This requires the application of individual marking. This technique is relatively simple when applied to most plants, vertebrates or mollusks. However, it seems impossible to apply it for the entire lifespan of insects and many other animals. Without individual marking one is unable to follow the survival and reproductive success of an individual, and therefore one fails to understand fully the selective advantage of its behavior. One example is the selective advantage of animal dispersion into suboptimum habitats.

Other difficulties are of a mathematical nature. Huston et al. (1988) see the prospects of applying the individual-based approach in ecology in the development of computers and the possibility of applying numerical simulations and numerical solutions. This suggests that we do not yet have mathematical tools good enough to allow analytical solutions. If this is the case one has to reject the possibility of building a new general theory of ecology based on the individual approach and supported by analytical solutions. One should not, however, give up too early. There are still possibilities that some new mathematical tools will be developed for the individual-based approach which, at least for some cases have analytical solutions.

There are two steps of mathematical analysis important for the individual-based approach. First, relying on the genetic, developmental and physiological properties of individuals, as well as their interrelations and characteristics of their habitat, one has to predict the distribution of their resource intake and, ultimately, the distribution of their reproductive success. Actually, we are able to study the distribution of body weight only, and we are concerned with the mechanisms determining this distribution. Second, from the distributions mentioned above we have to predict the dynamics of a population, its stability and, in further analysis, the stability of herbivore-plant, predator-prey and parasite-host systems. When we get this far, we can make further predictions concerning the stability of communities and ecosystems.

As discussed elsewhere (Łomnicki 1988), simple predictions about the character of distributions, based on some simple mathematical analysis of individual growth, disagree with many laboratory and field data. Recently, there has been considerable progress made in this field (Wyszomirski 1983, Uchmanski 1985, Kimmel 1986, Huston and DeAngelis 1987, Hara 1988) showing that determination of the body weight distribution from individual properties and interrelations among individuals is not as simple as expected. For example, some mathematical models show that competition implies skewness, and that skewness does not arise without competition, while other models and experimental data show that skewness may also arise when individuals are kept separately, unable to compete with each other. As the studies of the weight distributions become more detailed, numerical simulations become an important research tool for predicting empirical data.

If, based on theoretical and empirical investigations, one gets an idea what the distribution of individual resource intake or individual reproductive success looks like, the next step is to derive the dynamics of the population from these distributions. The idea which I first tried to develop more than a decade ago (Łomnicki 1978) was to derive population dynamics in discrete time units from individual resource intakes y which are assumed to be the function of individual ranks x. This idea does not require the assumption of social rank and social hierarchy. The function $y(x)$ can be derived from any distribution of individual resource intake; it is an inverse of the cumulative distribution of these resource intakes.

Since we do not have an explicit expression for the cumulative normal or lognormal distribution, we have been unable to apply the most common and most realistic forms of these distributions. What can be done is to find an expression for the function $y(x)$ which either approximates real distributions or describes some interesting forms of distributions like the one related to contest competition, or another form assuming equal resource partitioning. These expressions differ considerably from the real ones, which are found both in the field and in the laboratory. Therefore we can make only qualitative, not quantitative, predictions concerning population dynamics.

Commonly used distribution functions like normal or lognormal are only the descriptions of the net result of competition among individuals, usually expressed in body weight. From these distributions we may sometimes infer how the competition looks, but we can hardly predict how these distributions change when the amount of resources diminish or population size increases. However, some simple assumptions concerning the functions $y(x)$ allow such predictions.

There is an open question: shall we derive population dynamics using some realistic distributions like normal or lognormal, or shall we derive population dynamics directly from individual properties, ignoring the realistic distributions? If we intend to obtain analytical solutions, the models of population dynamics cannot be derived from the normal or lognormal distributions of individual resource intake or individual reproductive success.

Relations between Competing Individuals within a Single Populations: Some Concepts and Definitions

In applying the individual-based approach to ecology we cannot avoid describing intra-population variation and interrelations among population members. As in any new field which is initiated independently in several different places, however, one can expect different definitions of basic terms and frequent redefinitions of terms as the new data and concepts accumulate.

The reductionist approach in ecology is relatively easy to apply if we assume that population members are either identical or that they differ only by sex and age. For example, Schoener (1973) has developed models of population growth based on individual properties but assuming that all individuals are identical.

The models proposed and reviewed by Hassell and May (1985) are much more sophisticated in that the influence of individual behavior on spatial and temporal heterogeneity, and consequently on the dynamics of ecological systems, are analyzed. In these models, inequality among individuals is mainly due to local differences in population density, and the relations among individuals are not explicitly expressed.

The simplest and most commonly accepted concept of the individual-based approach is that of unequal resource intakes which can be expressed by variation in these intakes. Inequality and variation can be considered as synonyms, and they can be measured by the coefficient of variation or by the Gini coefficient, as proposed by Weiner and Solbrig (1984).

A term, which was very vaguely defined, is "asymmetry" of competition between individuals. For some authors (e.g., Begon 1984), unequal resource intakes or unequal sizes are sufficient reasons for the presence of asymmetry, so that asymmetry is used as a synonym for inequality and individual variation. Weiner (1990) has recently presented a more precise definition of asymmetry, in which the advantage of a larger individual is disproportionally higher than its larger size would indicate. Such a disproportionality may easily lead to increased variation, and further, to full monopolization of resources by stronger individuals.

While unequal resource partitioning may lead to the monopolization of resources by stronger individuals, these two terms—inequality and monopolization—cannot be equated. We require precise definitions of these terms. To understand interrelations among unequal individuals we have to answer the question of how a decrease in the amount V of resources for the entire population, and an increase in population size N, affect individual resource intake. There are two possibilities.

First, an individual resource intake y is related to differences in body size or to some other index or rank of differing competitive ability among individuals. If this relation is retained irrespective of the population density in relation to available resources (V/N), then

$$y = c(x)(V/N), \tag{1}$$

where $c(x)$ is constant for an individual of rank x. Such a relation does not change the proportion of resources taken by an individual, and should be called "ideal scramble competition." For this case one should not apply the terms asymmetry or hierarchy in spite of the presence of unequal resource partitioning. The ideal scramble competition may also include the case of equal resource partitioning among individuals.

Second, one can imagine that stronger individuals are absolute winners, so that if two individuals meet, the stronger one obtains all the available resource it needs, while the weaker one may take only what is left. If the individuals are ranked from the strongest to the weakest by rank x, then individual intake y is determined by individual rank x and the amount V of resources for the entire population. Several analytical models for such a case can be given (Łomnicki 1988), but an important point here is that individual resource intake y remains independent of the total number N of individuals, but dependent only on those of higher rank. This can be represented by the general expression

$$y = f(x, V). \tag{2}$$

Such a resource partitioning can be called "ideal contest competition." Models of population dynamics based on such a competition have been reported in my earlier papers (Łomnicki 1978, 1980) and in the paper by Gurney and Nisbet (1979).

Definitions of terms proposed here differ from those given elsewhere (Łomnicki 1988) where ideal contest competition was called "contest," while all others were called "scramble competition." Now I think that the terms contest and scramble should be used to describe how close a given relationship approaches ideal contest and ideal scramble competition, in a similar way to that in which the concept of despotic and free distributions (Fretwell 1972) are used. Ideal contest competition can also be called a competition due to monopolization of resources by stronger individuals.

A good example of ideal contest competition is the competition of plants for light, in which a shorter plant may not affect resource intake of a taller one. A similar situation may occur among animals which are able to defend their territories. Nevertheless, a shorter plant may have an effect on the mineral resource intake of a taller one, and territorial animals may obtain less resources due to the presence of many non-territorial individuals. Such a situation should be called "contest," not "ideal contest."

Perspectives

The individual-based approach to ecology creates a chance for a new paradigm. This chance can be exploited if we know the problems we have to solve. Is the individual-based approach indeed necessary, or is its application nothing but a different and more complicated way of creating already established theory? Are there some ecological phenomena which we do not yet understand, but suspect to be explainable by applying an individual-based approach? Are there any parts of ecological theory which should be revised, if individual properties are taken into account? Since we are just beginning to apply the individual-based approach, no clear answers to these questions can yet be provided.

My personal interest in the individual-based approach was first aroused when I was trying to explain animal dispersal into suboptimal habitats in a way compatible with natural selection theory. However, the models of population dynamics based on the individual approach seem to be able to explain much more than animal dispersion only. An obvious relation which these models help to understand is that between individual variation, contest competition and population stability. I think this relation is crucial for the individual-based approach to ecology.

As mentioned earlier, since we have not usually been able to determine the distribution of individual resource intake or individual reproductive success, real progress in the study of individual variation is made when studying mechanisms which determine the distribution of body size (for reviews see: Uchmanski 1985, Hara 1988). We already know that, in most cases, with increasing density in relation to available resources and with time, there is an increase in variation and in the skewness of the size distribution. We can therefore expect that when a sudden decrease in the amount of resources occurs, this may affect a population with low individual variation, and consequently, this may lead to its extinction.

What are other circumstances under which the variance in body weight will be extremely low? For plants, two such conditions are their simultaneous germination and homogeneity of their habitat—both phenomena common in agriculture. We do not have empirical evidence that homogeneity of the habitats of animals brings about similar effects. Preliminary experiments made by Ciesielska (1985) suggest that the heterogeneity of the medium in which the larvae of flour beetles, *Tribolium confusum,* are kept does not increase individual variation in body weight. This was probably due to free distribution (Fretwell 1972) of the larvae in the medium. More experiments are required on other species to find out whether habitat heterogeneity may increase individual variation among non-territorial animals.

Does variation in individual resource intake, but under ideal scramble competition, enhance population stability and persistence? As shown theoretically elsewhere (Łomnicki and Sedziwy 1989), this is not the case. If there is any effect of such variation on stability and persistence, it is so small that it appears virtually insignificant. This theoretical result was confirmed by experiments on flour beetles, *Tribolium castaneum,* displaying scramble competition (Korona and Łomnicki 1988). It can be concluded that individual variation itself is not sufficient for population stability and persistence, and that contest competition among individuals is also a necessary condition.

One cannot claim that individual variation itself is of no importance for population dynamics, because without an inequality among population members, contest competition is not possible. Nevertheless, the distribution of resource intakes alone is not sufficient to explain population dynamics. Detailed studies on relations among competing individuals are also needed. Natural and laboratory populations should be studied in order to determine how much competition among individuals is contest and how much of it is scramble. In population ecology of animals no individual approach is possible without taking into account animal behavior, especially that related to competition for limited resources. This raises another question of great practical importance: how to enhance or diminish the contest character of competition and consequently population stability and persistence?

If the stability and persistence of a population is caused by contest competition, then we can expect that contest competition will also enhance stability of a community of many competing species. A simple theoretical model by Jones (1979) suggests that variation in individual performance of either predator or prey stabilizes a predator-prey system. Unfortunately we have no empirical data to confirm these predictions.

I have suggested (Łomnicki 1988) that contest competition may explain animal emigration into suboptimal habitat, and therefore that it can further enhance stability of natural communities. More generally we can expect that any kind of habitat spatial heterogeneity may enhance stability of an ecological system. Micro-heterogeneity may act by increasing the variance of individual performance, and macro-heterogeneity may act by creating local habitats and the hostile space outside them in which emigrants are reduced in number. There is also a theoretical possibility that temporal heterogeneity enhances individual variation by the so called "coin flipping strategy," as defined by Cooper and Kaplan (1982). These are only theoretical suggestions. There is still a shortage of more precise theoretical models and of appropriate experiments to confirm these expectations.

At the end we should ask whether some spectacular or economically important increases in animal numbers could occur due to low intra-population variability, scramble competition, and consequently to low population stability. It is also interesting to know whether the extinction of some endangered species can be due to their low phenotypic variability and to scramble instead of contest competition. The individual-based approach suggests that it is theoretically possible, but it is also possible that the mechanisms proposed by this approach are less important than other commonly accepted mechanisms.

Conclusions

Application of the individual-based approach in ecology has allowed us to better understand some important ecological phenomena, especially contest and scramble competition, population stability, and animal dispersion into suboptimal habitat. Real progress has been made in our understanding of the physiological and ecological mechanisms which determine the distribution of body size. The present development of ecology is being influenced by the individual-based approach, but it is only of limited influence and does not yet represent any real breakthrough.

It is possible that the individual-based approach will lead to the development of alternative and superior ecological theories. The degree of success depends on to what extend this theory will be good enough to explain or predict those phenomena which the present theory has so far failed to predict. It is still an open question. We cannot beg the answer at this time.

Acknowledgements

I thank Ryszard Korona, Jan Kozłowski, and an anonymous reviewer for reading an earlier version of this manuscript and for helpful comments.

Literature Cited

Begon, M. 1984. Density and individual fitness; asymmetric competition. In *Evolutionary Ecology*, B. Shorrock (ed.), pp. 175–194. Blackwell, Oxford, England.

Braun-Blanquet, J. 1932. *Plant Sociology: The Study of Communities*. McGraw-Hill, New York.

Caplan, A. L. 1988. Rehabilitating reductionism. *Am. Zool.* **28**:193–203.

Ciesielska, M. 1985. Zroznicowanie fenotypowe poczwarek *Tribolium confusum* w niejednorodnych siedliskach. M.Sc. thesis, Jagiellonian University, Krakow, Poland.

Clutton-Brock, T. H., F. E. Guinness, and S. D. Albon. 1982. *Red Deer, Behavior and Ecology of Two Sexes.* Chicago University Press, Chicago.

Cooper, W. S. and R. H. Kaplan. 1982. Adaptive "coin-flipping": a decision-theoretic examination of natural selection for random individual variation. *J. Theor. Biol.* **94**:135–151.

Fretwell, S. D. 1972. *Populations in a Seasonal Environment.* Princeton University Press. Princeton, N.J.

Gurney, W. S. C. and R. M. Nisbet. 1979. Ecological stability and social hierarchy. *Theor. Popul. Biol.* **14**:48–80.

Hamilton, W. D. 1964. The genetical evolution of social behaviour. *J. Theor. Biol.* **7**:1–16.

Hara, T. 1988. Dynamics of size structure in plant populations. *Trends Ecol. Evol.* **3**:129–188.

Hassell, M. P. and R. M. May. 1985. From individual behavior to population dynamics. In *Behavioural Ecology: Ecological Consequences of Adaptive Behavior,* R. Sibley and R. Smith (eds.) pp. 3–32, Blackwell, Oxford, England.

Huston, M. and D. DeAngelis. 1987. Size bimodality in monospecific plant populations: a critical review of potential mechanisms. *Am. Nat.* **129**:678–707.

Huston, M., D. DeAngelis, and W. Post. 1988. New computer models unify ecological theory. *Bioscience* **38**:682–691.

Jones, R. 1979. Predator-prey relationship with particular references to vertebrates. *Biol. Rev. Camb. Philos. Soc.* **54**:73–97.

Kimmel, M. 1986. Does competition for food imply skewness? *Math. Biosci.* **80**:239–264.

Korona, R. 1989a. Evolutionarily stable strategies in competition for resource intake rate maximization. I. The model. *Behav. Ecol. Sociobiol.* **25**:193–199.

Korona, R. 1989b. Evolutionarily stable strategies in competition for resource intake rate maximization. II. Oviposition behavior in *Tribolium confusum. Behav. Ecol. Sociobiol.* **25**:201–205.

Korona, R. and A. Łomnicki. 1988. Individual differences and survival during competition: a simple theoretical model and empirical evidence for flour beetles *Tribolium castaneum. Funct. Ecol.* **2**:331–334.

Łomnicki, A. 1978. Individual differences among animals and natural regulation of their numbers. *J. Anim. Ecol.* **47**:461–475.

Łomnicki, A. 1980. Regulation of population density due to individual differences and patchy environment. *Oikos* **35**:185–193.

Łomnicki, A. 1988. *Population Ecology of Individuals.* Princeton University Press, Princeton, N.J.

Łomnicki, A. and S. Sedziwy. 1989. Do individual differences in resource intakes without monopolization cause population stability and persistence? *J. Theor. Biol.* **136**:317–326.

May, R. M. 1973. Stability and Complexity in Model Ecosystems. Princeton University Press, Princeton, N.J.

Mayr, E. 1982. *The Growth of Biological Thought*. Harvard University Press, Cambridge, MA.

Metz, J. A. J. and O. Diekmann (eds.). 1986. *The Dynamics of Physiologically Structured Populations*. Lecture Notes in Biomathematics 68. Springer-Verlag, Berlin.

Rubenstein, D. I. 1981. Population density, resource partitioning and territoriality in the Everglade pygmy sunfish. *Anim. Behav.* **29**:155–172.

Schoener, T. W. 1973. Population growth regulated by intra-specific competition for energy and time: some simple representation. *Theor. Popul. Biol.* **4**:56–84.

Uchmanski, J. 1985. Differentiation and distribution of body weights of plants and animals. *Philos. Trans. R. Soc. Lond. Biol. Sci.* **310**:1–75.

Weiner, J. 1990. Asymmetric competition in plant populations. *Trends Ecol. Evol.* **5**:360–364.

Weiner, J. and O. T. Solbrig. 1984. The meaning and measurement of size hierarchies in plant populations. *Oecologia* **61**:334–336.

Wynne-Edwards, V. C. 1962. *Animal Dispersion in Relation to Social Behaviour*. Oliver and Boyd, Edinburgh, Scotland.

Wyszomirski, T. 1983. A simulation model of growth of competing individuals of a plant population. *Ekol. Pol.* **31**:73–92.

2

Individual-Based Models: Combining Testability and Generality

W.W. Murdoch, E. McCauley, R.M. Nisbet, W.S.C. Gurney and A.M. de Roos

ABSTRACT. **A dilemma in ecology is posed by the contrast between simple strategic models, which offer generality but are exceedingly difficult to test against the performance of particular real systems, and complex tactical models, which aim to account for and be tested against particular systems but which do not easily yield generality. We explore the possibility that linked sequences of individual-based models provide the means for developing a theory of population and community ecology that is both testable and general. Individual-based models are by their nature preadapted to incorporating the mechanisms that determine a system's dynamics. In their more detailed forms they are therefore highly testable. For the same reason, they lend themselves to a process of progressive removal of particular features, which should facilitate distinguishing the "essential" mechanisms from those that account for detail rather than major dynamic features. We illustrate these ideas with a theoretical and experimental program investigating the interaction between populations of the zooplankter *Daphnia* and algal food supply.**

Our aims are to draw together some themes of the conference and comment on them and on possible future directions that individual-based modeling might take. Our sources for these themes was a set of position papers prepared for the conference. It turned out that these issues overlap our own preoccupations, and we therefore begin by addressing what we see as the central challenge in that part of theoretical ecology concerned with the dynamics of populations and communities. We then review the themes, discuss some research that relates to them, and finally say something about future directions.

The Challenge: Testability with Generality

We believe ecological theory should strive to achieve two key characteristics: testability and generality. A central problem, however, is that the two goals are frequently, and perhaps typically, incompatible (a related problem was discussed by Levins 1966).

The dilemma is illustrated by the distinction between "strategic" and "tactical" models (Holling 1966, May 1974, Nisbet and Gurney 1982). The former traditionally have been simple models that try to reach past the tangle of unique detail associated with each particular ecological system to grasp and abstract the features that account for those aspects of its dynamics that are generic. Such a model tries to leave behind "details" that account for the particularity of that system's dynamic behavior while "capturing the essential dynamics" (May 1979) of the system; Generality can be attained because, one hopes, many systems have a similar "essence." Tactical models, by contrast, seek to explain and project the detailed dynamics of particular systems, as might be useful, for example, in managing a resource population.

A tactical model is in principle highly testable precisely because it is a description of the mechanisms operating in a particular system: it can be asked to predict the system's dynamics under well-defined new circumstances. But how do we test strategic models? Their parameters are often highly abstracted (consider the "competition coefficient" of a Lotka-Volterra model), they do not correspond to well-defined and measurable biological quantities, and they cannot be reliably estimated in any particular ecological system. It is therefore difficult to make precise predictions for that system, and the model is as a consequence difficult to test. We return to this problem below.

A possible alternative is to develop tests that rely on predicted changes in qualitative dynamic behavior. For example, some headway has been made in testing the ability of very simple (Lotka-Volterra) predator-prey models to predict the qualitative pattern of change in equilibrium densities of the zooplankter *Daphnia* and its algal prey in response to increases in the nutrient level in lakes (McCauley et al. 1988, Leibold 1989). But so far these models cannot predict the dynamic behavior of this system (see below).

In fact, strategic models have rarely been tested. Rather they have been accepted or rejected largely on the basis of plausibility; i.e. on whether they seem to provide a reasonable explanation for observed phenomena. The problem here is that, typically, several competing simple models can provide a plausible explanation of a particular phenomenon. For example, *Daphnia/* algal populations are sometimes cyclic (Murdoch and McCauley 1985). The cycles can be "explained" by both a "paradox-of-enrichment" Lotka-Volterra model (Rosenzweig 1972) and by a delayed logistic model (Lawton 1987): in each case the parameters of the model can be set at "reasonable" values to produce cycles of about the right period. Both models almost certainly have the mechanisms wrong, however (McCauley and Murdoch 1987, Nisbet, Gurney, Murdoch, and McCauley 1989).

In sum, while the Holy Grail of a simple highly testable yet broadly general model of, say, population dynamics may be just around the corner, the evidence so far is not encouraging. In particular, there has as yet been little rigorous analysis to determine when some strategic model has omitted the "essence" of the phenomenon rather than merely its superfluous detail. We believe individual-based models (which include consistently formulated age-, stage- and other structured models) can help resolve this central dilemma, and our aim is to illustrate an approach.

Conference Themes

Two kinds of themes expressed in the position papers can be distinguished: those addressing the goals and promise of individual-based models, and those addressing their difficulties. First, there was broad agreement on two main potential advantages of individual-based models.

(1) Individual-based models can greatly increase our understanding of the system under study because they allow us to include a description of the actual mechanisms, the processes, that determine the vital rates of the different classes of individuals that make up the interacting populations.

(2) The models promise to be highly testable, again because mechanisms at the individual level are incorporated: we can readily examine these mechanisms, and they lead to testable predictions at the population model. (To avoid misunderstanding, we stress that the population models themselves need to be tested; basing the population model on a model of individual behavior that has been thoroughly tested does not constitute a test of the population model.)

Precisely, because they describe individual properties and processes in detail, however, individual-based models tend to be complex. This is perhaps especially so for those that model large collections of individuals. A second set of themes therefore expressed concern that the capacity for detail can itself prevent individual-based models from fulfilling their promise, that complexity has the potential to undermine every aspect of improved understanding and testability.

(1). *Understanding:* Models, especially those that follow large numbers of individuals, can be so complex that we lose the ability to understand how a particular mechanism operates within the multitude of detail and interacting mechanisms portrayed in the model.

(2) *Testability:* The models can also be so complex as to be essentially untestable—any observation can be accommodated by the model simply by altering a few functions or parameter values.

(3) *Data requirements:* Because of their great detail, the models have a huge appetite for data that can exceed our ability to satisfy it. Again, the likely consequence of this problem is a loss of testability: the model is likely to contain a more detailed description of phenomena than can be matched by observation and experiment and, as a result, there is an excess of "free" parameters that can be used to fit observations.

(4). *Generality:* We would add to these difficulties associated with complexity the problem of obtaining generality. The testability of the model may depend upon its being highly detailed and system-specific to ensure that the functions and parameter values can describe actual processes, and can be well estimated. But this particularity may prevent us from using the model to abstract and generalize the insight we obtain from a particular system.

In the next two sections we want to illustrate how these problems might be resolved. We leave for the Discussion one other difficulty, namely that involved in "scaling up" from simple to more complex systems.

Combining Testability and Generality

The approach we advocate below to make theory simultaneously testable and general is to express theory in a linked sequence of individual-based models, rather than in a single model. The sequence of models begins at the tactical end of the spectrum and ends at the strategic end, and the linkage is achieved by (1) explicitly deriving the simpler models by stripping detail from the more complex models, and (2) by successively comparing these models against new data to determine the loss in predictive capacity (either in detail or accuracy, or both) at each step (Fig. 2.1).

Such a sequence can exploit two main advantages offered by individual-based models. First, they are potentially eminently testable (a) because they are natural vehicles for incorporating detailed mechanisms that can be examined experimentally, (b) because their parameters and functions typically can be estimated directly, and (c) because they make more detailed predictions of population behavior and there are therefore more points at which to compare the model against reality, in contrast with unstructured models. Second, individual-based models lend themselves to sequential removal of detail in a process of progressive simplification.

By exploiting these two properties of individual-based models we can hope to turn the claim "this general model captures the essence of the process" from a statement of faith into a hypothesis to be tested. We will illustrate the point by some of our research on the *Daphnia*/algal system, emphasizing the process rather than particular results.

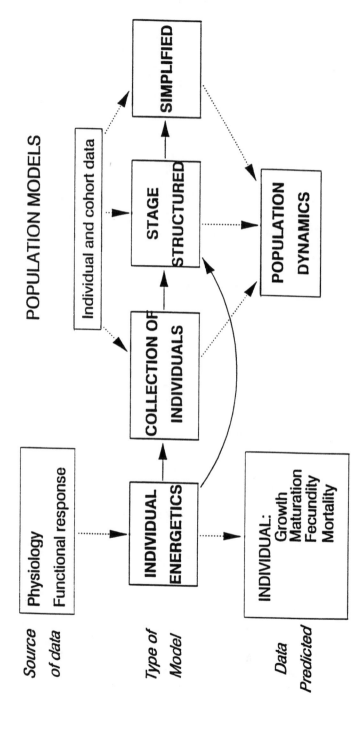

Fig. 2.1. Diagram of a research program to produce a linked sequence of experimentally tested models of an ecological system. The process begins with a model of individual properties, which serves as the basis for a sequence of progressively simpler individual-based population models. A key feature is that at each stage the model is tested against different data from those that are used to construct and parameterize it.

Testability

We turn first to testability. As discussed above, individual-based models are well adapted for incorporating the mechanisms that determine the system's dynamics. A population's dynamics are determined by the numbers recruiting, dying and moving, which are in turn determined by the properties of individuals, e.g. their behavior, feeding rates, growth rates, the rate at which they allocate assimilated energy to reproduction, and so on. These properties are of course influenced by the individual's state, e.g. its size. That is, the ultimate mechanisms of interest to population dynamics involve individual properties such as physiology and behavior, and how these properties change as the individual's environment changes. We stress that an individual's environment of course includes other individuals, and that, if variation in density affects individual properties, then it must be one of the "environmental" variables specifying individual performance.

If we wish to understand these basic mechanisms, the obvious starting point is therefore a model of the individual (e.g. of its physiology or behavior). But first we must ask: Which aspects of the individual should we model? The answer is: those properties thought to be important to the ecological (population-level) question being addressed, and here we emphasize the importance of basing the models in empirical observation, a point to which we will return.

The *Daphnia*/algal interaction is a particularly illuminating example. A very large body of observational and experimental work has been done on the population dynamics of this system in three situations: (1) in the laboratory (where the algae cannot grow and are regularly replenished by the experimenter), (2) in lakes, ponds, and reservoirs (where the algae are dynamic and there are many biological and physical complexities) and (3) in stock tanks (where the algae are dynamic, as they are in nature, but other complications such as predators are lacking, or are reduced—e.g. spatial heterogeneity). The stock tanks are thus an intermediate step between the laboratory and natural systems.

In a series of papers, McCauley and Murdoch have established a well-defined pattern of *Daphnia* dynamics that occurs in all three systems. The *Daphnia*/algal system has a predator-prey interaction in which *Daphnia* suppresses the algal food supply to a density that on average just allows *Daphnia's* fecundity to balance its death rate. The result is either stable population dynamics or cycles with broadly similar demographic patterns (Murdoch and McCauley 1985, McCauley and Murdoch 1987, 1990). Figure 2.2 shows, for example, that when cycles appear they have a remarkably consistent (small) amplitude over a large range of natural environments; cycles with very similar amplitude are seen in the laboratory and stock tanks (McCauley and Murdoch 1987). The patterns of cyclical change in age structure and

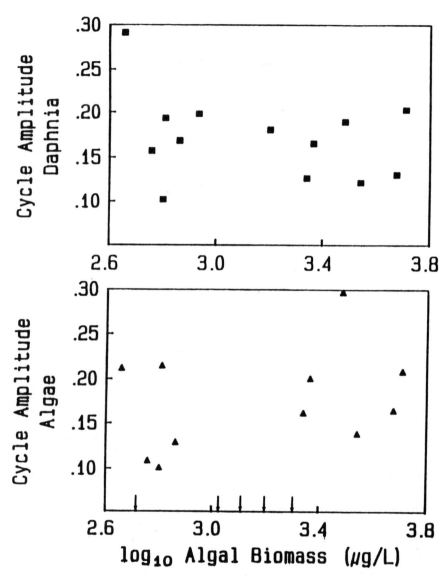

Figure 2.2. Amplitudes of cycles (in log units) observed for *Daphnia* (density) and algal (biomass) populations from lakes and ponds analyzed by McCauley and Murdoch (1987). There is no change in cycle period with algal biomass (which increases with nutrient status). Stable populations (indicated by arrows on the algal portion of the graph) occur over much of the observed range of nutrient levels. From McCauley and Murdoch (1990).

fecundity, which is a function of food supply, are also characteristic (McCauley and Murdoch 1987).

The major points, then, are (a) a well-defined and ubiquitous dynamic pattern has been established, and (b) we have shown that the underlying mechanism involves the interaction between *Daphnia* and its food supply. The individual *Daphnia* model must therefore focus on the factors affecting *Daphnia*'s rate of uptake of food. Foremost among these is *Daphnia* size, so the model must also account for *Daphnia*'s allocation of the assimilate among maintenance, growth and reproduction.

A further advantage of working with *Daphnia* is the enormous literature on individual energetics, which has allowed us to develop an individual model that predicts how Daphnia will grow, mature and reproduce as a function of its food intake. Other sets of data describe *Daphnia*'s individual performance (growth etc.) at different food levels, thus allowing the individual model to be tested on a data set entirely independent of those used in its development (Fig. 2.1). Our success in predicting *Daphnia*'s performance at different food levels is illustrated for size-at-maturity (Fig. 2.3), and other illustrations can be found in Gurney et al. (1990).

We now have an individual model successfully tested at all but the lowest food levels that have been studied in the laboratory (McCauley et al. 1990 examine lower food levels than have been looked at heretofore, and we are currently testing the model against these data). The next logical step (Fig. 2.1) is either a model that follows a large number of individual *Daphnia*, each obeying the behavioral and physiological rules embodied in the individual model (an *i*-state population model), or a very detailed "lumped" model, e.g. a stage- or physiologically-structured model that lumps individuals having the same or similar status (a *p*-state population model). Such a model might be tested first against existing or new detailed data on laboratory populations, and then against field population dynamics. Once again, the test data should be independent of the data giving rise to the model.

The idea is then to develop (and test at each step) a sequence of individual-based models of decreasing complexity. Our bias is towards lumped models rather than those that follow individuals explicitly. This bias is based on the premise that lumped models are more likely to lead to general insight, i.e. they can be more easily moved in the direction of strategic models.

In fact, for the present we have moved directly to a relatively simple stage-structured formulation. This model is based on the properties of the individual model, but it contains only three classes of individuals (young, immature adults and mature adults) and is perhaps too simple as a starting population model (see Fig. 2.4 and Nisbet, Gurney, Murdoch, and McCauley 1989). Murdoch et al. (in prep.) show that the model does quite well at predicting equilibrium values in some laboratory and field conditions.

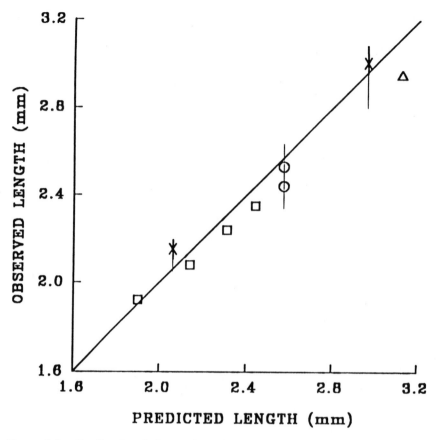

Figure 2.3. Predicted and observed maximum length attained by *Daphnia pulex* raised at different food levels. Data from Paloheimo et al. 1982 (△), Lynch et al. 1986 (○), Richman 1958 (□) and Taylor 1985 (x). Points along the 45° line would indicate a perfect match between observed and predicted values. Vertical lines are ±1 S.E. where available. Redrawn from Gurney et al. (1990).

However, in its original form it is unable to predict the small-amplitude cycles we have shown to be so widespread in natural environments and stock tanks (Fig. 2.2). These environments are characterized by much lower edible algal densities than those at which existing laboratory studies have examined *Daphnia*'s energetics. It is therefore possible that modifying the individual model in the light of new experiments at low food levels (McCauley et al. 1990) will lead to changes in the individual model and hence in the individual-based population model based upon it. That is, we need to recycle through the steps outlined in Fig. 2.1.

Population Model for *Daphnia*

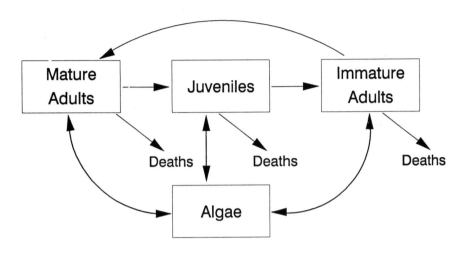

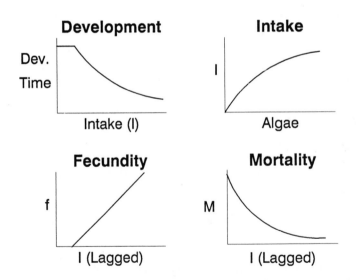

Figure 2.4. A stage-structured population model for interacting populations of *Daphnia* and algae. The relationships shown below the diagram of the model are empirically derived from laboratory studies.

The important lesson in this case concerns not these particular results, but the process: reworking assumptions about individual behavior should allow us once again to develop the population model independently of the data that will be used to test it. It is also possible, of course, that even when we correct the individual model, the population model will still be wrong. To take the simplest example, there might be density dependent interactions among *Daphnia* not included in our individual model. Formally, the way forward then is to recognize that other *Daphnia* are a component of the environment experienced by each individual *Daphnia*, and to further modify the individual model (Metz and deRoos, this volume). In practice, however, this modification will often amount to introducing new interaction terms in the population model. The take-home message is that individual-based population models must be tested directly; a model cannot be assumed to be "correct" on the grounds that the underlying individual model has been tested in certain specific contexts.

Generality

Individual-based models have two features that facilitate the creation of simultaneously testable and general theory. The first is that described above: because mechanisms at the individual level are incorporated directly into the population models, we can progressively strip details from the model and establish at each stage which feature(s) of the dynamics we then fail to account for. That is, we should be able to tell which simplifications drop mere "details," and which lose "essence."

Second, even quite simple lumped individual-based models may be more realistic than traditional strategic models. For example, compare the structure laid out in Fig. 2.4 with that contained in a standard analogous Lotka-Volterra model. Of course we might choose to complete the process of simplification by deriving an unstructured model; in the *Daphnia*/algal system, for example, this is a Lotka-Volterra type of model. But it is becoming clear that lumped models themselves are good candidates for strategic vehicles that abstract basic mechanisms.

Consider, for example, the growing number of stage-structured models that might provide generic explanations for cyclic population dynamics. These include single-species models for blowflies (Gurney et al. 1980, Stokes et al. 1988), moth populations in the laboratory (Gurney et al. 1983) (a related but differently-formulated model has been developed for cannibalistic Tribolium populations, Hastings and Constantino 1987), and for populations of marine kelp (Nisbet and Bence 1989), as well as two-species models for the *Daphnia*/algal system (Nisbet et al. 1989a), and for parasitoid-host interactions (Murdoch et al. 1987, Godfray and Hassell 1989 and Gordon et al. in press). Of these models, however, only the *Daphnia*/algal one rests upon an extensively tested linked sequence of models as proposed above.

The most obvious obstacle to achieving generality with individual-based models is the apparent need to develop new individual models for every species/environment combination under investigation. However even here the outlook is promising for two reasons. The most fundamental is that in many situations, energetic considerations play a major role in determining rates of growth and reproduction, and (more speculatively) survival. Modeling the vital rates then centers on modeling an individual's priority in allocating assimilated energy among growth, reproduction and maintenance. Similar allocation rules are to be expected in similar organisms, and experience gained in modeling one species can be valuable even when working with a very different organism. For example, many features of our individual *Daphnia* model (Gurney et al 1990) are used, sometimes with modification, in a model of growth and reproduction in marine mussels (Ross and Nisbet 1991); similarly, a model developed for *Daphnia* by Kooijman (1986) has been used to predict the pattern of reproduction in starved pond snails (Zonneveld and Kooijman 1989).

The second reason for optimism is that there is scope for substantial simplification of the population models. From experimental data on individuals, we can frequently identify development indices characterizing physiological age within each developmental stage, with the probability that an individual matures from one stage to its successor being determined by the value of this index (Gurney et al. 1986). A simple example of such an index for an insect in a population with no food limitation would be cumulative degree-days above some temperature threshold. More complex indices involving linear combinations of variables such as age and weight have proved successful in fitting data on *Daphnia* (Nisbet, Gurney, Murdoch, and McCauley 1989), and in studies of larvae of stored product insects (Gordon et al. 1988; Jones et al. 1991). Once a development index is defined, the population model can be recast, at least for the stages in question, in terms of a partial equation involving only one physiological variable—the development index. With additional assumptions, for example that all individuals within a stage have feeding rates that are identical or are related in some simple way to the development index, the population model may reduce to a relatively tractable set of ordinary or delay differential equations (Gurney et al. 1986).

The next task is to establish the degree to which population models from among those listed above can be generalized to "capture the essence" of many different ecological systems. That is, we need to extend the domain of some of them. A possible protocol for achieving this is to repeat in some other systems the process described above for the *Daphnia*/algal interaction, thus developing an array of sequentially linked and experimentally-tested models, in each case determining whether they reduce to the same strategic model.

Discussion

We have tried to show that individual-based models provide a basis for developing a truly testable yet general theory of population and community dynamics. Empirical ecologists have often been uneasy with strategic theory in population dynamics because so much of the biology is suppressed. Individual-based models portray that biology more clearly, and make more explicit what has been omitted. In their more detailed forms, they also lead to experimental tests in a much more direct way. They thus provide an excellent focus for collaboration between the experimentalist and the modeler, which we believe is a fruitful approach to developing improved theory.

There is, however, one particularly difficult area that we have not mentioned but that needs to be addressed: the problem of modeling many-species systems. In discussing the *Daphnia*/algal system, for example, we made no mention of other species, e.g. competitors and predators of *Daphnia*, that might influence the dynamics. In particular, it is likely to be difficult to deal with interactions across several trophic levels. Even though massive computer simulations following many individuals have sometimes included many species, these have typically belonged to a single guild such as forest trees, and even these "brute force" models have not tackled the several-trophic level problem. Models with several trophic levels are likely to be dauntingly complex. This may well reduce our understanding, and make testing more difficult and spurious fitting more likely, all concerns that have been expressed already with regard to existing individual-based models (see Conference Themes).

In many circumstances it may be possible to alleviate this problem by portraying many-species systems with few-species models. Real communities may at least sometimes consist of small sets of tightly-coupled species, each of which is only loosely-coupled to other species or sets of species. It may thus be possible to decompose the community into such subsets, and to treat other loosely-coupled sets as parts of the environment of the set of concern, thus simplifying the modeling problem (Fig. 2.5 and Murdoch and Walde 1989). Once again, the existence of loose-coupling between two species is a hypothesis that can be subjected to empirical test.

The *Daphnia*/algal system again provides an example. At least in some circumstances, we can describe the system's dynamics without specifying those of *Daphnia*'s predators (McCauley et al. 1988). Thus, fish and many invertebrate predators of *Daphnia* have long generation times relative to those of *Daphnia*, they typically feed on many other prey species, and they may feed on *Daphnia* during only a portion of the life history (Murdoch and Bence 1987). While these species may affect *Daphnia*'s rate of change, changes in their own density tend to be independent of short-term changes in *Daphnia*'s density, and it is such short-term change that we seek to ex-

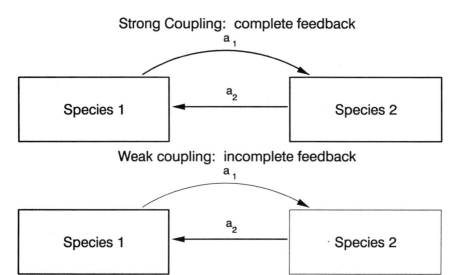

Strong Coupling: complete feedback

a_1

Species 1 a_2 Species 2

Weak coupling: incomplete feedback

a_1

Species 1 a_2 · Species 2

Species 2 becomes part of Species 1's environment

$$dN_1/dt = N_1 f_1(N_1) + a_1 N_1 g_1(N_2)$$
$$dN_2/dt = N_2 f_2(N_2) + a_2 N_2 g_2(N_1)$$

Figure 2.5. A diagrammatic representation of strong and weak coupling between two interacting populations. Strong coupling between two species occurs when species 1 strongly affects the short-term dynamics of species 2 (a_2 large), and the dynamics of species 1 is in turn strongly affected by species 2 (a_1 large). In weak coupling the interaction is weak in one or both directions (a_1 or a_2 small), and there is an incomplete feedback loop from species 1, through species 2, and back to species 1: no equations are needed describing the dynamics of species 2, whose effects on species 1 can instead be taken into account within the equations describing the dynamics of species 1. From Murdoch and Walde (1989).

plain in dynamic models of *Daphnia* populations (Fig. 2.6). The effect of these species on the *Daphnia*/algal system can therefore be described via the *Daphnia* death rate parameter or, perhaps, by making *Daphnia*'s death rate a function of *Daphnia*'s density (e.g. inversely density-dependent if the predator's functional response is type 2).

Stock-tank experiments using the backswimming bug *Notonecta* as the predator provide support for this approach in *Daphnia*/algal system (Murdoch unpub.). Orr et al. (1990), using data from the same experiment, show that the converse is also true: *Notonecta*'s dynamics are at most weakly coupled with those of *Daphnia* and can be described without reference to

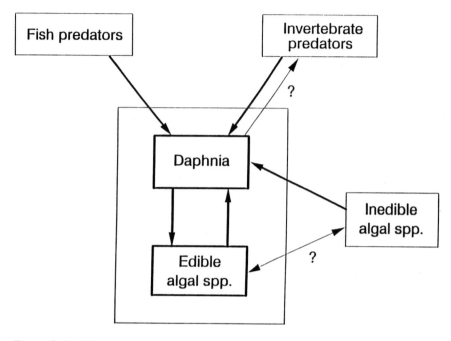

Figure 2.6. Diagram of possible types of coupling between the *Daphnia*/algal interaction and other species in the environment. From Murdoch and Walde (1989).

the dynamics of the *Daphnia*/algal system, even though the early stages of *Notonecta* feed mainly on *Daphnia*.

No doubt it will not always be possible to reduce the dimensionality of the system in this way, and we have no off-the-shelf alternative simplification schemes to offer. However it seems likely that the spirit of the approach advocated in this paper will survive, and that progress will involve careful study of the response of individuals to their (local) biotic and abiotic environment. This will lead to the development of population models incorporating the dynamics of all the trophic levels involved. Testing such complex models is one of the next major hurdles to be faced in the development of individual-based models.

Acknowledgements

The research reported here is based upon work supported by NATO to the authors and by NSF grant INT-9011084, and grant DE-FG03-89ER60885 from the Ecological Research Division. Office of Health and Environmental

Research, U.S. Department of Energy, to WWM. The latter support does not constitute an endorsement of the views expressed here.

Literature Cited

Godfray, H. C. J. and M. P. Hassell. 1989. Discrete and continuous insect populations in tropical environments. *J. Anim. Ecol.*, **58**:153–174.

Gordon, D. M., W. S. C. Gurney, R. M. Nisbet, and R. K. Stewart. 1988. A model of Cadra cautella growth and development. *J. Anim. Ecol.* **57**:645–658.

Gordon, D. M., R. M. Nisbet, A. M. deRoos, W. S. C. Gurney, and R. K. Stewart, 1991. Discrete generations in host-parasitoid models with contrasting life cycles. *J. Anim. Ecol.*, **60**:295–308.

Gurney, W. S. C., S. P. Blythe, and R. M. Nisbet. 1980. Nicholson's blowflies revisited. *Nature* **287**:17–21.

Gurney, W. S. C., R. M. Nisbet, and J. H. Lawton. 1983. The systematic formulation of tractable single species models involving age structure. *J. Anim. Ecol.* **52**:479–495.

Gurney, W. S. C., R. M. Nisbet, and S. P. Blythe. 1986. The systematic formulation of stage structured population models. In The Dynamics of Physiologically Structured Populations, J. A. J. Metz and O. Diekmann eds., pp. 474–494. Springer-Verlag, Berlin.

Gurney, W. S. C., E. McCauley, R. M. Nisbet, and W. W. Murdoch. 1990. The physiological ecology of *Daphnia*: a dynamic model of growth and reproduction. Ecology **71**:716–732.

Hastings, A. and R. F. Constantino. 1987. Cannibalistic egg-larva interactions in *Tribolium*: An explanation for the oscillations in population numbers. *Am. Nat.* **120**:36–52.

Holling, C. S. 1966. The strategy of building models of complex ecological systems. In Systems Analysis in Ecology, K. E. F. Watt (ed.), pp. 195–214. Academic Press, New York.

Jones, A. E. W. S. C. Gurney, R. M. Nisbet, and D. M. Gordon. 1991. Food degradation as a mechanism of intra-specific competition among stored product larvae. *Funct. Ecol.*, **4**:629–638.

Kooijman, S. A. L. M. 1986. Population dynamics on the basis of budgets. In The Dynamics of Physiologically Structured Populations, J. A. J. Metz and O. Diekmann eds., Springer-Verlag, Berlin.

Lawton, J. H. 1987. *Daphnia* population dynamics in theory and practice. *Trends Ecol. Evol.* **2**:233–234.

Leibold, M. A. 1989. Resource edibility and the effects of predators and productivity on the outcome of trophic interactions. *Am. Nat.* **134**:922–949.

Levins, R. 1966. The strategy of model building in population biology. *Am. Sci.* **54**:421–431.

Lynch, M., L. J. Weider, and W. Lampert. 1986. Measurement of the carbon balance in *Daphnia. Limnol. Oceanog.* **31**:17–33.

May, R. M. 1974. Stability and Complexity in Model Ecosystems (2nd ed.). Princeton University Press, Princeton, NJ.

May, R. M. 1979. Theoretical Ecology: Principles and Applications. Blackwell, Oxford, England.

McCauley, E. and W. W. Murdoch. 1987. Cyclic and stable populations: plankton as paradigm. *Am. Nat.* **129**:97–121.

McCauley, E., W. W. Murdoch, and S. Watson. 1988. Simple models and variation in plankton densities among lakes. *Am. Nat.* **132**:383–403.

McCauley, E. and W. W. Murdoch. 1990. Predator-prey dynamics in rich and poor environments. *Nature* **343**:455–457.

McCauley, E., W. W. Murdoch, and R. M. Nisbet. 1990. Growth, reproduction, and mortality of *Daphnia pulex*: life at low food. *Funct. Ecol.*, **4**:505–514.

Murdoch, W. W. and J. R. Bence. 1987. General predators and unstable prey populations. In Predation in Aquatic Communities: Direct and Indirect Effects. C. Kerfoot and A. Sih, eds., pp. 17–29. University Press of New England, Hanover, NH.

Murdoch, W. W. and E. McCauley. 1985. Three distinct types of dynamic behaviour shown by a single planktonic system. Nature **316**:628–630.

Murdoch, W. W., R. M. Nisbet, W. S. C. Gurney, and J. D. Reeve. 1987. An invulnerable age class and stability in delay-differential parasitoid-host models. *Am. Nat.* **129**:263–282.

Murdoch, W. W. and S. J. Walde. 1989. Analysis of insect population dynamics. In Toward A More Exact Ecology, J. Whittaker and P. J. Grubb (eds.), pp. 113–140. Blackwell, Oxford, England.

Nisbet, R. M. and J. R. Bence. 1989. Alternative dynamic regimes for canopy-forming kelp: A variant on density-vague population regulation. *Am. Nat.* **134**:377–408.

Nisbet, R. M. and W. S. C. Gurney. 1982. Modelling Fluctuating Populations. Wiley, New York.

Nisbet, R. M., W. S. C. Gurney, and J. A. J. Metz. 1989. Stage structure models applied in evolutionary ecology. In *Applied Mathematical Ecology*, S. A. Lewin, T. G. Hallam and L. J. Gross (eds.), pp. 428–449. Springer-Verlag, Berlin.

Nisbet, R. M. and W. S. C. Gurney, W. W. Murdoch, and E. McCauley. 1989. Structured population models: a tool for linking effects at individual and population level. *Biol. J. Linn. Soc.* **37**:79–99.

Orr, B. K., W. W. Murdoch, and J. R. Bence. 1990. Population regulation, convergence, and cannibalism in a backswimming bug. Ecology **71**:68–82.

Paloheimo, J. E., S. J. Crabtree, and W. D. Taylor. 1982. Growth model of *Daphnia.* Can. J. Fish. Aqua. Sci. **39**:598–606.

Richman, S. 1958. The transformation of energy by Daphnia pulex. Ecol. Monogr. **28**:273–291.

Nisbet, R. M., W. S. C. Gurney, and J. A. J. Metz. 1989. Stage structure models applied in evolutionary ecology. In *Applied Mathematical Ecology*, S. A. Lewin, T. G. Hallam and L. J. Gross (eds.), pp. 428–449. Springer-Verlag, Berlin.

Rosenzweig, M. L. 1972, Paradox of enrichment: destabilization of exploitation systems in ecological time. *Science* **175**:564–565.

Ross, A. H. and R. M. Nisbet. 1991. Dynamic models of growth and reproduction in the mussel Mytilus edulis. Funct. Ecol., in press.

Stokes, T. K., W. S. C. Gurney, R. M. Nisbet, and S. P. Blythe. 1988. Parameter evolution in a laboratory insect population. Theor. Popul. Biol. **34**:248–265.

Taylor, B. E. 1985. Effects of food limitation on growth and reproduction of *Daphnia*. Arch. Hydrobiol. Beih. **21**:285–296.

Zonneveld, C. and S. A. L. M. Kooijman. 1989. Application of a dynamic energy budget model to *Lymnaea stagnalis* (L.) Funct. Ecol. **3**:269–278.

3

From the Individual to the Population in Demographic Models

Hal Caswell and A. Meredith John

ABSTRACT. **Demographic population models are derived from descriptions of how individuals move through their life cycles. In this sense all demographic models are based on information about individuals. Individuals are described in terms of their *i*-states (sensu Metz and Diekmann), which provide the information necessary to specify the response of an individual to its environment. Age, size, and physiological state are typical *i*-state variables. The state of the population, or *p*-state, is derived from the *i*-states. If all individuals experience the same environment (we refer to this condition as mixing), the *p*-state is a distribution function over the set of *i*-states. Most demographic models assume this condition; we call these models *i*-state distribution models. When the mixing assumption fails, for example due to local interactions among individuals, each individual must be followed; we call these models *i*-state configuration models. We use this framework to examine examples of the relations between the individual and the population in demography, including multi-type branching processes and stable population theory, micro-simulations of reproduction and family structure, epidemic models and percolation theory, and hazard analysis.**

Introduction

The use of "individual-based models" (we will propose a different nomenclature in what follows) was urged by Huston et al. (1988) in a paper which led directly to this symposium. They claimed that most ecological models make assumptions that contradict two important properties of organisms. First, grouping individuals into categories violates the principle of the uniqueness of the individual. Second, by not distinguishing among the locations of individuals, the models violate the principle that interactions are inherently local.

The alternative that they propose is to use models based on explicit representation of individual organisms. Such models must be studied by numerical simulation, but recent increases in available computer power have made them more practical than they have ever been before. Huston et al. (1988) claim that such models escape some of the simplifying assumptions made by other ecological models. They can also be constructed directly from data on individuals, data which are routinely collected in studies of individual physiology and behavior.

In this paper we examine the special case of demographic population models. Some of the claims of Huston et al. (1988) underestimate the capabilities of traditional ecological and demographic models, and the extent to which they are already based on information about individuals. However, we will argue that the explicit inclusion of individual organisms is sometimes necessary, and more often convenient, and that such models can play an important role in theoretical ecology.

State in Population Models

Demography[1] is an approach to the study of populations which is based on the analysis of individuals. The word "demography" is derived from the Greek, loosely "writing about individuals," and all demographic models are, in a sense, individual-based models.

The simplest way to account for individuals is simply to count them, and the simplest population models are written in terms of the crude numbers of individuals and the crude rates by which those numbers change through time. But such models can have only limited success, and demographic models acknowledge that individuals differ among themselves in many important characteristics as they move through their life cycles. These differences are summarized in the state of the individuals.

It is impossible to predict the response of all but the very simplest natural systems from knowledge of current environmental stimuli alone. The problem is that the past of the system affects its response in the present. The state of a system is that information necessary to account for this history and determine, in combination with the present environment, the

---

[1]We use the word demography to identify an approach—one based on the characteristics of individuals—and not a subject species. Thus we will refer to "human demography" to denote that species-specific branch of the discipline concerned with a certain large primate. This usage will appear strange to some practitioners of that sub-discipline, but quite natural to demographers sensu lato.

"They tell me you're a demographer. Whatever happened
to Fred Biddlingmeyer?"

Figure 3.1. Cartoon by Peter Steiner, reprinted with permission.

system's behavior[2] (Zadeh 1969, Caswell et al. 1972, Metz and Diekmann 1986). Begin with the individual. Following Metz and Diekmann (1986), we define the state of an individual (the *i*-state) as the information needed to specify the response of the individual to its environment. Examples of *i*-state variables include age, size, marital status, instar, employment status, hunger, parity, lipid concentration, and geographical location.

The *i*-state encapsulates the information necessary to predict the individual's behavior. To move up to the level of the population requires a population state, the *p*-state of Metz and Diekmann. Since a population is a collection of individuals moving through their life cycles, responding to and influencing their environments, and because the additions to and losses from

---

[2]In deterministic systems it is the behavior that must be determined, in stochastic systems the probability distribution of behavior. The concepts are fundamentally similar in the two cases.

a population come from the reproduction, mortality, immigration and emigration of the individuals that comprise it, we can hope to derive the p-state from the information about i-state dynamics.[3]

The most direct and comprehensive way to construct a p-state is to recognize that a population is a system of interacting individuals and that the state of the population can thus be determined from the i-states of the individuals and a constraint function (Caswell et al. 1972) specifying the interactions among individuals. For reasons that will become clear later, we refer to this p-state as an i-state configuration. Consider, for example, a population of trees, where the i-state of a tree consists of its size and vigor. The i-state, together with the individual's environment (say, available light and water), provides all the information available on mortality, growth, and changes in vigor. An i-state configuration for a population of trees is a list giving the size and vigor of each individual, plus a constraint function specifying the environment of each individual as a function of the i-states of the individuals with which it interacts. If interactions are determined by spatial proximity, the i-state configuration would consist of the i-states and the spatial locations of each individual.

An important simplification occurs when all individuals in the population experience the same environment (we will refer to this as mixing). In this case, all individuals with the same i-state will have the same dynamics and can be treated collectively. The response of the population can then be predicted from the number of individuals in each i-state; we refer to this p-state variable as an i-state distribution. Given the assumption of mixing, age as an i-state leads to the age distribution as a p-state; size leads to the size distribution, and so on.

The i-state configuration contains much more information than the i-state distribution. Suppose individuals are grouped into discrete age classes 1, 2, 3, ..., k. The i-state distribution (the familiar age distribution) is given by the k-vector n, where n_i is the abundance of the ith age class. The state space is thus k-dimensional Euclidean space R^k (or perhaps the nonnegative cone of R^k). But suppose that the p-state is a configuration of age-classified individuals in two-dimensional space. The i-state configuration for N individuals is then a function mapping R^2 (the spatial coordinates of the individuals) to $\{1, 2, ..., k\}$ (the set of all i-states). The p-state space is the set of all possible such mappings, for all values of N (cf. Metz this volume).

---

[3]We suppose that the behaviors of interest at the population level can be specified in terms of the i-states. This may not be the case. Suppose that age is an adequate i-state variable for some species, and that population biomass is the response variable of interest. Age is an adequate p-state variable only if individual size is determined by age. If it is not, information on size must be included in the population model even though it is not strictly necessary at the individual level (Metz and Diekmann 1986).

The *i*-state distribution can be obtained from the *i*-state configuration, but not vice versa.

Construction of a population model requires a function specifying the dynamics of the *p*-state as determined by the current *p*-state and the environment. There are three main mathematical frameworks for such models. If the *p*-states and time are both discrete, the dynamics are modeled with population projection matrices (Caswell 1989). If the *p*-states are discrete but time is continuous, delay-differential equations are used (Nisbet and Gurney 1982). Finally, if both the *p*-states and time are continuous, the models are formulated as partial differential equations (Metz and Diekmann 1986). For each of these mathematical formulations, there are well-developed procedures for deriving the *p*-state population models from the *i*-state descriptions and powerful mathematical tools (differential equations, matrix algebra, stability theory, etc.) available for their analysis.

These models are what Huston et al. (1988) call "state variable," as opposed to "individual-based" models. The important distinction, however, is not between "individual-based" and "state-variable" models, because any population model must be based on a population state variable. And all demographic models are based, one way or another, on information about individuals. Instead, the important distinction is between two ways of obtaining population states from individual states—one using the complete individual configuration, the other reducing that configuration to a distribution function. The relation between our terminology and that of Huston et al. (1988) is:

This paper	Huston et al. (1988)
i-state configuration models	individual-based models
i-state distribution models	state variable models

The Individual in Demographic Models: Some Examples

We turn now to some examples, in no sense exhaustive, of demographic models and data analyses which display the transition from the individual to the population.

Branching Processes and Stable Population Theory

Stable population theory is used to project the population consequences of patterns of age- or stage-specific vital rates. Matrix population models (Leslie 1945, Caswell 1989) are one familiar version; a square nonnegative matrix A is used to project a stage-distribution vector n according to

$$n(t + 1) = An(t) \tag{1}$$

The matrix entries a_{ij} can be obtained from a directed graph representation of the movement of individuals through the stages of the life cycle. Once this individual information is incorporated into a matrix, it can be used to calculate the asymptotic rate of population growth, the stable population structure, reproductive value, transient behavior, etc. The entries of A can be made time-varying, stochastic, or density-dependent, albeit at the cost of increased analytical difficulty.

Equation (1) is an archetypal i-state distribution model; the matrix A explicitly maps the stage distribution n from t to $t + 1$. A closer look, however, reveals the i-state configuration model, in which each individual is followed, from which it is derived; such a model is called a multi-type branching process.

The simplest branching process model considers only one type of individual. It was one of the earliest stochastic population models, introduced over a century ago to study a problem in conservation biology—the extinction of surnames of English families (Watson and Galton 1874; for reviews of branching processes see Harris 1963, Athreya and Ney 1972).

Consider a semelparous organism with fixed generation length (this reduces the complexity of the i-state space to a minimum). All individuals are identical, and at reproduction each produces offspring, independently of the others, according to the probability distribution

$$P[\text{producing } k \text{ offspring}] = p_k \quad k = 0, 1, 2, \ldots \tag{2}$$

It would be easy to develop and program an individual-based simulation model of this population using the i-state configuration. In fact, what is now known as the Monte Carlo simulation method, which is precisely an i-state configuration simulation approach, was developed by Ulam and Von Neumann at Los Alamos in 1947 to study branching process models for the demography of neutrons—a situation in which the question of population explosion is literal, not figurative (Eckhardt 1989).

Standard treatments of branching processes begin by showing that, given the stated assumptions, the sequence of population sizes $N(t)$, $t = 0, 1, 2, \ldots$ is a Markov chain. That is,

$$P[N_{t+1} = k | N_t, N_{t-1}, \ldots, N_0] = P[N_{t+1} = k | N_t] \tag{3}$$

In other words, knowing N_t provides all the information there is to be known about N_{t+1}. This shows that the i-state distribution N_t is a p-state variable for this process.

N_t is a random variable. Its stochastic dynamics depend on the distribution of offspring number. Let $f(s)$ be the probability generating function of the distribution p_k:

$$f(s) = \sum_0^\infty p_k s^k, \quad |s| \le 1 \tag{4}$$

where s is a complex variable. Let $g_t(s)$ denote the probability generating function of N_t. This generating function satisfies a recursive relationship

$$g_t(s) = g_{t-1}(f(s)) \tag{5}$$

Given an initial condition N_0, (5) can be iterated to find the probability distribution of N_t for all t. The derivation of (5) involves writing the offspring produced at time t as a sum $\xi_1 + \xi_2 + \ldots + \xi_{N_t}$, where ξ_i is the number of offspring produced by the ith individual present at t. The crucial step is the assumption that the ξ_i are independently and identically distributed random variables from the distribution p_k. This is, in this case, the operational version of the mixing condition—that each individual experiences the same conditions as every other.

This approach can be generalized to time-varying environments, in which the p_i change with time, and to nonlinear models, in which the p_i are functions of N_t; in these cases the recursion relation (5) becomes a time-varying or nonlinear relation.

Explicit calculation of the generating function $g_t(s)$ by solving (5) is generally impossible. However, the expected population size can be found easily. Suppose that $f(s)$ does not change with time. Then

$$E[N_t] = g_t'(1) \tag{6}$$
$$= (f'(1))^t \tag{7}$$
$$= (R_0)^t \tag{8}$$

where R_0 is the mean number of offspring produced per individual and the primes denote differentiation with respect to s. Thus a constant environment produces exponential growth of the mean population at a rate given by the net reproductive rate. The extinction probability can also be calculated (Harris 1963, Athreya and Ney 1972), which is what Galton wanted in the first place.

Now consider the multi-type case, where the individuals are not identical. Instead, any finite number of types may be distinguished. The demography of an individual of type j is specified by a multivariate probability distribution $p^{(j)}(r_1, r_2, \ldots, r_k)$ which gives the probability that an individual of type j produces r_1 offspring of type 1, r_2 offspring of type 2, ..., and r_k offspring of type k. (The production of "offspring" may be the result of reproduction, of growth, of survival in the same state, etc.) All individuals are assumed to behave independently. The simplest assumption is that the probabilities $p^{(j)}$ are time-homogeneous, although this can be relaxed.

Again, it is not difficult to imagine an individual-based simulation of this population. Such a simulation is not necessary, however, since the properties of the population can be analyzed by straightforward generalizations of the analysis for the single-type process (although the notation does become more cumbersome). In particular, the sequence $n(t)$, $t = 0, 1, 2, \ldots$, where n is a vector giving the number of individuals of each type, can be shown to be a Markov chain. Thus the distribution of individuals into the different types is a p-state corresponding to the i-state model specified by the $p^{(j)}$. In terms of the probability generating functions

$$f^{(j)}(s) = \sum_{r_1} \sum_{r_2} \cdots \sum_{r_k} p^{(j)}(r_1, r_2, \ldots, r_k) s_1^{r_1} s_2^{r_2} \ldots s_k^{r_k} \qquad (9)$$

the mean process is given by

$$E[n(t + 1)|n(t)] = An(t) \qquad (10)$$

where the entries of the matrix A are the means

$$a_{ij} = \left. \frac{\partial f^{(j)}}{\partial s_i} \right|_{s=(1,\ldots,1)} \qquad (11)$$

Equation (10) is the familiar population projection matrix model (1); we have identified it as the expected value operator for a multi-type branching process.[4] The life cycle graph approach to constructing projection matrices (Hubbell and Werner 1979, Caswell 1982, 1989) is a simple way to specify directly the mean transition rates in (10).

Thus, lurking behind the well-known population projection models of demography is a family of stochastic descriptions of individuals. Given the mixing assumptions, those individual descriptions translate directly into the familiar i-state distributions without loss of information.

Micro-Simulation in Human Demography

In this section, we consider a tradition in human demography which uses explicit i-state configuration models, rather than relying on the reduction of these models to i-state distribution form. The emphasis in human demography since World War II has been on i-state transitions, following individuals through the life cycle (birth, sexual maturity, marriage, widowhood, migration, death), or through portions of the life cycle such as the female reproductive period (fecund, pregnant, post-partum infecundable). These studies are generally referred to as "micro-demography." They often involve

---

[4]A more elaborate and sophisticated treatment of these issues has been produced by Mode (1985) using age-dependent multi-type branching processes.

drawing population inferences from detailed data on individuals. Some micro-demographic models are merely particularly detailed *i*-state distribution models (e.g., Keyfitz 1977, Sheps and Menken 1973, Bongaarts 1977, John 1990), but human demographers have also made extensive use of *i*-state configuration models, which they refer to as micro-simulation models.

Since their introduction in the 1960's (Sheps 1967, Perrin 1967), micro-simulation models have been used extensively in the study of human reproduction and marriage. In these studies, reproductive histories of individual women are generated according to distribution functions defining transition probabilities between *i*-states. Once the individual reproductive histories have been generated, aggregation of the "data" yields the descriptive reproductive measures for cohorts (e.g., age specific or parity-specific fertility rates) and for the population (e.g., the total fertility rate or net reproduction rate).

> The actual simulation procedure for this model, written for the IBM-7094 computer, consists, of course, of putting an individual woman through the process for a given number of months, randomly selecting at the proper point and according to the specified distributions both the state to which the individual will pass next and the length of stay in that state. This process is repeated for a specified number of women . . . The output for the simulated cohort of females is designed to provide the usual information of demographic interest, together with some optional features. (Perrin 1967, p. 139)

REPMOD (Bongaarts 1977, Bongaarts and Potter 1983) is an example of a human reproduction simulation model, in which the possible *i*-states for a woman in a reproductive union are: fecundable, pregnant, post-partum infecundable, and sterile, possibly in combination with age and lactational history (Figure 3.2). The model is a semi-Markov process, in which the dynamics of the *i*-state distribution are determined by the waiting time distributions for each state transition. These distributions may be based on model output (e.g., the age at marriage is determined by the Coale (1971) marriage model), or hypothesized distribution functions (e.g., the waiting time to conceive has a geometric distribution, the duration of post-partum infecundability has a Pascal distribution, and the distribution of post-abortion infecundability has a geometric distribution). The addition of heterogeneity is straightforward: for example, the distribution of the waiting time in the fecund state may be heterogeneous with respect to woman's age and lactational history.

The output of REPMOD includes age-specific birth rates, birth interval distributions, and parity distributions. Since these rates and distributions are typical of the sort of data which can be collected from real populations, comparison of model aggregate output with data serves as an indication of how well the underlying (and often unmeasurable) individual processes are

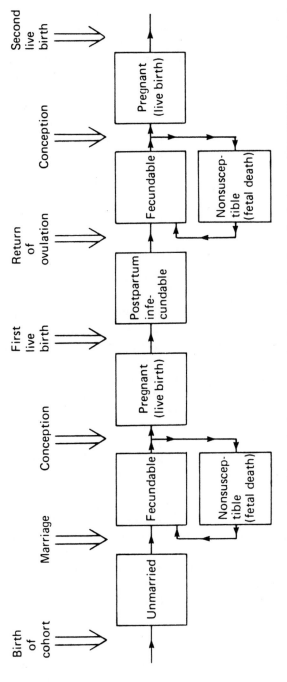

Figure 3.2. An *i*-state transition diagram for human reproduction from the micro-simulation model REPMOD (Bongaarts and Potter 1983).

captured by the model. A well-behaved micro-simulation model also permits the testing of the impact of individual behavioral variation on population aggregates; models of this sort have proven useful in attempts to predict the effect of changing age patterns of marriage or family building patterns (number and timing of births) on human reproduction patterns at the population level. For example, such models have demonstrated that China can achieve the same long-run target levels of population growth either by maintaining a one-child per couple policy, or by adopting a two-child policy with the mother's age at first birth relatively late and the children widely spaced, so that the mean generation length is increased.

Individuals in REPMOD proceed through their i-states without interaction. Thus the mixing condition is satisfied and the model can be analyzed as an i-state distribution model. In fact, the REPMOD simulation programs are written to permit either mode of analysis.

When interactions are important, individuals in the same i-state may experience different conditions, depending on their interactions. In such cases, micro-simulation models must use the complete i-state configuration. For example, Hammel et al. (1979) analyzed the demographic consequences of incest tabus in small human populations. They hypothesized that restricting marriage because of relatedness might significantly limit fertility, and that the effect at the population level should vary with population size. To test this hypothesis, they used a micro-simulation package (SOCSIM) in which individuals were characterized by age and sex and in which genealogical information was recorded to permit evaluation of relatedness. Marriages were generated in the model only between individuals eligible to marry under the incest rules being evaluated. Several hundred replicate simulations were performed for each of several population sizes and four patterns of incest tabu. The results showed that incest tabus could significantly restrict marital fertility, and that the effect decreased rapidly as population size increased.

MacCluer and Dyke (1976) used a similar approach to study the minimum viable size of human populations. Their model followed individual men and women characterized by age and relatedness. Marriages between siblings and between parents and offspring were prohibited, and restrictions were placed on the age differences between mates. Three mating patterns (monogamy, polygamy, monogamy with restrictions on first cousin marriage) and two levels of fertility and mortality were used. Populations were simulated for 200 years from each of several initial population sizes. Over this time scale, the minimum viable population size appeared to be fewer than three hundred, and perhaps fewer than one hundred individuals.

Epidemic Models

In models of aerogenically transmitted infectious disease, individuals move through a series of epidemiologic states (i-states) of variable duration. The distribution of individuals among these states at any point in time gives the *i*-state distribution. The dynamics of this distribution are derived from hypotheses about the rates of movement of individuals among the *i*-states.

In the simplest deterministic model (Kermack and McKendrick 1927, Hethcote 1989) the host population is divided into three classes: susceptible (*S*), infectious (*I*), and recovered, hence immune (*R*). The host population is assumed constant, *N*, so at all times $S(t) + I(t) + R(t) = N$. If *b* denotes the crude birth rate and μ denotes the crude death rate, then bN gives total births and μN gives total deaths. Since the population is stationary, the expectation of life at birth is $1/\mu$. These relationships constitute the demographic elements of the standard epidemiologic model.

It is usually assumed that the disease is propagated through a "catalytic" or "mass action" transmission process in which infections are acquired at a rate proportional to the number of encounters between susceptible and infectious individuals. The rate at which individuals pass from the infected class to the recovered class is constant and denoted γ. The model is written as a system of differential equations:

$$\frac{dS(t)}{dt} = \mu N - \mu S(t) - \beta S(t)I(t)$$

$$\frac{dI(t)}{dt} = \beta S(t)I(t) - (\mu + \gamma)I(t) \tag{12}$$

$$\frac{dR(t)}{dt} = \gamma I(t) - \mu R(t).$$

Summing the three equations yields $dN/dt = 0$, satisfying the assumption that population size is constant.

This simple model is typical of models of infectious disease transmission dynamics for developed countries where the assumptions of constant population size and negligible mortality until old age are credible, particularly if one is considering epidemics which occur on a shorter time scale than that required for significant changes in a slowly growing population.

The *i*-states in this model include no demographic detail. Expanding the *i*-state to include age, sex, or other demographic variables is an important problem. One approach (John 1990a,b, Tuljapurkar and John 1991) incorporates a demographic process and an epidemiologic process connected by a pair of link functions. The demographic system governs the dynamics of the age distribution $N(a, t)$, while the epidemiologic system governs transitions of individuals among the epidemiologic classes at each age: $S(a, t)$, $I(a, t)$ and $R(a, t)$. Thus the *p*-state of the model is a joint distribution of individuals among age and epidemiologic classes. In the absence of disease in the population, the simulation model reduces to a population projection model based on a Leslie matrix.

Analytical solution of these demographic-epidemiologic systems for the equilibrium population structure and the equilibrium force of infection is difficult. In addition, perturbation results for the effects of changes in the demography (i.e., the fertility or mortality schedules) or the epidemiology (e.g., changes in immunization schedules) cannot readily be found. Given these analytical difficulties, i-state configuration simulations might be an effective approach to the study of these complex models. Such simulations have been used by Ackerman et al. (1984) to study structurally simpler models of polio and influenza.

A second extension of the basic epidemiologic models is to situations in which infection spreads only to neighboring individuals. In such case i-state configuration models are a necessity, not a convenience, because the assumption that all individuals in a given i-state experience the same environment is obviously violated. Introduced by Bailey (1967) and Mollison (1977), and termed variously "contact epidemic processes" or "general epidemic models," these models have received more attention from mathematicians and physicists (e.g., Grassberger 1983, MacKay and Jan 1984, von Niessen and Blumen 1986, Durrett 1988, 1989) than from ecologists or epidemiologists.

Imagine an infinite d-dimensional integer lattice Z^d, at each point of which is an individual which may be healthy and susceptible (S), infected (I), or recovered (R). Let x denote a point on the lattice, let $\xi_t(x)$ denote the state of the individual at point x at time t, let k denote the number of immediate neighbors of x that are infected and let p denote the probability of transmitting the infection to a healthy neighbor. The dynamics are simple:

$$
\begin{aligned}
\xi_t(x) = R &\Rightarrow \xi_{t+1}(x) = R \\
\xi_t(x) = I &\Rightarrow \xi_{t+1}(x) = R \\
\xi_t(x) = S &\Rightarrow
\begin{cases}
P[\xi_{t+1}(x) = S|\xi_t(x)] = (1-p)^k \\
P[\xi_{t+1}(x) = I|\xi_t(x)] = 1 - (1-p)^k
\end{cases}
\end{aligned}
\tag{13}
$$

This formulation assumes that infection lasts for only a single time step and that there is no latent period; both assumptions may be relaxed, as may be the assumption of discrete time (Durrett 1989). In any case, Eqs. (13) is a stochastic, spatially localized version of the deterministic model (12).

The states of individuals in this model are clearly the same as those for the individuals in Eqs. (12). The p-state, however, is no longer the vector $[S(t), I(t), R(t)]$. Instead it is a function $\xi_t : Z^d \rightarrow \{S, I, R\}$ (Durrett 1988); that is, a list of the i-states of the individuals at each location in space. Thus this is an archetypal i-state configuration model.

Remarkably little is known about these systems, and little of that is known rigorously. Most of the available results rely on the equivalence of the contact epidemic model and percolation models[5] (Durrett 1989, Grimmett 1989). Thus there is a threshold value p_c for the probability of infection such that when $p < p_c$ the infection dies out with probability 1, and when $p > p_c$ a single infected individual gives rise, with probability 1, to an infection which spreads infinitely far. The value of p_c is known for only a few special cases.

There is a phase transition at $p = p_c$, and physicists have made much of analogies to statistical mechanics, noting that such quantities as the percolation probability, the mean size of the infection arising from a single individual, and the number of infected particles per individual scale allometrically with p in the neighborhood of p_c. They conjecture that the exponents describing these scaling relationships are universal; i.e., their value is independent of the details of the model, depending only on the dimensionality of the lattice. Mathematicians seem fond of pointing out that none of this is proven (Grimmett 1989).

Considerable attention has been devoted to theorems describing the asymptotic shape and growth rate of an infection beginning from a single infected individual (Durrett 1988, 1989). For $p > p_c$, the shape is convex and grows linearly in time, as $t \to \infty$. Computer simulations on a 2-dimensional lattice (Durrett 1988) show this clearly. As $p \downarrow p_c$, however, the smooth convex shape is very slow to appear. There seems to be a correlation distance, below which the spatial pattern of infection is fractal, with a dimension less than 2. This distance seems to scale roughly as $(p - p_c)^{-2}$. Thus, close to the threshold probability, spatial pattern s will remain irregular until the infection has spread over a large area.

The analysis of these models has challenged some outstanding mathematicians, the results are published in journals of probability theory, physics, and mathematics, and the model itself (identical individuals living on the points of a regular, infinite, and homogeneous lattice passing infection only to their immediate neighbors) is absurdly simple. Durrett (1988) notes that most of the available results are qualitative rather than quantitative, and that they do not lend themselves to computation. It is a sobering thought that such difficulties may be generic to *i*-state configuration models.

Interactions among individuals are not limited to spatial neighbors. More detailed *i*-state configuration models, along the lines of the micro-simulation models used to study human fertility, or of the epidemic models of Ackerman et al. (1984) would be appropriate for modeling diseases whose trans-

---

[5]Percolation models (Frisch and Hammersley 1963) originated as idealizations of fluid flow through porous media. Imagine a lattice of sites, with lines or "bonds" connecting the sites. Each bond is open with probability p and blocked with probability $1 - p$. A liquid introduced at a site(s) can flow only along open bonds. Percolation is said to occur if a liquid introduced at a single site spreads through the entire medium. This model is referred to as bond percolation; there is a corresponding site percolation model in which all bonds are open and sites may be blocked or not.

mission is influenced by interactions among household members or sexual partners (e.g., Barrett 1988).

Population Inference From Individual Data: Hazard Analysis and Related Topics

The construction of demographic models necessitates inferring p-state parameters from data on i-state transitions. Huston et al. (1988) emphasize the ease of collecting individual data as a benefit of the i-state configuration approach. However, the transition rates which form the basis of an i-state distribution model can also be obtained from data on individuals. One of the most powerful analytical tools is hazard analysis (e.g., Cox and Oakes 1984; for applications see Trussell and Hammerslough 1983, John et al. 1987, John 1988).

Individual data generally take one of two forms: cross-sectional (also called current status, or period) data, in which information about the individual's state at a single time is recorded, or prospective data, in which individuals are followed for an extended period and their movements among states recorded. For humans, there is a third type of data—retrospective history data— in which the life history of the individual for a specified period prior to the interview is reported at the survey interview. An example is the retrospective maternity history collected for individual women in the World Fertility Surveys (Cleland and Scott 1987). These surveys were conducted in 44 developing countries in the late 1970s and early 1980s; in each country a standardized survey interview was used to record reproductive histories for approximately 5,000 women.

The p-state parameters that one wishes to estimate from these data are rates or probabilities (of survival, of reproduction, of conception, etc.). The data consist of individual transitions. In essence, hazard models are regression models in which the dependent variable is the rate at which the transition occurs, known as the hazard. For example, for a reproduction model, the dependent variable may be the monthly rate of conception.

Formally, let $f(t)$ denote the probability of the event happening at time t and let $F(t)$ denote the cumulative probability of the event happening before time t; i.e., $F(t) = \int_0^t f(s)ds$. Then the hazard function is $f(t)/(1 - F(t))$; it gives the risk of the event occurring at t, given that it has not yet occurred. This formulation is familiar to ecologists in the context of life table analysis, where $f(t)$ is the distribution of age at death, $1 - F(t)$ is the survivorship curve, and the hazard is the mortality rate $\mu(t)$. However, we emphasize that the use of hazard analysis is not restricted to mortality; to illustrate we will use as an example the transition from fecund to pregnant.

There are several ways of formulating hazard models; their common feature is the expression of the hazard as a function of a linear combination of explanatory variables. Let λ denote the hazard, z the vector of variables, and β a vector of coefficients. Then the hazard is written as $\lambda(\beta'z)$; the coefficients in β describe the relationship between the covariates and the hazard (e.g., the relation between lactation and the risk of conception).

Let $\lambda_i(t)$ denote the hazard for individual i at time t (e.g., the probability of individual i conceiving at time t conditional on not yet having conceived by time t). The two most common forms relating λ_i to the covariates z_i are the loglinear form

$$\lambda_i(t) = e^{\beta'z_i} \tag{14}$$

and the logistic form, which uses the log-odds of the hazard as the dependent variable in a linear model

$$\ln \frac{\lambda_i(t)}{1 - \lambda_i(t)} = \beta'z_i \tag{15}$$

or equivalently,

$$\lambda_i(t) = \frac{e^{\beta'z_i}}{1 + e^{\beta'z_i}} \tag{16}$$

These forms always result in nonnegative hazards. If the hazard were modeled simply as a linear function of z_i, some combinations of β and z_i would yield an estimated hazard less than zero, which is inconsistent with the definition of a hazard. While (14) and (16) appear quite different, the estimated coefficients are similar when the hazard $\lambda_i(t)$ being estimated is small (Allison 1982, Foster et al. 1986).

The hazard model is a particularly powerful tool because it can incorporate censored data; i.e., data in which the fate of some individuals is not observed. For example, a group of fecund women may be observed either until they conceive or until the survey ends; at the end of the survey the only information available about some women is that their wait was at least as long as the period of observation. Excluding these women from the analysis biases the sample to those with the shortest conception waits. However, in the hazard model, these women can be included as censored observations, thus rendering the sample unbiased. The models can also incorporate time-varying covariates; e.g., a fecund woman may move from the lactating to non-lactating state while she waits to conceive, and this transition may have a significant impact on her hazard.

Hazard models yield estimates of population level transition rates, but they are estimated from event history data on individuals. The data consist of information on whether the event has happened, and on covariates describing the individual (e.g., age, sex) and her environment (e.g., population density). These individual results and the population-level hazard model are connected by the likelihood function. The likelihood of the model given the data is proportional to the probability of the data given the model. The parameters in the model are estimated by maximizing the likelihood function. This usually requires iterative calculation, but the necessary routines can be found in the survival analysis routines in such software as SAS, STATA, and BMDP.

Using hazard analysis, complex information on i-state transition rates can be obtained from individual data. John (1988), for example, used a logistic hazard model to examine the factors determining conception rate (following a birth) in a sample of Bangladeshi women. The data, obtained in a prospective study, consisted of monthly interviews with 403 individual women over a period of five years. Covariates included time since resumption of menses, age, religion, weight, season, proportion to the month the couple are separated, husband's occupation, and the duration and extent of breast-feeding. The results demonstrated that time since resumption of menses, age, and the proportion of the month separated all had significant effects, but that season, socio-economic variables, and body size had little effect. Lactation effects were of particular interest, and the analysis showed that they had significant effects, reducing the probability of conception. Not only did the results indicate the significance of the covariates, but also provided a set of equations from which the conception rates could be calculated as functions of the covariates.

When the *i*-State Distribution is not Enough

Dynamic i-state distribution models are the fundamental tools of demographic theory, which relies on them to link the properties of individuals and the dynamics of populations. They produce equations (ordinary differential, partial differential, or difference) which can be attacked by a variety of powerful mathematical tools. The attack may be resisted, but if successful its results are more powerful and more general than the simulation of i-state configuration models. For example, simulation of individuals following a multi-type branching process might suggest that population size eventually increases exponentially. Repeating the simulation many times might suggest that this always happens, and that the rate depends on the stage-specific vital rates. But branching process theory provides powerful general results that go beyond those obtained from simulations (e.g., if the dominant eigenvalue of the expected value matrix A is greater than one, the population converges with probability 1 to a random vector proportional to the dominant right eigenvector w_1 of A). Simple matrix algebra applied to A reveals the conditions (primitivity of the expected value matrix) under which this asymptotic growth pattern appears, what happens when those conditions are not fulfilled, and how to calculate the eventual rate of increase. These conclusions could not be reached by simulation alone, no matter how elaborate or sophisticated. This power is not something to be lightly discarded, even though it may require approximations in the development of the models.

Be that as it may, there are at least three situations in which i-state distribution models become intractable or impossible, and in which a direct attack on the i-state configuration may be the only possible tactic. These situations are: complicated i-states, small populations, and local interactions.

As the complexity of the i-state increases, the number of dimensions in the i-state distribution may become inconveniently large. Simulation of an i-state configuration model may become easier than numerical solution of a complicated i-state distribution model. There is a trend in population ecology toward incorporating more detailed individual information in models (e.g., the work of the inter-continental *Daphnia* modeling conspiracy, the usual suspects in which include Kooijman (1986), Hallam et al. (1990), McCauley et al. (1990), and Gurney et al. (1990)). Combined with increased computer power, this trend will probably lead to increased use of i-state configuration simulations. The difficulties encountered by John (1990a,b) and Tuljapurkar and John (1991) in the analysis of age-classified epidemic models suggests that an i-state configuration analysis might be useful here as well.

It is widely appreciated that the dynamics of small populations may be strongly affected by demographic stochasticity. Because population projection models are expected value operators for underlying multi-type branching process models, their accuracy declines when population size is small. The underlying branching process models, which include demographic stochasticity, are i-state distribution models, just as the expected value models are. However, their analysis may be difficult enough that i-state configuration simulations become an attractive approach to their study. Interest in demographic analysis of threatened or endangered species, and questions of the viability of small populations, will likely lead to increased use of i-state configuration simulations in this context.

These uses of i-state configuration models are practical conveniences rather than theoretical necessities. There are, after all, no theoretical limits on, say, the size of a population projection matrix. The situation is different when the assumption of mixing fails. In this case, i-state configuration models play a fundamental role. If the failure of mixing results from simple spatial heterogeneity, where individuals in different locations experience different environmental conditions, it can be remedied by enlarging the i-state to include location. But if the heterogeneity results from local interaction of individuals, the entire configuration of the population is fundamentally required to predict its dynamics. The interactions may be in geographical space, as in the generalized epidemic models and their percolation theory counterparts, or in some more abstract space, as in the kinship-based micro-simulation model of Hammel et al. (1979). In either case, Huston et al. (1988) are correct to emphasize the importance of local interactions as a problem leading to *i*-state configuration models.

Farmer (1990) draws a potentially useful distinction between fixed and dynamic interaction structures in a recent paper on 'connectionist' models, which

> "consist of elementary units which can be 'connected' together to form a network. The form of the resulting connection diagram is often called the architecture of the network . . . Since the modeler has control over how the connections are made, the architecture is plastic. This contrasts with the usual approach in dynamics and bifurcation theory, where the dynamical system is a fixed object whose variability is concentrated into a few parameters . . . Dynamics occurs on as many as three levels, that of the states of the network, the values of connection strengths, and the architecture of the connections themselves." (Farmer 1990, p. 154)

Farmer cites neural networks, classifier systems, immune networks, and autocatalytic chemical reaction networks as examples. These are differentiated from models like cellular automata, in which the units interact, but the interactions are "fixed, completely regular, and have no dynamics." *i*-state configuration models in which the interactions themselves evolve (e.g., due to relatedness, household structure, or pair formation) may share some properties with other connectionist models.

Parting Thoughts

We conclude with some thoughts about future uses and future problems with *i*-state configuration models.

First, *i*-state configuration models may be useful for estimating the parameters in *i*-state distribution models. Suppose that the set of *i*-states is large and the rules for transitions among them complicated. An *i*-state configuration model could provide "data" from which the waiting time distributions and transition probabilities could be estimated using hazard models. Or, the probabilities calculated from the individual-based simulation could be used to derive maximum likelihood estimates of parameter values (A. Solow, pers. comm.).

Second, most analyses of most models are interested in asymptotic behavior. Since *i*-state configuration models are usually stochastic, analyzing and interpreting asymptotic behavior will require extra care because results on convergence of stochastic processes are generally more subtle than the results for the convergence of deterministic series.

Stochastic matrix population models provide an excellent example. The asymptotic behavior of time-invariant models is unambiguously described by the dominant eigenvalue of the projection matrix. The demographic strong ergodic theorem guarantees that every initial population will eventually grow at this rate. If the vital rates vary stochastically, however, population size is a random variable, and one must choose how to describe its asymptotic properties. One can calculate the asymptotic growth rate of the mean population size

$$\ln \mu = \lim_{t \to \infty} \frac{1}{t} \ln E[N(t)] \tag{17}$$

or the mean population growth rate

$$\ln \lambda_s = \lim_{t \to \infty} \frac{1}{t} E[\ln N(t)] \tag{18}$$

where $N(t) = \Sigma_i n_i(t)$ denotes total population size. The interesting part is that these two rates are in general not equal; $\ln \lambda_s \leq \ln \mu$, with strict inequality in general. Which rate to choose? The answer is provided by the ergodic theory of random matrix products (Furstenburg and Kesten 1960, Cohen 1976a, Tuljapurkar and Orzack 1980; see reviews by Tuljapurkar 1989, 1990), which proves that, under a wide array of conditions,

$$\lim_{t \to \infty} \frac{1}{t} \ln N(t) = \ln \lambda_s \tag{19}$$

with probability 1. That is, except for a set of measure zero, every realization of this process eventually grows at the rate $\ln \lambda_s$, even though the mean population size is growing at the faster rate $\ln \mu$. Thus it is perfectly possible to have the mean population size increasing exponentially ($\ln \mu > 0$) while the probability of extinction is approaching 1 ($\ln \lambda_s < 0$). Without the guidance of the ergodic theory, it is doubtful that simulation analysis (e.g., Boyce 1977) would uncover the true nature of the asymptotic behavior of the model.

For an individual-based example, consider a model of the gregarious settlement of larval benthic invertebrates. Suppose there is a finite number of patches in which larva may settle, and that the attractiveness of a patch is directly proportional to the number of settled individuals in that patch. Begin with all patches equal (say, with a single individual in each) and follow the settlement of a population of larvae. Figure 3.3a shows the results of a simulation of 1000 individuals for the simple case of two patches. The occupancy of the two patches converges to proportions of 0.6334 and 0.3666.

Since this is a stochastic simulation, we had better replicate it. Figure

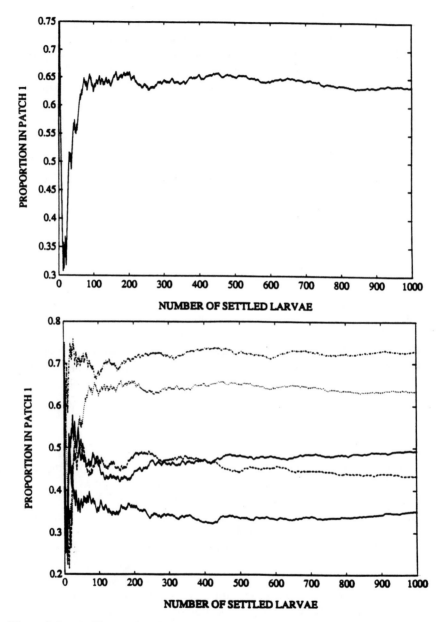

Figure 3.3.   (a) The results of a simulation of gregarious settlement into two patches by individual planktonic larvae of a benthic invertebrate. There are two patches; the proportion of larvae settling in Patch 1 is plotted against time. (b) The results of five replicates of the experiment in (a).

3.3b shows the results of five replicates. Each converges to a fixed proportion of settled larvae in each patch, but the proportion is different each time. This model, introduced to population biologists by Cohen (1976b), is an example of a Polya urn scheme (Eggenberger and Polya 1923). It has the surprising property that, with probability 1, every realization converges to fixed proportions in each patch, but that the proportions are "nondegenerate limit" random variables, uniformly distributed between 0 and 1 (or, for k patches, uniformly distributed on the unit simplex). See Arthur et al. (1987) for an application to the problem of industrial firms choosing locations; the analogy to larval settlement is not coincidental.

These examples underscore the likelihood that study of the asymptotic properties of stochastic i-state configuration models will require concepts and methods that go beyond the stability analyses so useful in studying similar deterministic models.

Conclusions

The individual plays an essential role in demographic models. Choosing a p-state variable is the crucial step in constructing a population model from individual information. Demographic theory has been developed using the i-state distribution as a p-state variable. There are powerful methods, including hazard models, for estimating the parameters in these models from data on individual organisms, and equally powerful analytical techniques for analyzing the models. However, when the i-state space is very complicated, simulation of i-state configuration models may be an alternative to the solution of the dynamic equations for an i-state distribution model. More fundamentally, when local interactions (in either geographical or some other, more abstract space) are important, the mixing assumption required for i-state distribution models fails. In these cases, i-state configuration models are a necessity, not a convenience. As models of individual organisms become more detailed, and computers become more powerful, i-state configuration models will become increasingly useful in demography.

Acknowledgements

We are grateful for numerous comments from other participants in the symposium, which helped to clarify the issues discussed here. The importance of the ideas of J. A. J. Metz should be obvious, but deserves special acknowledgment here. This research was supported by NSF Grants OCE-8900231 and BSR-8704936 and DOE Grant DE-FG02-89ER60882 to H.C. and NIH Grant AI29418 to A. M. J. Woods Hole Oceanographic Institution Contribution 7415.

Literature Cited

Ackerman, E., L. R. Elveback, and J. P. Fox. 1984. *Simulation of Infectious Disease Epidemics*. Charles C. Thomas, Springfield, IL.

Allison, P. 1982. Discrete time methods for analyses of event histories. In *Sociological Methodology*, S. Leinhardt (ed.) pp. 61–98, Jossey-Bass, San Francisco.

Arthur, W. B., Y. M. Ermoliev, and Y. M. Kaniovski. 1987. Path-dependent processes and the emergence of macro-structure. *Eur. J. Operat. Res.* **30**:294–303.

Athreya, K. B. and P. E. Ney. 1972. *Branching Processes*. Springer-Verlag, New York.

Bailey, N. T. J. 1967. The simulation of stochastic epidemics in two dimensions. *Fifth Berkeley Symp. Math. Stat. Probab.* **4**:237–257.

Barrett, J. C. 1988. Monte Carlo simulation of the heterosexual selective spread of the human immunodeficiency virus. *J. Med. Virol.* **26**:99–109.

Bongaarts, J. 1977. A dynamic model of the reproductive process. *Popul. Stud.* **31**:59–73.

Bongaarts, J. and R. Potter. 1983. *Fertility, Biology and Behavior: an Analysis of the Proximate Determinants*. Academic Press, New York.

Boyce, M. S. 1977. Population growth with stochastic fluctuations in the life table. *Theor. Popul. Biol.* **12**:366–373.

Caswell, H. 1982. Stable population structure and reproductive value for populations with complex life cycles. *Ecology* **63**:1223–1231.

Caswell, H. 1989. *Matrix Population Models: Construction, Analysis, and Interpretation*. Sinauer Associates, Sunderland, MA.

Caswell, H., H. E. Koenig, J. A. Resh, and Q. E. Ross. 1972. An introduction to systems science for ecologists. In *Systems Analysis and Simulation in Ecology, Vol. 2.* pp. 3–78, B. C. Patten (ed.), Academic Press, New York.

Cleland, J. and C. Scott. 1987. *The World Fertility Survey: an Assessment*. Clarendon Press, Oxford, England.

Coale, A. J. 1971. Age patterns of marriage. *Popul. Stud.* **25**:193–214.

Cohen, J. E. 1976a. Ergodicity of age structure in populations with Markovian vital rates, I: countable states. *J. Am. Stat. Assoc.* **71**:335–339.

Cohen, J. E. 1976b. Irreproducible results and the breeding of pigs (or nondegenerate limit random variables in biology). *BioScience* **26**:391–394.

Cox, D. R. and D. Oakes. 1984. *Analysis of Survival Data*. Chapman and Hall, New York.

Durrett, R. 1988. Crabgrass, measles, and gypsy moths: an introduction to interacting particle systems. *Math. Intell.* **10**:37–47.

Durrett, R. 1989. *Lecture Notes on Particle Systems and Percolation*. Wadsworth and Brooks/Cole, Pacific Grove, CA.

Eckhardt, R. 1989. Stan Ulam, John von Neumann, and the Monte Carlo Method.

In *From Cardinals to Chaos*, N. G. Cooper (ed.), pp. 131–143. Cambridge University Press, Cambridge, England.

Eggenberger, F. and G. Polya. 1923. Uber die Statistik verketteter Vorgange. *Zeit. Angew. Math. Mech.* **3**:279–289.

Farmer, J. D. 1990. A Rosetta stone for connectionism. *Physica D* **42**:153–187.

Foster, A., J. A. Menken, A. K. M. A. Chowdhury, and J. Trussell. 1986. Female reproductive development: a hazard model analysis. *Soc. Biol.* **33**:183–198.

Frisch, H. L. and J. M. Hammersley. 1963. Percolation processes and related topics. *SIAM J.* **11**:894–918.

Furstenberg, H. and H. Kesten. 1960. Products of random matrices. *Ann. Math. Stat.* **31**:457–469.

Grassberger, P. 1983. On the critical behavior of the general epidemic process and dynamical percolation. *Math. Biosci.* **63**:157–172.

Grimmett, G. 1989. *Percolation*. Springer-Verlag, New York.

Gurney, W. S. C., E. McCauley, R. M. Nisbet, and W. W. Murdoch. 1990. The physiological ecology of Daphnia: a dynamic model of growth and reproduction. *Ecology* **71**:716–732.

Hallam, T. G., R. R. Lassiter, J. Li, and L. A. Suarez. 1990. Modeling individuals employing an integrated energy response: application to Daphnia. *Ecology* **71**:938–954.

Hammel, E. A., C. K. McDaniel, and K. W. Wachter. 1979. Demographic consequences of incest tabus: a microsimulation analysis. *Science* **205**:972–977.

Harris, T. E. 1963. *The Theory of Branching Processes*. Springer-Verlag, Berlin.

Hethcote, H. W. 1989. Three basic epidemiological models. In *Applied Mathematical Ecology*, S. A. Levin, T. G. Hallam and L. J. Gross, (eds.), pp. 119–144, Springer-Verlag, Berlin.

Hubbell, S. P. and P. A. Werner. 1979. On measuring the intrinsic rate of increase of populations with heterogeneous life histories. *Am. Nat.* **113**:277–293.

Huston, M., D. DeAngelis, and W. Post. 1988. New computer models unify ecological theory. *BioScience* **38**:682–691.

John, A. M. 1988. Lactation and the waiting time to conception: an application of hazard models. *Hum. Biol.* **60**:873–888.

John, A. M., J. A. Menken, and A. K. M. A. Chowdhury. 1987. The effects of breastfeeding and nutrition on fecundability in rural Bangladesh: a hazards-model analysis. *Popu. Stud.* **41**:433–446.

John, A. M., J. A. Menken, and J. Trussell. 1988. Estimating the distribution of interval length: current status versus retrospective history data. *Popu. Stud.* **42**:115–127.

John 1990a. Transmission and control of childhood infectious diseases: does demography matter? *Popul. Stud.* **44**:195–215.

John 1990b. Endemic disease in host populations with fully specified demography. *Theor. Popul. Biol.* **37**:455–471.

Kermack, W. O. and A. G. McKendrick. 1927. A contribution to the mathematical theory of epidemics. *Proc. R. Stat. Soc. Lon. A Math. Phys. Sci.* **115**:700–721.

Keyfitz, N. 1977. *Applied Mathematical Demography*. Wiley, New York.

Kooijman, S. A. L. M. 1986. Population dynamics on the basis of budgets. In *The Dynamics of Physiologically Structured Populations*, J. A. J. Metz and O. Diekmann (eds.), pp. 266–297. Springer-Verlag, Berlin.

Leslie, P. H. 1945. On the use of matrices in certain population mathematics. *Biometrika* **33**:183–212.

MacCluer, J. W. and B. Dyke. 1976. On the minimum size of endogamous populations. *Social Biology* **23**:1–12.

MacKay, G. and N. Jan. 1984. Forest fires as critical phenomena. *J. Phys. A: Math. Gen.* **17**:L757–L760.

McCauley, E., W. W. Murdoch, R. M. Nisbet, and W. S. C. Gurney. 1990. The physiological ecology of Daphnia: development of a model of growth and reproduction. *Ecology* **71**:703–715.

Metz, J. A. J. and O. Diekmann. 1986. *The Dynamics of Physiologically Structured Populations*. Springer-Verlag, Berlin.

Mode, C. J. 1985. *Stochastic Processes in Demography and their Computer Implementation*. Springer-Verlag, New York.

Mollison, D. 1977. Spatial contact models for ecological and epidemic spread. *J. R. Stat. Soc. Ser. B* **39**:283–326.

Nisbet, R. M. and W. S. C. Gurney. 1982. *Modelling Fluctuating Populations*. Wiley, New York.

Perrin, E. B. 1967. Uses of stochastic models in the evaluation of population policies. II. Extension of the results by computer simulation. *Fifth Berkeley Symp. Math. Stat. Probab.* **4**:137–146.

Sheps, M. C. 1967. Uses of stochastic models in the evaluation of population policies. I. Theory and approaches to data analysis. *Fifth Berkeley Symp. Math. Stat. and Probab.* **4**:115–136.

Sheps, M. C. and J. A. Menken. 1973. *Mathematical Models of Conception and Birth*. University of Chicago Press, Chicago.

Trussell, J. and C. Hammerslough. 1982. A hazards model analysis of the covariates of infant and child mortality in Sri Lanka. *Demog.* **20**:1–26.

Tuljapurkar, S. D. 1989. An uncertain life: demography in random environments. *Theor. Popul. Biol.* **35**:277–294.

Tuljapurkar, S. D. 1990. *Population Dynamics in Variable Environments*. Springer-Verlag, New York.

Tuljapurkar, S. D. and A. M. John. 1991. Disease in changing populations: growth and disequilibrium. *Theor. Popu. Biol.*, in press.

Tuljapurkar, S. D. and S. H. Orzack. 1980. Population dynamics in variable environments I. Long-run growth rates and extinction. *Theor. Popul. Biol.* **18**:314–342.

von Niessen, W. and A. Blumen. 1986. Dynamics of forest fires as a directed percolation model. *J. Phy. A: Math. Gen.* **19**:L289–L293.

Watson, H. W. and F. Galton. 1874. On the probability of the extinction of families. *J. Anthrop. Inst. GB Irel.* **4**:138–144.

Zadeh, L. A. 1969. The concepts of system, aggregate, and state in system theory. In *System Theory*, L. A. Zadeh and E. Polak, eds., pp. 3–42. McGraw Hill, New York.

Part II

Techniques of Individual-Based Modeling

Section Overview

An important aspect of this symposium/workshop was the identification and assessment of different specific approaches within the umbrella of individual-based modeling. Two broad methodologies have developed among those who model the internal structure (sizes, ages, and other characteristics) of populations and communities. The first methodology (referred to generally in this volume as the i-state distribution approach, following Metz and Diekmann 1986; see Chapter 3 for ref.) relies primarily on analytic tools such as Leslie matrix models and partial differential equations that can deal with distributions of characteristics in populations. These methods originated a few to several decades ago, but have been undergoing a rapid increase in sophistication in recent years (e.g., Metz and Diekmann 1986, Caswell 1989, Tuljapurkar 1989; see Chapter 3 for refs.), including the use of numerical methods when analytic solution is impossible. The second methodology (referred to in this volume as the i-state configuration approach) is based on the simulation of numerous interacting individual organisms and relies on high speed computers to carry out the computation and present results in comprehensible forms.

Within the two general methodologies are numerous specific techniques. This is especially true of the burgeoning field of computer simulations. For this reason, an overview of available techniques and discussions of the probable future trends is as important at this time as is a review of applications, which constitutes Sections III and IV. The papers in this section perform a variety of functions. Some papers illustrate specific techniques of individual-based modeling by applying these techniques to particular problems, some evaluate a given technique and discuss its applicability, and some compare the relative merits of different approaches.

DeAngelis and Rose explicitly attempt to compare and contrast i-state configuration models and *i*-state distribution models. The authors make the point that the methodologies are conceptually quite similar. However, their regimes of applicability differ. The *i*-state configuration models should be more accurate descriptions of populations that are likely to have strong stochastic components and to be subject to highly localized interactions. On the other hand, for many situations where these factors are not considerations, the economy of formulation of *i*-state distribution models and the body of analytic and numerical methods available to solve them favors their use. However, much also depends on individual preferences of modelers, as well as the purposes of model construction. General theoretical conclusions might be suggested by extensive simulations of an *i*-state configuration model, but are much more readily obtainable from *i*-state distribution models.

Metz and de Roos also survey the uses and limitations of both *i*-state configuration and distribution models. They note that the adequacy of deterministic partial differential equation models (the so-called physiologically structured models) depends in general on the assumptions that all individuals experience the same environment, that their effect on the environment is additive, and that the number of individuals is sufficiently large. How large is sufficient depends on the details of the model. Further assumptions of local mass action (effects arise from the sum of individuals in a local neighborhood) and sufficiently slow local environmental variation in space are needed for describing populations in space by the *i*-state distribution approach. Given this framework, such models can be used very effectively in a number of problems in population ecology. In particular, examination of partial differential equation physiologically structured models can reveal further legitimate simplifications that can be made and numerous general properties that can then be used to guide more detailed study of specific systems. Metz and de Roos are optimistic about the use of their individual-based approach, though dealing with systems with strong short-range interactions remains a difficulty for this approach.

The general considerations of the first two papers are followed by the analysis of a specific size-structured population model (Cushing and Li). The approach used is a discrete-stage model of a general population with one juvenile stage and two adult classes, large and small adults. The relative proportion of large to small adults in a particular cohort varies inversely with the density (intraspecific competition) that the cohort experienced in the juvenile stage. This very simple model leads to some remarkable results, one being that in populations with either very low or very high inherent net reproductive rates, the type of intraspecific competition contained in this model can be destabilizing, causing oscillatory or "chaotic" dynamics.

Haefner reviews current and potential applications of parallel computers for individual-based models. Both *i*-state distribution and *i*-state configuration approaches may benefit from conceptual and hardware aspects of parallel computing. For example, i-state configuration models have analogies in finite-state automata (FSA), in which an individual is represented as being in one of a sequence of discrete states (for example, age). However, FSAs would need to be generalized to describe continuous variables (e.g., size) as well. While Haefner points out that most single species i-state distribution models (partial differential equations with size structure, for example) are not so large as to require parallel processing, multipopulation models may benefit from both vector methods and multiprocessor methods of computation.

In view of the difficulties that i-state distribution models have in describing strong local interactions of individuals, it is well worth noting that a class of models that are specifically designed to mimic systems with strong local interactions, cellular automata, have been in existence for four decades. Phipps (this volume) presents some lessons from these models that may be of relevance to ecological modeling. These models consist of a two- or three-dimensional array of "cells," each of which can adopt up to m different discrete states, and each of which can undergo transitions during a time step according to rules that take into account the states of neighboring cells. These rules can be either purely deterministic or have some stochasticity. Different sets of rules can lead to quite different behaviors of the cellular automata, including a rich array of complex spatiotemporally organized behaviors. A number of generalizations have been derived from cellular automata that, despite the relative simplicity of these models, may have relevance to ecological systems. One of these generalizations is the neighborhood coherence principle (NCP), by which cooperative cell networking may in some cases tend to maintain local homogeneity in the states of cells in spite of the tendency of individual cells to deviate. This may be particularly important in the analysis of certain plant populations.

Cellular automata have the property that individual entities change from one state to the next through time in parallel or concurrently, which is true as well of entities in nature. However, Palmer (this volume) argues that the uniform time step of cellular automata and many other model types is only a special case of a rich array of possible temporal organizations which can include different spatial and temporal scales. To implement these ideas, Palmer develops an object-oriented programming approach using the Smalltalk computer language. An object-oriented program is a system of logical entities, called objects, that can communicate with each other. An ecological system can be represented as such a system of objects that send and receive messages and change in response to these messages. The entities act somewhat independently, but are constrained by their places in the hierarchy of objects (for example, individual organisms of a given species may be subordinate hierarchically to the object "species niche," which may itself be subordinate to "location." Palmer's discussion suggests that the concept of concurrency on several spatial and temporal scales will play an increasingly important role in the conceptual foundations of individual-based models.

4

Which Individual-Based Approach Is Most Appropriate For a Given Problem?

D. L. DeAngelis and K. A. Rose

ABSTRACT. **Two general approaches exist for modeling populations with internal structure; *i*-space distribution models, such as Leslie matrix models or McKendrick-von Foerster and Sinko-Streifer partial differential equations, and *i*-space configuration models, such as computer simulation models of large numbers of individuals. The basic similarities and differences of these approaches are discussed. Each model type has certain advantages and disadvantages, depending on the nature of the population being modeled, the types of questions the model is supposed to address, and the type of data available. Small populations, populations that are subject to a high degree of temporal stochasticity in the environment, and populations in which environmental exposure and encounters with other individuals are likely to vary greatly within the population, are usually best described by *i*-space configuration models, which require computer simulation. Examples are presented showing that *i*-space distribution models may not accurately represent some of the subtleties of population behavior under such circumstances. Large populations in which all individuals are relatively similar are more likely to be satisfactorily modeled by *i*-space distribution models. The possibility of analytic solution of such models or of applying well-known numerical techniques means that the results of such models are more easily checked and are in forms that are more general and more convenient to use than those of *i*-space configuration models.**

Introduction

Metz and Diekmann (1986) provided a useful distinction between different population modeling approaches, reiterated by Caswell and John (this volume). They distinguished between *p*-states and *i*-states in structured populations. The *p*-states represent states characterizing a whole population, such as population size and the average physiological size (e.g., average weight or length) of individuals in the population. The *i*-states represent states of individual organisms within the population, such as size, age, energy reserves, degree of satiation, aggressiveness, sex (male or female), and maturity (sexually mature or not). The spaces of such states are called *p*-space and *i*-space.

Within the *i*-state characterization, Metz and Diekmann (1986) noted two general types, *i*-state configuration models and *i*-state distribution models. An *i*-space configuration model follows individuals as discrete entities. An aggregation of the set of *i*-states for the individuals in the population results in the *p*-states. For example, for the *i*-space of sizes, S, the set of sizes (S_1, S_2, ..., S_{n-1}, S_n) for all n members of the population is the *i*-space configuration and the related *p*-state variable of mean size is $\sum_{i=1}^n S_i/n$. An *i*-space distribution approach for this example may use a continuous function, the size-frequency distribution, $f(S)$, that approximates the *i*-state configuration. It may, alternatively, use a discrete distribution function, such as a vector, f, of numbers of organisms in a number of discrete size classes.

The *i*-space distribution and configuration approaches each have some advantages and disadvantages for studying certain problems. We have become accustomed to the usefulness of the *i*-state distribution approach because of its success in many areas of physics. For example, many of the properties of gases can be predicted by starting with the assumption of a Maxwellian distribution of velocities of gas atoms. Whether biological populations can be effectively modeled by the same sort of *i*-space distribution approaches that have proven useful for many problems in the physical sciences is an important question. Biological populations do not consist of identical individuals in the way that gases consist of identical molecules, and do not have population sizes in the range of typical gases (10^{23}).

The distribution approach has been used successfully for several decades to give insights into biological populations with age and/or size structure (e.g., Leslie matrix models, the McKendrick-von Foerster equation). However, configuration approaches have also been applied over the past two decades to a variety of problems and have been very successful in increasing the understanding of certain types of phenomena, such as forest succession (e.g., Shugart 1984). We know of no systematic attempts to compare these two approaches and to try to establish guidelines for situations under which one approach may be preferable over the other. Here we will use three basic criteria to determine whether it will be advantageous to use one approach or the other in a given circumstance. These criteria concern (1) the importance of population size and of variation within populations in making predictions, (2) elegance and economy of formulation, and (3) the appropriateness of the approach with respect to modeling objectives, the availability of data, and the ability to corroborate the model.

Parallels and Differences Between i-Space Configuration And i-Space Distribution Approaches

Although based on different ways of representing individual organisms in a population (as discrete entities and continuous or discrete distributions, respectively), the *i*-state configuration and *i*-state distribution approaches have underlying parallels. Consider the situation in which the population size, $n(t)$, and the individual physiological sizes are kept track of through time, t. What follows will be limited to the deterministic situation with no stochasticity in growth and survival. The alternative representations are as follows.

Size and Growth of Individuals

Continuous distribution: The changes at one point along the size continuum variable, S, due to growth in size alone is described by

$$dS(t)/dt = G_d[S, E(t)]. \tag{1}$$

Configuration: The growth in size, S_i, of *i*th individual is described by

$$dS_i(t)/dt = G_c[S_i, E(t)] \quad [i = 1, 2, \ldots, n(t)], \tag{2}$$

where G_d is the growth rate at point S along the continuum of sizes and G_c is the growth rate of the *i*th individual as a function of its current size, S_i. These growth rates are also functions of time and all relevant environmental factors [lumped together as a time-dependent function $E(t)$]. In the special case where all individuals of the same size in a population go through the same experiences, the equality $G_c = G_d$ holds. Otherwise, G_c may vary among individuals, even those having the same size. The total number of individuals, $n(t)$, may change through time.

Population Size Through Time

Continuous distribution: $f(S, t)$, where $f(S, t)\Delta S$ = number of individuals at time t between sizes S and $S + \Delta S$:

$$M_d[S, E(t)] = \text{mortality rate per unit size class per}$$
$$\text{unit time,}$$
$$\int_{S_0}^{\infty} f(S, t)dS = n(t). \tag{3}$$

where S_0 is the size of the smallest individual in the population. The change through time of the population size-frequency distribution due to mortality alone is

$$df(S, t)/dt = -M_d[S, E(t)]f(S, t). \qquad (4)$$

Configuration: The probability of mortality per unit discrete time step, Δt, for each individual can be defined as $M_c[S_i, E(t)]$, such that Δt is small enough that $M_c[S_i, E(t)]\Delta t \ll 1$. This mortality rate is dependent on size of the individual. Choose a random number, R, for each individual at each time step Δt. If $0 < R < M_c[S_i, E(t)]$, then the ith individual dies during that interval. The population at time $t + \Delta t$ is $n(t + \Delta t) = n(t) - n_d$, where n_d is the number of individuals that died in the interval $(t, t + \Delta t)$.

Reproduction

Continuous distribution:

$$B_d[S, E(t), t] = \text{reproduction rate per unit size class per unit time.}$$

The term,

$$\int_{S_A}^{\infty} B_d[S, E(t), t]f(S, t)dS = df(S_0, t)/dt$$

(where S_A is the size of the smallest individual capable of reproduction) is a time-dependent boundary condition for the size-frequency distribution, setting the flux of new organisms at the lower size limit, S_0, into the system. The direct temporal dependence of reproduction on t represents species-specific reproductive patterns in time.

Configuration: $B_c[S_i, m, E(t), t]$ can be defined as the probability that an individual of size, S_i, reproduces m offspring during the time period $(t, t + \Delta t)$. Choose a random number, R, for each individual i at Δt. If $\sum_{j=1}^{m} B[S_i, j, E(t), t] < R < \sum_{j=1}^{m+1} B[S_i, j, E(t), t]$, then the ith individual produces m offspring during that interval. The population size at time $t + \Delta t$ is incremented by the number of offspring, m, produced during the interval $(t, t + \Delta t)$.

Whole Model

Continuous distribution: The components of the distribution approach can be combined into a partial differential equation,

$$\partial f(S, t)/\partial t + (\partial/\partial S)\{G_d[S, E(t), t]f(S, t)\} = -M_d[S, E(t), t]f(S, t) \qquad (5)$$

plus the boundary condition,

$$df(S_0, t)/dt = \int_{S_A}^{\infty} B_d[S, E(t), t] f(S, t) dS. \qquad (6)$$

This model is a version of the Sinko-Streifer equation for size dependence in a population (Sinko and Streifer 1967). It is a special case of a more general equation, the McKendrick-von Foerster equation, which includes age as an independent variable as well.

Configuration: The configuration approach does not yield such a compact representation. Instead, one keeps track of S_i (from Eq. 2) through time for each living organism. The set of living organisms is kept track of through bookkeeping of deaths and births. This is well-suited to implementation in a computer simulation.

The above is a highly simplified model to illustrate the parallels and differences between the two approaches. It is possible to incorporate into either approach many more i-state variables.

Choice of Approaches: Factors Related to Population Size and Variability

When the model population size, $n(t)$, is very large the configuration and distribution approach models should yield similar predictions. By "similar" we mean that, as $n \to \infty$, when the simulated individuals of the configuration model in each size step $(S, S + \Delta S)$ are added together at time t, their numbers should be approximated by $f(S, t)$ calculated from Eqs. (5) and (6). One further condition besides large population size that must be satisfied is that every individual in the configuration population is effectively exposed to the same environment, $E(t)$, through time. We will examine some of the effects that can arise when either or both of these conditions are not met.

Sensitivity to Initial Conditions Due to Small Population Size

The larger the population modeled by both a configuration and a distribution model, the more likely the two will produce results that are similar. The question "How large?" is not answerable in general because it is highly conditional. An example will be used to show that different finite populations drawn at random from the same population may undergo very different dynamics. This results from very small differences in the initial conditions of the populations resulting from the fact that they consist of finite numbers of individuals.

Consider the results of a series of laboratory experiments on largemouth bass (DeAngelis et al. 1979). In these experiments, 50-gallon aquaria were stocked with young-of-the-year bass (approximately 250 individuals in each) selected at random from a large population in a holding tank. The fish were fed, but could also cannibalize each other if size differences between individuals within the cohort were large enough.

Suppose one tried to model the subsequent dynamics of the two initial size-frequency distributions of largemouth bass in two aquaria (shown in the top frames of Figs. 4.1a and 4.1b) using the i-state distribution approach. Since the populations of the two tanks were drawn from the same larger population and since the size-frequency distributions look superficially similar, one might be tempted to fit the same continuous function to both of these distributions, in which case analysis of a distribution-type model (e.g., an elaboration of the Sinko-Streifer model to include cannibalism) would predict the same outcome for the two aquaria. However, the outcomes of the laboratory experiments proved to be highly sensitive to initial conditions. The size-frequency distributions evolved within 48 days to the two very different size distributions shown in the bottom of Figs. 1a and 1b. One explanation for this was that a larger fish could cannibalize a smaller one if it exceeded the size of the smaller one by a sufficient ratio. The largest of the fish introduced into the aquaria were able to cannibalize on the smallest individuals through much of the experiment. These experimental results were analyzed using a configuration model (DeAngelis et al. 1979). Computer simulations of the configuration model indicated that a slightly smaller size (a few millimeters smaller) of the largest few fish in Fig. 4.1b can cause the difference indicated in the experiments.

Simulation of these dynamics using a continuous distribution model with slight differences in the initial size-frequency distribution has not been attempted. Certainly such a model could be built. However, it is unlikely that the continuous distribution approach could easily duplicate the great difference in dynamics that resulted from initial size differences involving only a few individual fish in a finite population. The distribution approach will usually be ineffective when dealing with small populations. For very large populations and populations that are not highly sensitive to slight differences in size (no cannibalism or dominance based on size), or differences in other i-state variables, the distribution approach may be appropriate.

Stochasticity

The distribution approach assumes that all individuals at the same point in state space encounter the same environment and, therefore, the population changes at a deterministic rate given by Eq. (5). In reality, $B_d[S, E(t)$ and $M_d[S, E(t)]$ are not deterministic rates but fluctuate because they are based on stochastic events experienced by each individual. Thus, the population number, $n(t)$, fluctuates rather than following a deterministic curve (Fig. 4.2). Fluctuations may have little consequence (not push the population to extinction) in large populations, if the fluctuations are small relative to the

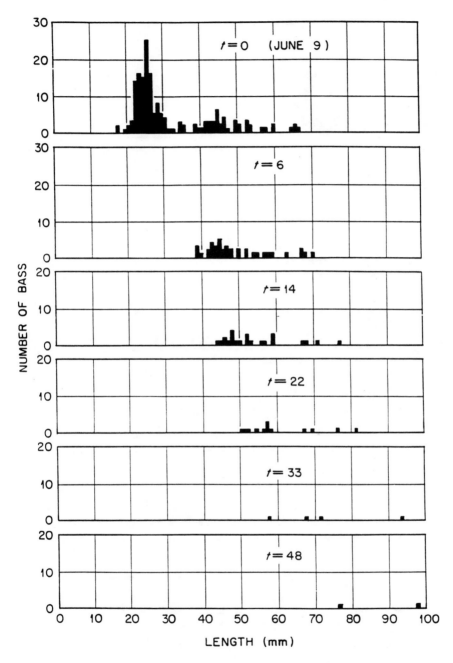

Figure 4.1. (a) Initial and subsequent size-frequency distributions of young-of-the-year largemouth bass over 48 days in a laboratory aquarium.

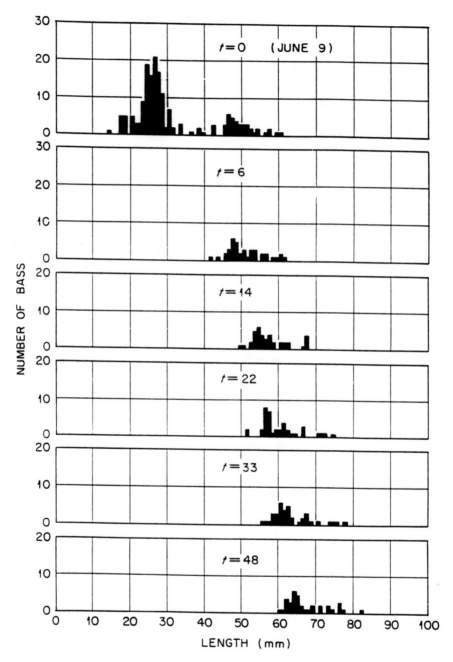

Figure 4.1. (b) Size-frequency distribution of largemouth bass in a second laboratory aquarium. (DeAngelis et al. 1979.) Time (*t*) is in days.

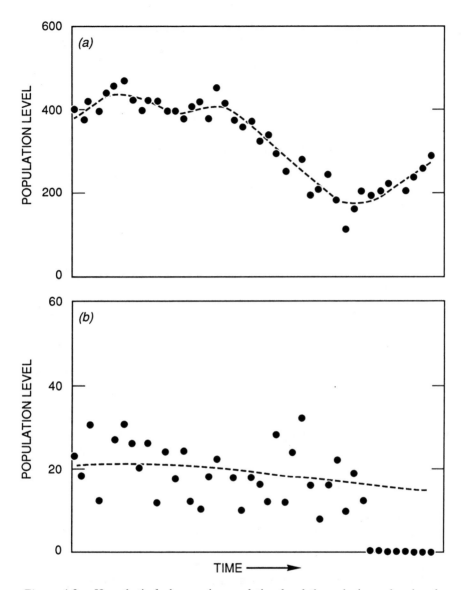

Figure 4.2.   Hypothetical changes in population level through time, showing demographic stochasticity around deterministic means. (a) In this case, the fluctuations in population are relatively small compared with the mean population level. (b) The fluctuations are large compared with the mean population. Extinction occurs through a random downward fluctuation in population, though the deterministic solution predicts population survival.

short-time scale average population size at low density (Fig. 4.2a). However, cases where stochasticity can have large consequences include:

1. Small populations, where fluctuations can make a profound difference. Note that following the distribution approach, the smooth curve tracking the time-average population may remain greater than zero even though the population has become extinct (Fig. 4.2b).

2. Both large and small populations, where fluctuations in sizes or other i-state variables of a small number of individuals in the populations are amplified relative to sizes or other i-state variables of other individuals, leading to dominance by a few individuals. For example, Fig. 4.3 shows a changing size distribution of smallmouth bass modeled by a computer simulation configuration approach (DeAngelis et al., in press). In the model, each fish added a stochastic increment of length each day. The size of this increment was positively related to initial size and random encounters with prey. Since the fish competed with each other for prey, success of some fish penalized the rest. A few fish in Fig. 4.3 managed by stochastic "good luck" to achieve a large enough size early in their life history to continue growth at a much faster rate than the others.

Rice et al. (submitted) provide a computer simulation that shows how differences in size distributions can develop when stochasticity in growth rates is taken into account. They modeled a cohort of first-year bloaters (*Coregonus hoyi*) over a sixty day period during which the bloaters were exposed to predation that decreased in intensity with increasing bloater size. Each bloater was assigned an initial growth rate, v, from the distribution

$$F_v(v) = N_0 \exp[-(v - v_0)^2/b^2]/(2\pi)^{1/2}b, \tag{7}$$

where N_0 is the initial number of bloaters, v_0 is the mean growth rate, and b is the standard deviation of growth rates. All fish growth rates that were assigned values less than zero by Eq. (7) were adjusted to 0, and all growth rates assigned values greater than one were set to 1.

In one set of simulations each bloater kept the same growth rate from day to day. In this case, it is possible to describe the size distribution, $f(S, t)$, by a special form of the Sinko-Streifer equation,

$$\partial f/\partial t + v\partial f/\partial S = -\alpha(2.7/S - 0.066)f, \tag{8}$$

where S is size and $\alpha(2.7/S - 0.066)$ is the size-dependent mortality rate due to predation.

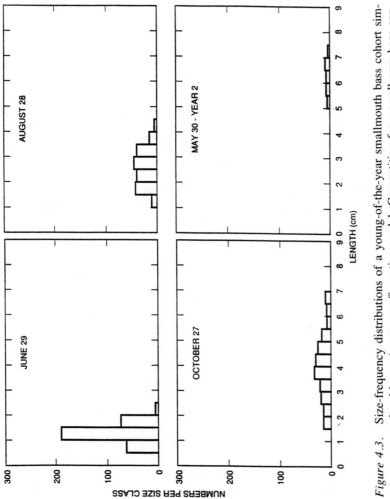

Figure 4.3. Size-frequency distributions of a young-of-the-year smallmouth bass cohort simulated by an *i*-space configuration model. Competition for a small prey base constrains the growth of the majority of bass, but a few, by chance, are able to grow large enough to exploit large prey and to continue to grow at a rapid rate. Only the fast growers escape winter starvation mortality.

Computer simulations of the configuration model equivalent to Eq. 8 run for sixty days, both with and without predation, yielded the relative size frequency distributions shown by the shaded histograms in Fig. 4.4a,b. These results agree very well with the solution of the i-space distribution model shown in the appendix of Rice et al. (submitted).

Rice et al. (submitted) also performed computer simulations of the configuration model in which the individual growth rates started at values assigned by Eq. (7), but increased or decreased randomly each day by a small increment, though staying within the range of 0 to 0.8 mm d^{-1}. The resultant size-frequency distributions after sixty days, both with and without predation, are shown in the unshaded histograms of Fig. 4.4a,b. There is clearly a significant contribution by the random component to the spread in sizes. Because size can be highly positively correlated with survival probability (Shuter et al. 1980), realistic representation of growth, and thus size, is critical. This day-to-day stochastic element in growth can be a crucial determinant of population structure, and it cannot easily be incorporated into i-space distribution models such as the Sinko-Streifer model. Stochastic partial differential equations based on the Sinko-Streifer model can be formulated, but are extremely difficult to analyze compared with the ease of simulating the configuration model.

Population Mixing

Closely related to stochastic effects on populations are the effects of imperfect "population mixing." The assumption of perfect population mixing means that every member of a population that is in the same point in i-space (e.g., same age, size, and condition) is exposed to the same environment (both the external abiotic and biotic environment and the environment created by others in the population). This is certainly not true for plants, and this lack of mixing has important consequences for plant population and community dynamics. Consider Fig. 4.5a, which shows seedlings scattered on a plot. The local environment of each seedling, in terms of the crowding effect of other seedlings, is different from that of all others, and the resultant pattern of growth of the population (Fig. 4.5) is highly dependent on the precise initial pattern (hence, this effect has some relationship to "sensitivity to initial conditions").

This imperfect mixing within populations is even more important for plant communities consisting of many species populations. Each plant is exposed to a unique set of circumstances, a factor that has implications for the preservation of diversity in natural communities.

Whereas plant populations are commonly assumed to exhibit a high degree of spatial variation, populations of mobile animals tend to be viewed as more mixed. The latter assumption is probably erroneous in most cases. Take, for example, larval fish feeding in an apparently homogeneous pelagic

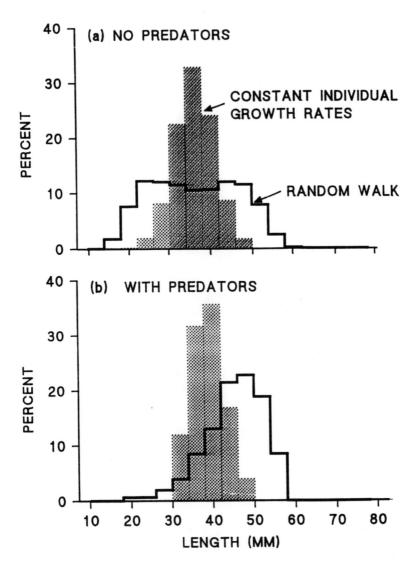

Figure 4.4.   Size-frequency distributions (in percent per size class), of young-of-the-year bloater populations modeled by *i*-space configuration models. (a) No predation mortality occurs and (b) predation mortality occurs. The shaded histograms [which are nearly identical to eyeball aproximations of exact solutions of Eq. 8], the Sinko-Streifer, show simulations in which the growth rates of individuals remain constant, whereas the unshaded histograms represent computer simulations in which there are random day-to-day changes in growth increments of individuals, so that an *i*-space distribution model of the Sinko-Streifer sort cannot be used. (From Rice et al., submitted.)

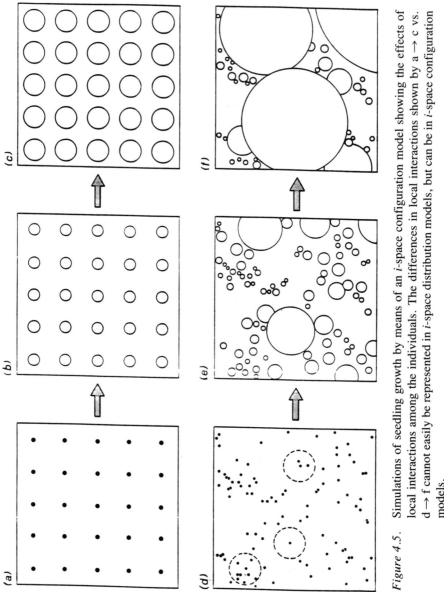

Figure 4.5. Simulations of seedling growth by means of an *i*-space configuration model showing the effects of local interactions among the individuals. The differences in local interactions shown by a → c vs. d → f cannot easily be represented in *i*-space distribution models, but can be in *i*-space configuration models.

environment of a lake. Most larval fish encounter prey as individuals during discrete feeding events. Even in a perfectly uniform environment, each individual fish would have a different history of encounters, even over a relatively short time period of a few days. Stochastic elements in the behavioral ecology of predator-prey interactions will undoubtedly yield differences in prey consumptions rates, growth, survivorship, etc. The different histories of encounters of the fish may be partly simply a result of stochasticity. However, other sources of imperfect population mixing also occur. For example, in the early life stages of many fish species, fish tend to form conspecific schools of similarily sized fish and sometimes of related individuals (e.g., siblings). Thus, systematic differences in individual histories of encounters with prey, predators, and other elements of the environment are expected. Assortative mating within populations is another source of imperfect population mixing, with potential influence on population dynamics.

Perfect mixing of larval fishes would only be approximated if all the larvae had enough time to sample enough of the environment, so that they all experienced essentially the same environment. This is unlikely to be the case in a natural situation over the short time intervals that can be crucial to larval growth and survival. The survival of larval fish will, therefore, be highly dependent on the details of variance in the experiences of the individual fish over a short time period.

Choice of Approach: Factors Related to Elegance and Economy of Formulation

Simplicity and elegance of formulation of a model, and the ease with which it can be analyzed, are often important criteria for the acceptance of models. This is partly true because models that are simple enough to be studied analytically yield solutions that are general and for that reason of potential theoretical utility. It is much easier to understand the behavior of a model from an analytic solution than from a large number of numerical solutions or computer simulations of special cases.

There is perhaps also some tendency to believe, at least in the physical sciences, that simpler models may be more "correct" than more complex models. The fact that a large variety of physical phenomena can be explained from Newton's three laws of motion, Maxwell's equations of electrodynamics, the three fundamental laws of thermodynamics, the Dirac equation, and a handful of other simple and aesthetically pleasing equations has produced this bias towards elegance. However, in some cases the desire for elegance can be misleading. Pearl's infatuation with the logistic curve, and his belief that it must represent a basic "law" of biological growth, is an instructive example of just how misleading such a bias can be (see Kingsland 1981). The logistic curve, the Lotka-Volterra equations, and other models based on them, have simplicity and elegance and also the capacity to produce dynamic behavior that looks realistic. However, they are not laws in any sense. At best, they are mathematical representations that may be good for describing some general types of behavior, but that seldom adequately accounts for actual population behavior.

Convenience of formulation and ease of solution of models may be important considerations in selecting a modeling approach. Equation (5) is a much more compact formulation than the equivalent configuration model, which is a computer simulation. A great deal of mathematical analysis has been performed on models of the form of Eq. (5) and so can be brought to bear on any new problem formulated by that approach. Also, numerical simulation packages for such equations are available.

The advantages of the Sinko—Streifer or McKendrick—von Foerster models begin to dissipate, however, when the environment of the population becomes more complex. For example, when G_d, M_d, and B_d are strongly time-dependent, analytic solution is generally not possible, though numerical solutions may still be fairly easy (de Roos 1989). The situation becomes difficult even for numerical solution when the following sorts of complications are added: (1) spatial heterogeneity and movement of individuals, and (2) importance of differences among individuals with respect to multiple properties, such as degree of satiation (stomach contents), aggression, cannibalism, and levels of lipids, toxicants and stress. Additional properties (or i-state variables) can in principle be included in Eq. (5) by generalizing $(\partial/\partial S)\{G_d[S, E(t), t]f(S, t)\}$ to

$$\sum_{j=1}^{K} (\partial/\partial X_j)\{G_d, j[X, E(t), t]f(X, t)\} \qquad (9)$$

where X is a vector of properties in the model and K is the number of properties accounted for.

Despite the compactness of Eq. (9), even numerical solution becomes very difficult as K increases beyond 3. An additional problem with the distribution formulation in such cases is that, unless the size of the total population is very large, the approximation of continuity inherent in $f(X, t)$ over the ranges of all K X_i's is likely to be violated for at least one of the i-state variables; that is, some i-state classes are likely to be very small.

These added complications do not make the use of i-space configuration models appreciably more difficult. Each individual i in the population now carries K properties, including spatial location, that must be individually followed in the computer simulation. However, as long as submodels can be formulated to update all of these for each individual in the simulation from time step to time step, there are no technical difficulties. The information demands for such a formulation are heavy (as they would be for the distribution formulation as well). While the tools for analysis are not as standardized as for distribution models, the means for exploring the consequences of such models are available.

Choice of Approach: Factors Related to Data Availability and Model Corroboration

An essential aspect in deciding what modeling approach to apply has to do with the objective of the analysis. If the objective is to produce qualitative predictions without much concern for the quantitative details, then the *i*-space distribution approach has advantages. In this situation, the details of the model formulation are likely to be flexible enough to allow general solutions of the model and permit the use of the many mathematical techniques available for analyzing distribution models. As the objectives of the analysis tend toward quantitative predictions with emphasis on mechanistic interpretations of model behavior, the *i*-space configuration approach becomes the preferred approach. *I*-space configuration models sacrifice analytic tractability to allow inclusion of process-level information at the appropriate level of detail. *I*-space configuration models also tend to be more tightly coupled to empirical information and thus also offer an advantage if evaluation of the quality and quantity of available information is an important objective. The process of constructing a detail-rich configuration model requires the quantitative synthesis of available information. This synthesis and linking together of the different empirical sources of information is arguably the most effective method for identifying potentially critical areas where information is lacking.

The time scales of the predictions of interest can also affect the selection of a modeling approach. Predictions can be categorized as short-term (within a generation) and long-term (multiple generations). Data on size and/or age distributions through time are typically available for most populations of interest. Both *i*-space distribution and *i*-space configuration models provide predictions of size and age distributions on short-term and long-term time scales. An *i*-space configuration is likely to be more appropriate for short-term predictions at which fine-scale detail (sensitivity to initial conditions, imperfect mixing, and stochasticity) can play an important role. Long-term data generally consist of highly aggregated quantities (e.g., means, variances) collected at relatively long intervals (e.g., annually) making the *i*-space distribution approach more appropriate.

In discussions at the workshop from which this book evolved, data availability arguments have been invoked to justify the use of both distribution and configuration approaches. In our opinion, neither the *i*-space distribution nor the *i*-space configuration approach offers obvious advantages in the ease with which available information is used to formulate the model. The information required to formulate comparable *i*-space distribution and configuration models is generally similar. This was illustrated above for the situation in which population number and physiological size were tracked (see "Parallels and Differences Between *i*-Space Configuration and *i*-Space Distribution Approaches"). For the situation illustrated, application of both approaches would require specification of the functional forms of G, M, and B, and their dependences on the *i*-state variables and on the environment. In general, such extensive information as how process rates (e.g., G, M, and B) simultaneously vary as functions of *i*-space variables and the environment are rarely available. Furthermore, many applications of population models are site-specific, making the likelihood of extensive data being available to completely specify all model inputs practically impossible.

One can imagine situations where available data are clearly more compatible with either an *i*-space distribution or configuration approach. Some populations, especially ones that are small and easily observed (e.g., breeding colonies of birds, troops of baboons, small populations of vertebrates in relatively bounded areas), are studied empirically by keeping track of individuals. In such cases, quite a bit of information is usually available on detailed behaviors of individuals and interactions among individuals. What is often desired in this case is a highly mechanistic model that predicts population properties based on a detailed knowledge of individuals. The *i*-space configuration model may be the more appropriate in such cases.

At the other extreme, very large populations that are hard to observe as individuals, so that information is generally available only at the aggregate level, are probably best modeled by the *i*-space distribution approach.

In most actual cases either approach may be applicable. In such cases any intrinsic advantages that one formulation may have over the others are probably outweighed by the abilities that the modeler may have in one type of approach or the other. This is because proper model formulation and specification of model inputs by means of either approach is non-trivial and relies heavily on the cleverness and skills of the investigators developing the models and their ability in properly synthesizing and using information collected from a variety of populations and systems.

Model corroboration consists of comparing model predictions of size and/or age distributions to measured distributions of these variables. Both the *i*-space configuration and distribution models generate distributions of size, age, and other *i*-state variables in the model. An advantage of the *i*-space configuration approach for model corroboration is that many *i*-space variables can be included in the model; the appropriate *i*-space variables for each individual in the model (computer code) are simply tracked. Including many *i*-space variables in a distribution model is possible, but difficult. The *i*-space distribution models can require different solution techniques and even become intractable as the number of *i*-space variables increases. Thus, data in addition to the basic size and age distributions, such as diet as a function of size, condition indices, and relative (%/day) and absolute (e.g., mm/day or mg/day) growth rates in terms of length and weight, can also easily be compared to predictions from an *i*-space configuration model. Having the ability with the *i*-space configuration approach to compare many *i*-space variables to available data permits more constraints to be placed on the combinations of inputs that result in realistic model simulations, and thus can result in more rigorous model corroboration.

The *i*-space configuration models also offer an advantage in corroboration when, as mentioned earlier, information is available on the time history of specific individuals. In some fish species, daily records of growth are preserved as otolith rings (Campana and Neilson 1985). Individual fish can be sampled and their birth dates and growth histories from birth to capture reconstructed. Because *i*-space configuration models specifically follow a sample of discrete individuals, there is perfect congruence between model predictions of growth and otolith-derived histories.

There are possible disadvantages of using the *i*-space configuration approach in cases of populations that undergo drastic numerical declines over the time scale of interest. Corroboration of model predictions requires a sufficient number of survivors at the end of simulation runs to permit meaningful analysis of values of survivors' *i*-space variables. Survival rates in early life stages of marine fish are commonly ≪1% (Bailey and Houde 1989) making those individuals that ultimately survive early life extremely rare. Under such high mortality, many individuals must be followed initially to ensure a sufficient number of survivors at the end of the simulation. This problem of too few survivors at the end of the simulation does not arise with the *i*-space distribution approach, which produces predictions of the entire probability distribution of all *i*-space variables throughout the simulation.

Conclusions

There are many advantages associated with using an *i*-space distribution approach for population modeling. This type of approach has a long, rich history of application, resulting in the accumulation of a large body of knowledge on how to interpret and analyze their behavior. This approach will undoubtedly continue as a vital part of population modeling. However, when assumptions underlying the *i*-space distribution approach are not met (i.e., when fine scale detail within the population becomes important) or their practical limitations are exceeded (e.g., there are many *i*-space variables to be kept track of), the configuration approach becomes more appropriate.

The distribution and configuration approaches appear superficially to be quite different and unrelated types of models. We have attempted to show that, in fact, these two approaches have quite parallel formulations. The logic of model development should initially be the same in both cases and should diverge only as the particulars of the situation under study and the specific questions to be asked of the model point towards the use of one approach over the other.

We believe that the use of the configuration approach will increase dramatically with time. However, we envision many situations in which both approaches will be used together; a configuration approach for certain life stages or time periods coupled to a distribution approach to permit long-term predictions. Care must be used as we deviate from the sound foundations and rigorous mathematics underlying the distribution approach and pursue the fuzzy world of the configuration approach. We should not abandon the distribution approach, but use it as a starting point for developing the more detail-rich configuration models that may often be needed. The boundary region between the areas of applicability of the two approaches should be a rich area for study.

Acknowledgements

The authors thank L. J. Gross, L. W. Barnthouse, and K. O. Winemiller for their useful comments on an earlier draft of this paper. This research was sponsored by the Electric Power Research Institute under Contract No. RP2932-2 (DOE No. ERD-87-672) with the U.S. Department of Energy under Contract No. DE-AC05-840R21400 with Martin Marietta Energy Systems, Inc. Environmental Sciences Division Publication No. 3833.

Literature Cited

Bailey, K. M. and E. D. Houde. 1989. Predation on eggs and larval of marine fishes and the recruitment problem. *Adv. Mar. Biol.* **25**:1–83.

Campana, S. E. and J. D. Neilson. 1985. Microstructures of fish otoliths. *Can. J. Fish. Aquat. Sci.* **42**:1014–1032.

DeAngelis, D. L., D. C. Cox, and C. C. Coutant. 1979. Cannibalism and size dispersal in young-of-the-year largemouth bass: experiments and model. *Ecol. Modell.* **8**:133–148.

DeAngelis, D. L., L. Godbout, and B. J. Shuter. 1991. An individual-based approach to predicting density-dependent dynamics in smallmouth bass populations. *Ecol. Modell.*, **57**:91–115.

de Roos, A. 1989. Daphnids on a Train: Development and Application of a New Numerical Method for Physiologically Structured Population Models, Ph.D. Dissertation, University of Leiden, The Netherlands.

Kingsland, S. E. 1985. *Modeling Nature*. University of Chicago Press, Chicago.

Metz, J. A. J. and O. Diekmann (eds.) 1986. *The Dynamics of Physiologically Structured Populations. Lecture Notes in Biomathematics 68.* Springer-Verlag, Berlin.

Rice, J. A., T. J. Miller, K. A. Rose, L. B. Crowder, E. A. Marschall, A. S. Trebitz, and D. L. DeAngelis. 1991. Growth rate variation and larval survival: Implications of an individual-based size-dependent model. *Canadian Journal of Fisheries and Aquatic Sciences,* submitted.

Shugart, H. H. 1984. *Theory of Forest Dynamics: The Ecological Implications of Forest Succession Models.* Springer-Verlag, New York.

Shuter, B. J., J. A. McLean, F. E. J. Fry, and H. A. Regier. 1980. Stochastic simulation of temperature effects on first-year survival of smallmouth bass. *Trans. Am. Fish. Soc.* **109**:1–34.

Sinko, J. W. and W. Streifer. 1967. A new model for age-structure for a population. *Ecology* **48**:910–918.

5

The Role of Physiologically Structured Population Models Within a General Individual-Based Modeling Perspective

J. A. J. Metz and A. M. de Roos

ABSTRACT. **In this paper we outline a framework for dealing in a reasonably general manner with individual-based population models. As the resulting mathematical structures are necessarily stochastic, we spend some effort delineating the cases which allow clear-cut deterministic approximations. The most important exceptions are those cases where, at some time or other, the number of individuals in a key category is bound to be low, or in which interactions among individuals are extremely local and yet large numbers of individuals are coupled in a chain- or net-like fashion.**

As a second topic, we review some general results for the class of physiologically structured population models, to wit the conditional linearity principle (conditional on the time course of the environmental conditions, the population process is linear in the initial state), the calculation of invasion criteria, a potpourri of model simplification principles, and some useful and relatively straightforward numerical and analytical techniques for the spatially homogeneous case.

As a third topic, we consider a general modeling philosophy which explicitly tries to account for the inherent conflict between our wish for comprehension and the large amounts of detail present in any concrete situation. Within this framework we discuss possible strategies for ascertaining that conclusions from simple models have some wider applicability, and the corresponding need for appropriate higher order concepts.

Finally we provide some pointers to important, but also very difficult, open problems.

Introduction

(i) Populations by definition are made up of individuals. Therefore, it may not come as a surprise that the classic models of ecology (the Pearl-Verhulst logistic and Volterra-Lotka differential, Nicholson-Bailey difference, Kermack-von Foerster partial differential, and Lotka integral equation models, the Galton-Watson branching process, etc.) already allow individual-based interpretations. In fact, most of them were derived originally from individual-based considerations. Their main fault is that they deal only with very restricted, and therefore unrealistic, types of individuals and environments.

(ii) Our communal insight increases by and large by accretion at the margins. It is of little use to try establishing footholds too far out at sea: one should dare but not drown.

(iii) No model will ever embody the final truth. Neither does any field study or any lab experiment tell once and for all what makes the world tick. Individuals come in many different kinds, and reality always shows additional underlying complexity when we look more closely.

Below we try to provide a general perspective for individual-based modeling based on these three considerations. We concentrate in particular on the link between deterministic, differential equation oriented models, and models in which each individual is represented as a separate entity. Of course physiologically structured population models as set forth in Metz and Diekmann (1986) appear prominently. We refrain, however, from giving yet another exposition of their methodology, and consider only some of its restrictions, as well as results, in order to illustrate possible directions for individual-based modeling in general. Moreover, we unashamedly use the examples with which we are most familiar.

A (Fairly) General Framework for Individual-Based Population Models

1. On the Basic Laws of Population Biology and the Necessity for Always Considering Classes of Models

Population biology proper has only two really basic laws of its own:

(i). Individuals do not materialize spontaneously; they are born from other individuals.

(ii). The number of individuals in a closed region can decrease only as a result of the deaths of some of them.

On a more refined but still fairly general level, the laws of genetics tell us which sorts of individuals are born from which other ones.

What remains are particulars, though at various levels of specialization. It is the task of the population biologist to work out these special cases. How do the properties of individuals and environment interact in various specific instances to determine whether the individuals give birth or die, and, if the environment is not everywhere the same, what are the rules which govern their movement between the various places? Moreover, at the theoretical end we have the question of how the various potential rules at the individual level translate into the behavior of populations as a whole.

Notice that the latter question cannot be answered by experiment alone. Experiments can only be used to test whether our theoretical deliberations indeed capture the essentials of a particular field or lab situation.

The theoretical question of how rules at the individual level relate to whole population phenomena is necessarily a question about classes of models. The most we can conclude from a finite number of simulation runs is that under particular conditions certain mechanisms can generate certain phenomena. Even this is not easy, as there is always the snag that we may be dealing with artifacts of the particular implementation of our conceptual constructs. Usually we want a little more, and it is therefore important to have a good conception about which classes of models allow stronger types of inference. Of course, the stronger the methods the narrower the class of models. In this paper we have set ourselves as a first task the tentative delineation of the largest possible classes of models that are amenable to particular technical means.

2. On the Concept of State (of Individuals and of a Population) and an Appropriate Definition of Environment

In order to chart the temporal development of a population it pays to revert to a state representation.

By definition, a state description should be such that

(i). given the present state and the intervening input (environmental history) the future states are fully determined, and

(ii). given the present state and the present input value (condition of the environment) the present output (behavior) is fully determined.

Here fully determined should be interpreted in a stochastic (Markovian) sense.

The relevant behavior of individuals consists on the one hand of giving birth and dying, and on the other hand impinging on the environment. In the case of a population, all the population statistics that have our interest count as behavior, together with, again, any influences which the population may exert on the environment.

Let Ω denote the state space of an individual (i-state space) and let S denote physical space. The state of the population at any particular time can always be expressed as the ordered n-tuple of the states of its n members together with some indicator of the selection made by each individual for its input from among the variables which are collectively called the condition of the environment. (Notice that here "the condition of the environment" may stand for something very grand, like the values of a great many variables as a function of the physical space in which the organisms live.) Usually the location of an individual in space can be used as a sufficient indicator, so that the population state space can be represented as:

$$\bigcup_{n=0}^{\infty} (S \times \Omega)^n$$

where n is population size, $(S \times \Omega)^0 = 0$, $(S \times \Omega)^1 = (S \times \Omega)$ and $(S \times \Omega)^n = (S \times \Omega) \times (S \times \Omega)^{n-1}$ for $n > 1$.

The minimal description of the environment at a particular place and time may have to contain quantities such as temperature, nitrogen availability, food density, or predation pressure.

The essential element in the above deliberations is that any interaction between individuals is always couched in terms of intervening variables which are considered as components of the environment.

3. A First Go at Deterministic Models: Large Number Limits

If and only if

(i)   all individuals experience the same environmental input, and

(ii) the output to the environment can be calculated by adding all individual contributions,

we can collapse the population state space to just a frequency distribution over Ω. We call this distribution the p-state following Metz and Diekmann (1986).

If and only if in addition to (i) and (ii)

(iii) the number of individuals is sufficiently large, we can use a law of large numbers limit (in the style of Kurtz 1981, or Ethier and Kurtz 1986) to arrive at a deterministic model of the type discussed extensively in Metz and Diekmann (1986), and, somewhat more summarily, in Metz et al. (1988): the so-called physiologically structured population models.

What "sufficiently large" means depends on the details of the model. As a rule-of-thumb the number of individuals in any key category, for example the fertile females, should never fall below ten, or a hundred if we are conservative. The delineation of key categories in turn depends on the smoothness of the coefficient functions. For example, if everybody behaves almost the same there is but one key category, to wit the whole of Ω.

Notice that the use of "p-state" is predicated upon some rather strong assumptions (compare the distinction between i-state configuration and i-state distribution, or p-state, made by Caswell and John, this volume). Metz and Diekmann (1986) used (i) to define a population. They refer to (ii) as the mass action principle or law of mass action *sensu lato*.

4. On Large Number Limits Taking Account of Spatial Differentiation

If and only if

(i) we have local mass action, i.e. locally the population output can be approximated by adding the contribution of all individuals "in the neighborhood," and

(ii) the numbers of individuals in each neighborhood over which the environment stays approximately the same is sufficiently large, we can again use a deterministic approximation, but now with the meta-population state a distribution over $S \times \Omega$.

Notice that in the case of a continuous S we assume implicitly that the environment as perceived by the individuals does not fluctuate too wildly over space. This is an intricate point to which we shall return more than once. A second point is that an appropriate specification of continuous space deterministic limit models catering to any but the most straightforward movement patterns is far from easy! Some good general references are Okubo (1980, 1986) and Turchin (1989).

Van den Bosch et al. (1990) review an alternative, more immediate route from microscopic specifications to macroscopic phenomena involving space, which can be very effective provided some fairly strong restrictions are satisfied (see also section 10).

Further good, but relatively more phenomenologically oriented, general sources are Edelstein-Keshet (1988) and Murray (1989). Avnir (1989) contains a number of reviews about possible as yet largely unexplored alternative ways of accounting for the fine structure of ecological space. For our own pleasure we point to the epidemiological literature for some beautiful combinations of data analysis and modeling of spatial phenomena (e.g., Bacon 1985, Cliff and Hagget 1988).

5. On Moving Versus Fixed Individuals

Which mechanisms lead to (local) mass action and/or spatially smooth environmental variation? (Notice that due to our definition of environment these need not be two separate questions.)

At least two mechanisms lead to the required result:

(i) The individuals are fixed but the environment can be characterized wholly by continuous quantities such as temperature or nitrogen availability, which are smoothed by a fast diffusion mechanism.

(ii)   The individuals themselves roam about so that the number of individuals with which they potentially interact is large.

The latter assumption is implicit in the definition of quantities like predation pressure: We do not consider the environment as consisting of a number of particular predators and their positions relative to the individual under consideration, but in a way we average over the local predator distribution.

We do not yet have any mathematical proofs that (i) and particularly (ii) work. However, from the success of deterministic models based on local mass action type assumptions in physics and chemistry, where systems are more accessible experimentally, one gets the feeling that we need not bother (see e.g., Keizer 1987, for a nicely explicit discussion of these issues).

6. *A Final Go at Determinism: Generalized Individuals*

The basic requirement for the "law-of-large-numbers" approximation to hold is the availability of entities which are coupled by some form of mass action. In sections 3 and 4 we considered the case where these entities were the individuals themselves, but this is not the only sort of entity that may be usefully introduced.

For the sake of argument, we shall call any assemblage of individuals which stays intact for a time of order unity on the time scale of our model, a generalized individual. Conditional on their i-state and the condition of the environment, the production of offspring by ordinary individuals living in the same environment automatically conforms to the mass action principle. Generalized individuals may be "born" in any odd manner, for example by being assembled from (parts of) generalized individuals of various other kinds. Therefore we need to add as a separate requirement for our generalized individuals that the rate at which they are "born" can be expressed in mass action type population statistics. Notice that the "death"-rate automatically satisfies the mass action principle due to our definition of i-state.

One obvious example of generalized individuals is married couples, and single males and females, in demographic models, if we make the usual assumption that, given the i-state and the locality of the partners, marriages occur at random.

A second type of generalized individuals are the "sites" in safe site and lottery type models (see e.g., Geritz et al. 1988, and the references therein). In these models only one individual is assumed to remain per empty site from the propagules arriving from everywhere in, say, fall, while individuals in different safe sites are assumed not to compete. (Notice that in this class of models the distinction between ordinary and generalized individuals is only minimal).

A final example is provided by the prey-predator-patch models of Diekmann et al. (1988, 1989) (see also Metz and Diekmann 1986), in which the patches acted as generalized individuals having the states of the local prey and predator populations for their i-states. The aerial plankton from which the founders of the local populations derive acts as the law of mass action type coupling agent.

7. On Local Interactions

Unfortunately there exist an abundance of ecological systems in which many short range interactions, connecting many individuals in a chain- or net-like fashion, form an important part of the picture. In particular, sessile organisms like corals or plants tend to compete only with their immediate neighbors. In that case it is not possible to arrive at any clearcut deterministic model by letting the number of individuals go to infinity.

Some prototype models exhibiting the problems one may run into when interactions are only local are: the Harris contact process (cf. Liggett 1985; an example of an application to individual-based population modelling can be found in Verboom et al., 1991), the nearest neighbor epidemic (cf. Mollison 1977, 1986, Durret 1988a, 1988b Grasberger 1983, Cox and Durrett 1988), or more generally, stochastic cellular automata (cf. Hogeweg 1988, and also Toffoli and Margolis 1987, and Wolfram 1986).

Notice that contrary to the examples in physics, where we usually have either long-range or short-range interactions, in population biology we frequently have a mixture of interaction types. Plants compete with their neighbors but are grazed by moving herbivores and their propagules spread over large distances. As yet we do not have any good comprehensive methodology for coping with these classes of problems, though the approximate semi-descriptive modeling techniques advocated by Pacala and Silander (1985, Pacala 1986a, 1986b, 1987, Silander and Pacala 1990) provide one step in the right direction.

8. First Finale: On the Two-Way Traffic Between the Population and Organismic Levels

Until now we have concentrated on the step from the properties of individuals to population phenomena. However, the fact that reproduction is almost, but not totally faithful, inevitably leads to evolution (in the biological sense), except in the most contrived scenarios. Therefore there exists a two-way traffic between the properties of individuals and of populations: The mechanisms of population change ultimately reside in the individuals comprising it. Conversely, only those properties of individuals which, under the then current conditions, gave their bearers an advantage in the race to contribute to future generations have left their mark in the present.

Subject to generally unknown but often guessable constraints (deriving ultimately from detailed physics or chemistry, either directly or through the grand unknown of the developmental process as molded by the preceding evolutionary path), individuals are *grosso modo* adapted to the environmental conditions under which they live. Notice, however, that these environmental conditions are not extrinsically determined constants. First of all, your environment contains not only the individuals of the species with which you interact but also your conspecifics. Secondly, environmental conditions may fluctuate on various spatial and temporal scales. The currency relative to which we should evaluate adaptiveness, customarily called fitness, should ultimately be derived from detailed population dynamical considerations, though in particular cases it may reduce to simple (e.g. energetic) measures. In our opinion it is one of the prime tasks of the theoretical population biologist to elucidate the precise nature of "fitness" for ever more generalized/ realistic environmental scenarios. The theory of evolution is the core theory of biology, and we should not hesitate to spend considerable effort in refining it. This should provide us with at least one general goal for our deliberations in the next set of sections.

Some General Results for Physiologically Structured Models

At present all results for individual-based models having some semblance of generality pertain to the deterministic models resulting from a "law-of-large-numbers" approximation. In the following sections (9 to 13) we shall assume that such models can be applied without impunity.

9. The Conditional Linearity Principle

In this and the next section, we shall concentrate on deterministic models phrased in terms of ordinary as, opposed to generalized, individuals. Such models share the following useful property:

When the course of the environment is given,
the dynamics of the *p*-state is linear

(but time-varying except in a constant environment). This "conditional linearity principle" suggests the following promising line of attack for further mathematical development of the theory of deterministic individual-based population models:

  (i) Consider the existence and uniqueness problem for general time varying linear population problems.

  (ii) Derive appropriate smoothness results for the resulting input-output relations (from the environment through the population "box" to the environment).

  (iii) Treat autonomous population problems, resulting from closing environmental feedback loops, as equations among population input-output relations using fixed point arguments.

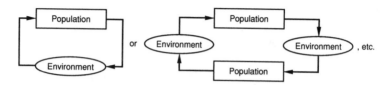

(iv)   Within this framework consider linearized stability, bifurcation theory, etc. This approach fits in nicely with our model building toolbox. As yet we are still far from completing such a program, but some clear progress is being made by Odo Diekmann and coworkers (see Diekmann in press, for a survey).

10.   On Invasion Problems

An interesting corollary to the conditional linearity principle pertains to the invasion problem. If a species is rare, the environment is set by the other species. Whether invasion is possible depends on whether the dominant Lyapunov exponent of the invading species is positive or negative. (If the environment is constant the dominant Lyapunov exponent reduces to the dominant eigenvalue of the population differential generator. A good discussion about the properties and the approximate calculation of dominant Lyapunov exponents for discrete-time stage-structured populations in stochastic environments can be found in Tuljapurkar 1989, 1990). This linearity principle can be exploited to study the possible addition of species to an existing assemblage (some imaginative estimates for dominant Lyapunov exponents can be found in the work of Peter Chesson and coworkers, see Warner and Chesson 1985, Chesson 1986, 1988, 1989, Chesson and Huntly 1988, 1989, and references therein) or the ability of a disease to gain foothold in a population (cf. Diekmann et al. 1990).

Unfortunately, the extinction problem is not the immediate mirror image of the invasion problem. On the one hand there may be large excursions leading to lower and lower returns, and eventually to extinction due to stochastic effects, even though at each low the deterministic trend is for the population to eventually increase. On the other hand there is the possibility for the coexistence of an internal attractor in combination with one or more boundary attractors, as in the cannibalism model of Van den Bosch et al. (1988). No invasion does imply extinction in the relatively common cases where any other environment is always worse than the virgin one, as in most resource limited single species models (including most epidemic models).

The "linearity when rare" principle can also be used very effectively to calculate the speed of spatial spread of a species newly introduced into a virgin territory (see Van den Bosch et al. 1990).

If we replace "species" in the preceding discussion by mutant and "other species" by resident type, we are fully set up for the calculation of Evolutionarily Stable life history Strategies (ESS). Notice that here fitness apparently has to be equated with the dominant Lyapunov exponent! (Metz et al., in submitted; compare also Tuljapurkar 1989, 1990).

11. On Invasion and the Generalized Individual

Now consider deterministic models phrased in terms of generalized individuals. If all "birth" rates are linear in the p-state, as was the case for ordinary individuals, the considerations from section 10 extend without change.

However, even for birth rates which are nonlinear in the p-state, the invasion problem tends to be linear. Consider, for example, the dynamics of a sexually transmitted disease in the context of a marriage model. When the disease is still rare an unmarried ill individual will always marry a healthy partner. The density of healthy partners is at least initially independent of the dynamics of the disease; only in later stages of the epidemic does it become necessary to account for the probability that two ill individuals may together initiate a new partnership. Therefore, we can argue in terms of a linear dynamics on the space of frequency distributions over the possible illness- and partnership-states. (However, we do have to account for the possibility that an ill individual is still married to an ill individual by whom (s)he has been infected or whom (s)he has infected him/herself.) More details may be found in Diekmann et al. (in press).

To tackle ESS life history problems for population models phrased in terms of infinitely many groups, each consisting of a relatively small number of ordinary individuals, it usually pays to revert to inclusive fitness arguments (see e.g., Taylor 1988a, 1988b, 1989, 1990). Conversely, most inclusive fitness arguments are implicitly predicated upon local demographic stochasticity, as one's relatives will otherwise become too diluted to be "recognizable," while explicit calculations usually assume a spatial structure of the generalized individual type.

12. On Some Simplification Principles

Individual-based models have the unfortunate tendency to be very complicated. Therefore it is of the greatest importance to have tool boxes for systematic model simplification. At this moment we know of only four general principles which typically pertain to the population level (as opposed to simplifications of the underlying i-model; see Heijmans and Metz 1989, for a discussion about the connection between i- and p-level simplifications).

    (i)   If (a) all individuals are born equal (in the probabilistic sense) and (b) the i-state develops independent of the environment, it is possible to revert to population dynamically equivalent age-representations (see Metz and Diekmann 1986, chapter IV). (This principle allows extensions to (1) physiological age and (2) multiple ages; think, for example, of the times since the arrival of the prey and predator foundresses in the prey-predator-patch models introduced in section 6). The important point is that for sufficiently large time the i-state distribution necessarily factorizes into an age-dependent component which varies with time, times a fixed distribution for each age over the remaining component of the i-state space. In the particular case that the environment is constant, (b) is trivially fulfilled. Therefore age-representations are particularly useful tools both for calculating equilibria and for dealing with invasion problems.

(ii) Under certain conditions on the i-dynamics the necessarily infinite dimensional evolutionary systems describing p-behavior, symbolically represented by a partial integro-differential equation, allow finite dimensional counterparts, corresponding to a representation in terms of ordinary differential equations, which faithfully generate the population input-output relation. This property has been dubbed Linear Chain Trickability (LCT). In Metz and Diekmann (1991, in prep.) various (sufficient, necessary and sufficient) conditions are given for physiologically structured population models to be LCT.

(iii) If the environment makes itself felt only in an additional i-state independent death rate term, any environmental changes will only influence the temporal development of the overall density of the population, but not its composition. Therefore the i-state distribution will converge to a stable form, as in constant environments, thus allowing, asymptotically, a one dimensional representation of the p-state. This result still applies if, in addition, the i-state dependent component of the death rate, the birth rate, and the rate of i-state movement are all modulated in a multiplicative manner with one and the same strictly positive environmentally dependent constant (see Metz and Diekmann 1986, sections I.4.5, II.14, IV.5.1).

(iv) Sometimes it is possible to use time scale arguments of various sorts. (For a rigorous approach one would need infinite dimensional variants of Tikhonov's singular perturbation theorem [Tikhonov et al. 1985, VII.2].) As one example consider a population which for all environmental conditions grows only very slowly, i.e. for which the dominant eigenvalues of the p-differential generators for the various environmental conditions are all close to zero. Assume, moreover, that the environmental condition itself is just a continuous linear functional of the *p*-state. If the remaining parts of the spectra of the *p*-differential generators are uniformly bounded away from zero, we may distinguish a short time-scale process in which the *p*-state converges to a stable *i*-state distribution, and a long time-scale process of population growth in which the stable *i*-state distribution just waxes or wanes without changing shape.

13. On Numerical and Analytical Techniques

André de Roos (1988, 1989, De Roos et al., in press, De Roos and Metz, 1991) has developed an efficient numerical method for dealing with general nonspatial physiologically structured population models, based on an interesting combination of biological and mathematical insights.

Numerically solving even the equations describing the deterministic versions of individual-based models tends to be a large task, unless one can make use of special properties. The Glasgow team of Roger Nisbet, Bill Gurney and Steve Blythe have specialized in representing ever more general classes of physiologically structured population models as delay differential equations (Gurney and Nisbet 1982, Nisbet and Gurney 1983, Gurney et al. 1983, Nisbet et al. 1985, Gurney et al. 1986) which can be solved very quickly with the locally developed Solver package. Their techniques provide a considerable extension of the Linear Chain Trickery program discussed in the previous section. If one still has a certain freedom of choice in choosing the coefficient functions of one's model, it pays to opt for a linear chain trickable combination.

The analytical approach to deterministic individual-based models proceeds in the usual manner: First calculate equilibria and then determine their linearized stability through a mixture of numerical and analytical techniques. In practice this is only feasible if everybody is born equal (or, more generally, if everybody at some point of its life has to pass through one of an a priori given finite set of *i*-states, and, when we are dealing with a spatial model, space either consists of a finite number of homogeneously mixed patches, or else the combination of the movement process and the spatial domain allows a separable coordinate system; see Morse & Feshbach, 1965). Otherwise it is not possible to separate out a characteristic equation for closer study. Some good examples of this approach may be found in the work of the Glasgow group and their collaborators on insect population dynamics (e.g., Nisbet and Gurney 1984, Gurney and Nisbet 1985, Murdoch et al. 1987), in the work of Charles Godfray on parasitoid-host interaction (Godfray and Hassell 1989), and in the Daphnia work of André de Roos (1989, De Roos et al. 1990).

Some Modeling Principles

14. On the Metaphorical Nature of Models

As scientists we should always at least aim at providing a coherent and general picture of the relationship between mechanisms and phenomena as opposed to the consideration of particular cases only. This view leads to the following modifications of May's scheme of tactical versus strategic modeling, which should illuminate the essentially metaphorical nature of models:

General and Encompassing	**Strategic models**	
Special cases	Tactical models with a strategic goal mathematically as simple as possible, constructed to uncover potential generalizations	Tactical models with a practical goal constructed for prediction/testing, usually incorporate lots of technically awkward detail

The idea is that we should specify our strategic models in as general a manner as we can achieve, constrained only by a few basic laws on the one hand, and on the other hand some special assumptions representing the particular biological mechanism we wish to single out for special consideration. Of course we hope that there exist emergent phenomena exhibited by the strategic class as a whole, or else that the emergent phenomena allow a consistent classification. Moreover, we hope that any such generalizations/classifications may already be discovered by looking at specific examples.

In this view very simple examples living only in our imagination stand on equal footing with complicated simulation models purportedly mimicking concrete situations. The connection between the two is through the encompassing class of strategic models. The simpler models usually provide a better route to understanding, but only if they capture "the essentials" of the mechanism under consideration. Whether the latter is the case can only be judged by embedding them in a sufficiently rich spectrum of models all belonging to the same strategic class.

15. On Candidate Phenomena

In the linear realm, that is, when individuals do not influence their environment directly or indirectly, we have a pretty good general idea about the behavioral possibilities of population models: In the long run, an ergodic environment leads to either overall exponential growth or extinction. In the spatial case there is the added feature that in a homogenous terrain, the area in which the population density is above any particular sufficiently high level eventually becomes circular, with a radius which expands at a fixed rate. The main open questions are quantitative: what are the probabilities of either extinction or overall exponential growth, and how fast is that growth?

In the nonlinear realm we can rarely do better than study carefully chosen prototype problems. For an appropriate use of prototype models and the concomitant formulation of robustness principles (or conjectures; see next section) we should have good theoretical categories for classifying the potential population dynamical phenomena.

Obvious candidates from a mathematical viewpoint are the various possible attractor structures of dynamical systems theory (see e.g., Bergé et al. 1984, Schaffer 1985, Thompson and Stewart 1986, Holden 1986, Schuster 1988, Ruelle 1989a), the various generic bifurcation patterns for solutions of parametrized equations (see e.g., Guckenheimer and Holmes 1983, Golubitsky and Schaeffer 1985, Golubitsky et al. 1988, Ruelle 1989b), and the various scaling laws of statistical physics (see e.g., Ma 1976, Burckhardt and van Leeuwen 1982, Stauffer 1985, Grimmet 1989, and Grasberger 1983).

A second, more relevant (but often related) type of candidate phenomena may be derived from established, if overly simplified, ecological theory. Examples are the competitive exclusion principle and the paradox of enrichment (*sensu* Rosenzweig, 1971). Incidentally, the competitive exclusion principle in the form given to it by Levin (1970) almost immediately extends to fairly large classes of physiologically structured population models. A discussion of the paradox of enrichment can be found in De Roos (1989, see also De Roos et al. 1990). This case is slightly more involved, as the population structure may already generate limit cycle behavior all by itself. Paradox of enrichment type population cycles can dominate, however, in large areas of parameter space.

A third class of phenomena again has to do with the spatial structure of populations, like population waves (see e.g., Okubo 1980, Murray 1989, Van den Bosch et al. 1990, Hadeler and Rothe 1975, Shigesada and Kawasaki 1986) and spatially mediated coexistence (see e.g., Yodzis 1989, and the references therein, and Tilman 1988).

A fourth class of phenomena refers to characteristics of the p-state. An example is provided by the question of whether the distribution of the sizes is unimodal or bimodal.

Clearly this list can be extended. Our main message, though, is that it is all too easy to get lost in the riches of our conceptual or simulated worlds unless we adhere to rather strict rules for organizing our results.

16. On Robustness, Invariance Principles and Universality Conjectures

Prototype problems are worth their salt only if the inferred model properties are robust, or better still, have some kind of "universality". (The latter is presently a very hot topic in statistical physics.) The problem is one of ascertaining that this is indeed the case. Some possible indicators are:

(i) The model has been derived through a more or less rigorous limit argument from a class of more complicated models, and the conclusions concern phenomena which survive the limit procedure (compare the "invariance principles" from probability theory). Examples are provided by the introduction of deterministic models in sections 3 and 4, and some of the model simplification procedures from section 12.

(ii) The phenomena belong to a class which in some, usually more restricted (e.g. an ODE) context, has been found to be structurally stable. Examples are the well established bifurcation patterns of equilibria and periodic attractors (see e.g., Golubitsky and Schaeffer 1985, Ruelle 1989b).

(iii) The conclusions turn out in hindsight to allow a direct mechanistic interpretation. An example is provided by the work of Van den Bosch et al. (1988) on cannibalism models. These authors were able to interpret the analytical criteria for the occurrence of various bifurcation patterns in terms of energetic considerations which clearly transcended their specific model assumptions.

(iv) The model allows some approximate self-similarity on rescaling, i.e., the behavior of different sized aggregates of (generalized) individuals can be considered as approximately the same, possibly after one or more additional rescalings, e.g. of time. In that case the associated phenomena satisfy corresponding scaling laws, and more importantly, are independent of the details of the assumed spatial structure. Many of the universality principles of statistical physics belong to this category. This idea applies in particular to many problems with local interactions. As such, it is of considerable interest, as it provides at least some means of organizing the extensive simulation efforts in this area.

(v) ? (More rules of thumb comparable to the preceding ones would be most welcome!)

Some Open Problems

17. On Some Mathematical Loose-Ends and/or Open Problems

This section is first of all a short apology for everything which we presented as a mathematical fact, though strictly speaking it was only a conjecture inferred from a number of examples backed up by heuristic arguments. For biologists like us this may suffice, but our mathematical colleagues might think differently.

For one thing, a lot of mathematical finery has still to be added before our very general statement about convergence to deterministic models in sections 3 and 6 can be considered a theorem in the strict sense attached to that word by modern probabilists (the same holds true for the spatially distributed case discussed in section 4, only more so). One approach which suggests itself follows the lines of attack for the existence and uniqueness problem for the deterministic models set forth in section 9.

Apart from those cases where we overstepped a little, there are the cases where we indicated clear open problems of a mathematical nature, like the derivation of the law of mass action from assumptions about movement patterns (section 5), the proof of existence and uniqueness theorems for fairly general physiologically structured population models (section 9), or the proof of Tikhonov's singular perturbation theorem, and extensions of it to infinite time intervals, for the dynamical systems from structured population models (section 12 (iv)).

Last but not least, we more or less implicitly indicated a number of open mathematical problems of various sorts. Examples are: finding ways for quickly estimating dominant Lyapunov exponents, the study of persistence (cf. Hofbauer and Sigmund 1988) for infinite dimensional dynamical systems, and the study of linear chain trickability for more complicated *i*-state spaces.

The problems discussed in the next two sections are premathematical, i.e. they have not yet been clarified to a stage that we can state precise conjectures. Rather, each section represents a hesitant attempt at finding the adequate framework for a better problem definition.

18. On the Disentangling of Temporal and Spatial Scales and the Essential Role of Stochasticity on the Individual Level

The environment of individuals varies on various temporal and spatial scales. Very fast or very local fluctuations can be done away with at the *i*-level by some appropriate averaging, as was implicit in the application of the (local) law of mass action. Very slow or wide-spread variations can be incorporated explicitly into the population model structure as time-varying parameters. The problem is how we should deal with intermediate scale inhomogeneities.

The problem of medium-scale environmental fluctuations gains in importance from the following observation. Physiologically structured population models which do not incorporate any stochasticity in the timing of offspring production by an individual show extremely singular behavior. Apparently some amount of individual stochasticity is necessary to smooth out irregularities deriving either from the initial condition or accumulated over the course of time. Usually we just fudge the necessary stochasticity by smearing out the production of offspring in an artificial manner (cf. Metz and Diekmann 1986, example II.6.3.1, and Heijmans and Metz 1989). It would be nice if we would have at least an inkling of how to proceed in a methodologically more rigorous and/or biologically more sound manner.

The continuous time formulation calls for a more detailed description of the processes at the i-level and hence the problem of how to realistically incorporate individual stochasticity in i-state movements and/or birth moments rears its head in a very explicit manner. Discrete time models are necessarily somewhat more descriptive, and in the case of matrix models (Caswell 1989) also more approximate, as the underlying continuous time processes have been lost sight of. Therefore the problem appears less troublesome in the discrete time case.

19. On Mechanism Versus Description

In our, admittedly biased, opinion we should always formulate our models as mechanistically as we can (see sections 14 and 16 for some reasons why). In fact, we prefer a model based on mechanistic considerations which is a bit of the mark, over a much better fitting purely descriptive model.

There are cases where it is possible to arrive at particular descriptive models in a rigorous manner starting from mechanistic considerations. An example is provided, again, by age representations. If the environment is constant and everybody is born equal, an age representation will always tell the whole population dynamical story, even though at the i-level we deal only with a description in terms of cohort averages. Knowledge that such a descriptive approach should work, and more importantly, a good idea of when it should work and when not, is a great asset. This means that we know when the underlying mechanism allows us to revert, without loss of mechanistic interpretability, to a modeling framework partially couched in terms of more easily obtainable descriptive statistics. It would be nice to know whether there exist other examples of descriptive models allowing a ready mechanistic justification, and, if such examples exist, to have a catalogue of all possible types.

More generally, we need a better insight into the balance between mechanism and description in various modeling contexts.

Final Remarks

In this paper we tried to convey three messages:

(i)   In sections 1 to 8 we gave a general heuristic outline of what we consider to be the proper mathematical framework for considering individual-based models. Sections 1 and 2 were more axiomatic, concentrating on conservation laws and the proper definitions of environment and population, while sections 5, 18 and 19 gradually concentrated on the hidden biological intricacies.

(ii) In sections 14 to 16 we sketched a modeling philosophy which, in our opinion, accords particularly well with the individual-based approach. We should never consider a model on its own. Within a more encompassing framework we can try to prove theorems of greater or lesser generality, and we can arrive at conjectures for such theorems through the diligent study of, preferably simple, prototype problems. If at all possible we should derive these prototype problems through explicit limit procedures since this provides us with implicit information about the range of applicability of the model results. Of course prototype problems are usually, again, whole families of models to which the previous principles apply.

(iii) Last but not least we concentrated on a particular class of prototype problems characterized by the assumption that individuals have sufficiently large interaction ranges. For these models we described in sections 9 to 13 some of the more general results, while their position relative to general individual-based models was outlined in sections 3 to 7.

What one calls results of course depends on one's perspective. A question can only take shape when the concepts needed to put it into words have been developed. The development of such concepts can be considered a result in itself. In this sense (i) to (iii) were all concerned with results. A second theme of this contribution has been the delineation of problem areas.

In section 8 we formulated what we consider scientifically to be the most challenging goal from a long term perspective. In sections 10 and 11 we indicated some pertinent partial results. Other goals are possible, however. We hope that our deliberations have helped you to put some of your own goals into a sharper focus.

The most important gray area in our research methodology has been indicated in section 7: We do not have available any off-the-shelf techniques for dealing somewhat more generally with purely short range interactions.

And the situation is even worse for problems in which interactions of different types and ranges are mixed.

Finally, sections 18 and 19 indicate some potential weak points in the whole individual-based modeling approach. Of course no approach is without faults. Our feeling is that a reductionist attitude has at least the great advantage of providing links between many disparate areas of understanding, and individual-based modeling is the ultimate reductionist strategy available to us ecologists!

Acknowledgements

This paper owes its existence to the willingness of many people to engage in extended discussions on its subject matter. The following people certainly deserve explicit recognition for their contributions as well as their patience: Odo Diekmann, Stephan Geritz, Bas Kooijman, Roger Nisbet and Wim van der Steen.

Literature Cited

Avnir D. (ed.). 1989. *The Fractal Approach to Heterogeneous Chemistry; Surfaces, Colloids, Polymers.* Wiley, Chichester.

Bacon, P. J. (ed.). 1985. *Population Dynamics of Rabies in Wildlife.* Academic Press, London.

Bergé, P., Y. Pomeau and Ch. Vidal. 1984. *Order Within Chaos: Towards a Deterministic Approach to Turbulence.* Wiley, New York.

Burkhardt, T. W. and J. M. J. van Leeuwen. (eds.). 1982. *Real-space Renormalization.* Springer-Verlag, Berlin.

Caswell, H. 1989. *Matrix Population Models.* Sinauer Associates, Sunderland, MA.

Chesson, P. L. 1986. *Environmental variation and the coexistence of species.* In *Community Ecology,* J. Diamond and T. Case. (eds.), pp. 240–256. Harper & Row, New York.

Chesson, P. L. 1988. Interactions between environment and competition: how fluctuations mediate coexistence and competitive exclusion. In *Community Ecology,* A. Hastings. (ed.), pp. 51–71, Lecture Notes in Biomathematics 77. Springer-Verlag, Berlin.

Chesson, P. L. 1989. A general model of the role of environmental variability in communities of competing species. In *Models in Population Biology.* A. Hastings. (ed.), pp. 97–123, Lectures on Mathematics in the Life Sciences 20. Springer Verlag, Berlin.

Chesson, P. L. and N. Huntly. 1988. Community consequences of life-history traits in a variable environment. *Ann. Zool. Fenn.* **25**:5–16.

Chesson, P. L. and N. Huntly. 1989. Short-term instabilities and long-term community dynamics. TREE. **4**:293–298.

Cliff, D. and P. Haggett. 1988. *Atlas of Disease Distributions*. Blackwell, Oxford.

Cox, J. T. and R. Durrett. 1988. Limit theorems for the spread of epidemics and forest fires. *Stochastic Processes Appl.* **36**:171–191.

de Roos, A. M. 1988. Numerical methods for structured population models; the escalator boxcar train. *Num. Meth. Part. Differ. Eq.* **4**:173–195.

de Roos, A. M. 1989. Daphnids on a Train: Development and Application of a New Numerical Method for Physiologically Structured Population Models. PhD Thesis, Leiden University, Leiden, the Netherlands.

de Roos, A. M., J. A. J. Metz, and O. Diekmann. Studying the dynamics of structured population models: a versatile technique and its application to Daphnia *Am. Nat.*, in press.

de Roos, A. M., J. A. J. Metz, E. Evers, and A. Leipoldt. 1990. A size dependent predator-prey interaction: who pursues whom? *J. Math. Biol.* **28**:609–643.

de Roos, A. M. and J. A. J. Metz. Towards a numerical analysis of the escalator boxcar train. In *Differential Equations with Applications in Biology, Physics and Engineering*. J. A. Goldstein, F. Kappel and W. Schappacher, (eds.), pp. 91–113, Marcel Dekker, New York. 1991.

Diekmann, O. 1991. Dynamics of structured populations. In *Journées de la Theorie Qualitative d'Equations Differentielles*. Marakesh,

Diekmann, O., K. Dietz, and J. A. P. Heesterbeek. The basic reproduction ratio R_O for sexually transmitted diseases. *Math. Biosci.*, in press.

Diekmann, O., J. A. P. Heesterbeek, and J. A. J. Metz. 1990. On the definition and the computation of the basic reproduction ratio R_O in models for infectious diseases in heterogeneous populations. *J. Math. Biol.* **28**:365–382.

Diekmann, O., J. A. J. Metz, and M. W. Sabelis. 1988. Mathematical models of predator/prey/plant interactions in a patch environment. *Exp. & Appl. Acarol.* **5**:319–342.

Diekmann, O., J. A. J. Metz, and M. W. Sabelis. 1989. Reflections and calculations on a prey-predator-patch problem. *Acta Appl. Math.* **14**:23–35.

Durrett, R. 1988a. Crabgrass, measles, and gypsy moths: an introduction to modern probability. *Bull. Am. Math. Soc.* **18**:117–143.

Durrett, R. 1988b. *Lecture notes on particle systems and percolation*. Wadsworth & Brooks/Cole, Pacific Grove, Ca.

Edelstein-Keshet, L. 1988. *Mathematical Models in Biology*. Random House, New York.

Ethier, S. N. and T. G. Kurtz. 1986. *Markov Processes: Characterization and Convergence*. Wiley, New York.

Geritz, S. A. H., J. A. J. Metz, P. G. L. Klinkhamer, and T. J. de Jong, 1988. Competition in safe sites. *Theor. Popul. Biol.* **33**:161–180.

Godfray, H. C. J. and M. P. Hasell. 1989. Discrete and continuous insect populations in tropical environments. *J. Anim. Ecol.* **58**:153–174.

Golubitsky, M. and D. G. Schaeffer. 1985. *Singularities and groups in bifurcation theory I*. Springer-Verlag, New York.

Golubitsky, M., I. Stewart, and D. G. Schaeffer. 1988. *Singularities and groups in bifurcation Theory II*. Springer-Verlag, New York.

Grasberger, P. 1983. On the critical behavior of the general epidemic process and dynamical percolation. *Math. Biosci.* **63**:157–172.

Grimmett, G. 1989. *Percolation*. Springer-Verlag, New York.

Guckenheimer, J. and P. Holmes. 1983. *Nonlinear Oscillations, Dynamical Systems, and Bifurcations of Vector Fields*. Springer-Verlag, New York.

Gurney, W. S. C. and R. M. Nisbet 1983. The systematic formulation of delay-differential models of age or size structured populations. In *Population Biology*, M. I. Freeman and C. Strobeck. (eds.), pp. 163–172, Lecture Notes in Biomathematics 52. Springer-Verlag, Berlin.

Gurney, W. S. C. and R. M. Nisbet. 1985. Fluctuation periodicity, generation separation, and the expression of larval competition. *Theor. Popul. Biol.* **28**:150–180.

Gurney, W. S. C., R. M. Nisbet, and S. P. Blythe. 1986. The systematic formulation of models of stage-structured populations. In *The Dynamics of Physiologically Structured Populations*. J. A. J. Metz and O. Diekmann. (eds.), pp. 474–494, Lecture Notes in Biomathematics 68. Springer-Verlag, Berlin.

Gurney, W. S. C., R. M. Nisbet, and J. H. Lawton. 1983. The systematic formulation of tractable single species models incorporating age structure. *J. Anim. Ecol.* **52**:479–496.

Hadeler, K. P. and F. Rothe. 1975. Travelling fronts in nonlinear diffusion equations. *J. Math. Biol.* **2**:251–263.

Heijmans, H. J. A. M. and J. A. J. Metz. 1989. Small parameters in structured population models and the Trotter-Kato theorem. SIAM *J. Math. Anal.* **20**:870–885.

Hofbauer, J. and K. Sigmund. 1988. *The Theory of Evolution and Dynamical Systems*. Cambridge University Press, Cambridge, England.

Hogeweg, P. 1988. Cellular automata as a paradigm for ecological modelling. *Appl. Math. Comput.* **27**:81–100.

Holden, A. V. (ed.). 1986. *Chaos*. Manchester University Press, Manchester.

Keizer, J. 1987. *Statistical Thermodynamics of Nonequilibrium Processes*. Springer-Verlag, New York.

Kurtz, T. G. 1981. *Approximation of Population Processes*. Society for Industrial and Applied Mathematics, Philadelphia.

Levin, S. 1970. Community equilibria and stability, and an extension of the competitive exclusion principle. *Am. Nat.* **104**:413–423.

Liggett, T. M. 1985. *Interacting Particle Systems*. Springer-Verlag, New York.

Ma, S. 1976. *Modern Theory of Critical Phenomena*. Benjamin, Reading, Ma.

Metz, J. A. J. and O. Diekmann. (eds.). 1986. *The Dynamics of Physiological Structured Populations*. Lecture Notes in Biomathematics 68. Springer-Verlag, Berlin.

Metz, J. A. J. and O. Diekmann. 1991. Exact finite dimensional representations of models for physiologically structured populations I: the abstract foundations of linear chain trickery. In *Differential Equations with Applications in Biology, Physics and Engineering*. J. A. Goldstein, F. Kappel and W. Schappacher. (eds.), pp. 269g–289g. Marcel Dekker, New York.

Metz, J. A. J., R. M. Nisbet, and S. A. H. Geritz. How should we define "fitness" for general ecological scenarios? *Trends in Evol. Ecol.* (accepted for publication).

Metz, J. A. J., A. M. de Roos, and F. van den Bosch. 1988. Population models incorporating physiological structure: a quick survey of the basic concepts and an application to size-structured population dynamics in waterfleas. In *Size-structured Populations*. B. Ebenman and L. Persson. (eds.), pp. 106–126. Springer-Verlag, Berlin.

Mollison, D. 1977. Spatial contact models for ecological and epidemic spread. *J. Roy. Statis. Soc. Ser. B* **39**:283–326.

Mollison, D. 1986. Modelling biological invasions: chance, explanation, prediction. *Phil. Trans. R. Soc. Lond. B Biol. Sci.* **314**:675–693.

Morse, P. M. and H. Feshbach. 1965. *Methods of Theoretical Physics. I*. McGraw Hill, New York.

Murdoch, W. W., R. M. Nisbet, S. P. Blythe, W. S. C. Gurney, and J. D. Reeve. 1987. An invulnerable age class and stability in delay-differential parasitoid-host models. *Am. Nat.* **129**:263–282.

Murray, J. D. 1989. *Mathematical Biology*. Springer-Verlag, Berlin.

Nisbet, R. M. and W. S. C. Gurney. 1982. *Modelling Fluctuating Populations*. Wiley, Chichester.

Nisbet, R. M. and W. S. C. Gurney. 1983. The systematic formulation of population models for insects with dynamically varying instar duration. *Theor. Popul. Biol.* **23**:114–135.

Nisbet, R. M. and W. S. C. Gurney. 1984. "Stage-structure" models of uniform larval competition. In *Mathematical Ecology*. S. A. Levin and T. G. Hallam, (eds.), pp. 97–113. Lecture Notes in Biomathematics 54. Springer-Verlag, Berlin.

Nisbet, R. M., S. P. Blythe, W. S. C. Gurney, and J. A. J. Metz. 1985. Stage-structure models of populations with distinct growth and development processes. *IMA J. Math. Appl. Med. Biol.* **2**:57–68.

Okubo, A. 1980. *Diffusion and Ecological Problems: Mathematical Models*. Springer-Verlag, Berlin.

Okubo, A. 1986. Dynamical aspects of animal groupings: swarms, schools, flocks and herds. *Adv. Biophys.* **22**:1–94.

Pacala, S. W. 1986a. Neighbourhood models of plant population dynamics. II: multispecies models of annuals. *Theor. Popul. Biol.* **29**:262–292.

Pacala, S. W. 1986b. Neighbourhood models of plant population dynamics IV: single and multispecies models of annuals with dormant seeds. *Am. Nat.* **128**:859–878.

Pacala, S. W. 1987. Neighbourhood models of plant population dynamics III: Models with spatial heterogeneity in the physical environment. *Theor. Popul. Biol.* **31**:359–392.

Pacala, S. W. and J. A. Silander Jr. 1985. Neighbourhood models of plant population dynamics I: Single-species models of annuals. *Am. Nat.* **125**:385–411.

Rosenzweig, M. L. 1971. Paradox of enrichment: destabilization of exploitation ecosystems in ecological time. *Science* **171**:385–387.

Ruelle, D. 1989a. *Chaotic Evolution and Strange Attractors*. Cambridge University Press, Cambridge, England.

Ruelle, D. 1989b. *Elements of Differentiable Dynamics and Bifurcation Theory*. Academic Press, Boston.

Schaffer, W. M. 1985. Can nonlinear dynamics elucidate mechanisms in ecology and epidemiology? *IMA J. Math. Appl. Med. Biol.* **2**:221–252.

Schaffer, W. M. 1988. Perceiving order in the chaos of nature. In *Evolution of Life Histories*, M. Boyce. (ed.), pp. 313–350. Yale University Press, New Haven.

Schuster, H. G. 1988. *Deterministic Chaos: an Introduction*. VCH, Weinheim.

Shigesada, N. and K. Kawasaki. 1986. Travelling periodic waves in heterogenous environments. *Theor. Popul. Biol.* **30**:143–160.

Silander, J. A. Jr. and S. W. Pacala. 1990. The application of plant population dynamic models to understanding plant competition. In *Perspectives on Plant Competition*. J. B. Grace and D. Tilman (eds.), pp. 67–91. Academic Press, San Diego.

Stauffer, D. 1985. *Introduction to Percolation Theory*. Taylor and Francis, London.

Taylor, P. D. 1988a. Inclusive fitness models with two sexes. *Theor. Popul. Biol.* **34**:145–168.

Taylor, P. D. 1988b. An inclusive fitness model for dispersal of offspring. *J. Theor. Biol.* **130**:363–378.

Taylor, P. D. 1989. Evolutionary stability in one-parameter models under weak selection. *Theor. Popul. Biol.* **36**:125–143.

Taylor, P. D. 1990. Allele-frequency change in a class-structured population. *Am. Nat.* **135**:95–106.

Thompson, J. M. T. and H. B. Stewart. 1986. *Nonlinear Dynamics and Chaos*. Wiley, Chichester.

Tikhonov, A. N., A. B. Vasil'eva and A. G. Sveshnikov. 1985. *Differential Equations*. Springer-Verlag, Berlin.

Tilman, D. 1988. *Plant Strategies and the Dynamics and Structure of Plant Communities*. Princeton University Press, Princeton, N.J.

Toffoli, T. and N. Margolus. 1987. *Cellular Automata Machines: A New Environment for Modelling*. MIT Press, Canbridge, MA.

Tuljapurkar, S. 1989. Uncertain life: Demography in random environments. *Theor. Popul. Biol.* **35**:227–294.

Tuljapurkar, S. 1990. *Population Dynamics in Variable Environments*. Lecture Notes in Biomathematics 85, Springer-Verlag, Berlin.

Turchin, P. 1989. Beyond simple diffusion: models of not-so-simple movement of animals and cells. *Comments Theor. Biol.* **1**:65–83.

van den Bosch, F., A. M. de Roos and W. Gabriel. 1988. Cannibalism as a life boat mechanism. *J. Math. Biol.* **26**:619–633.

van den Bosch, F., J. A. J. Metz and O. Diekmann. 1990. The velocity of spatial population expansion. *J. Math. Biol.* **28**:529–565.

Verboom, J., K. Lankester and J. A. J. Metz. Linking local and regional dynamics in stochastic metapopulation models. *Biol. J. Linn. Soc.* 1991. **42**:3g–55.

Warner, R. R. and P. L. Chesson. 1985. Coexistence mediated by recruitment fluctuations: a field guide to the storage effect. *Am. Nat.* **125**:769–787.

Wolfram, S. ed. 1986. *Theory and Applications of Cellular Automata*. World Scientific, Singapore.

Yodzis, P. 1989. *Introduction to Theoretical Ecology*. Harper & Row, New York.

6

The Dynamics of a Size-Structured Intraspecific Competition Model with Density Dependent Juvenile Growth Rates

J. M. Cushing and Jia Li

ABSTRACT. The dynamics of a size-structured population in which adult fertility correlates with body size, and in which adult body size at maturation is dependent (through competitive effects) upon population density during juvenile growth, are studied. A simple discrete model for a population with one juvenile size class and two adult size classes, one consisting of larger and more fertile individuals than the other, is derived and analyzed. The competitive effects on juvenile growth are separated into those due to competition from other juveniles and those due to adults. Parameter regions are determined in which the dynamics equilibrate, approach 2-cycles, or result in chaotic oscillations. The results suggest that intra-specific competition of this kind between juveniles and adults is destabilizing for either small or very large inherent net reproductive rates, strong competitive effects tending to result in a synchronous 2-cycle in which juveniles and adults avoid competition. For intermediate values of the inherent net reproductive rate, however, intra-specific competition has a stabilizing influence, promoting equilibration where there would otherwise be chaotic oscillations.

Introduction

Biological populations have a natural propensity for exponential growth which in the long run, of course, must be held in check. Nonlinear models of population growth do this by incorporating "density effects" which serve to decrease either fertility or survival. The majority of models for population growth describe highly aggregate population level variables, such as total population size or biomass, and as a result can account for intraspecific competition due to increased population density in only a very qualitative manner. In order to better account for the mechanisms causing intraspecific competition, it is necessary to use a structured population model in which relevant physiological or behavioral characteristics of individual organisms are distinguished.

There are many means by which individuals of a species might compete for common resources, and by which such competition might express itself. One type of intraspecific competition which has received recent attention is that which can occur between juvenile and adult members of the population (May et al. 1974, Bellows 1982, Tschumy 1982, Ebenman 1987, Ebenman 1988a, Cushing and Li 1989, Cushing and Li 1991, Cushing 1991). The possibility of niche overlap between juveniles and adults exists for species with relatively "simple" life cycles, as opposed to those that undergo significant metamorphosis or otherwise experience radical ontogenetic niche shifts during their development. This includes most reptiles, fish, mammals,and hemimetabolous insects.

The dynamical consequences of juvenile and adult competition has been investigated by means of simple age-structured models by several authors with the conclusion that such competition is usually destabilizing (May et al. 1974, Bellows 1982, Tschumy 1982), but that under certain circumstances can be stabilizing (Ebenman 1987, Ebenman 1988a, Cushing and Li 1989, Cushing and Li 1991, Cushing 1991). In these age-structured models the effects of juvenile vs. adult competition result in either reduced juvenile survival or adult fertility.

Another important effect of competition experienced by juveniles, which is not included in any of these age-structured models, is that of slowed growth. Body size is often a more important physical attribute than is chronological age (Werner and Gilliam 1984, Ebenman and Persson 1988). Body size, not age, is often the key factor in determining vital rates such as fertility, survival, and individual growth rates as well as susceptibility to environmental hazards (such as predation and cannibalism), metabolic demands, etc. Thus, intraspecific competition can slow juvenile growth, reduce size at maturation, and consequently affect population growth by reducing fertility (Wilbur 1980, Botsford 1981, Prout and McChesney 1985, Ebenman and Persson 1988).

In an attempt to study the consequences of slowed juvenile growth due to intraspecific competition, Ebenman (1988b) placed, in a rather *ad hoc* manner, a time delay in his age-structure difference equation model and concluded that strong competition of this sort has a destabilizing effect on the population dynamics. As pointed out by Ebenman, however, it would be more appropriate to analyze this phenomenon by means of a size-structured model. In a later paper, Ebenman (1988b) extended his model to include variable adult size. This extension, however, is sufficiently complicated that virtually no analytical results are attainable and it must be analyzed numerically.

In this paper we consider a model of size-structured, intraspecific competition in which increased competition during juvenile growth reduces size at adulthood and thereby reduces adult fertility. Our goal here is to derive a model which is simple enough to be as analytically tractable as possible, and yet still capture these essential features. We wish to understand the asymptotic dynamics of the model and to draw some conclusions about the stabilizing or destabilizing effects of this kind of intraspecific competitive interaction.

Model Equations

We wish to write down a model for the dynamics of a population in which an adult individual's fertility is correlated with its body size, which in turn is dependent upon the amount of competition experienced during juvenile growth. In this paper we will consider only the simplest version of the kind of discrete model we have in mind, which nonetheless captures these basic features. We will ignore density dependent fertility and survival rates and focus on the effects of density on growth rates. Such extensions of the model will be studied in a future paper.

Imagine a population in which (surviving) juveniles mature at one of two possible adult sizes after a fixed unit of time $t = 1$. The fraction $\phi(W)$ of surviving juveniles that mature at the larger adult size is dependent upon the amount of competitive pressure experienced during juvenile growth, which is measured here by a weighted total population size $W = J + \beta_1 A_1 + \beta_2 A_2$. Here J is the number or density of juveniles, A_1 is the number of smaller adults, and A_2 is the number of larger adults. Thus the competition coefficients $\beta_i \geq 0$ measure the effect that an adult individual of size i has on juvenile growth, relative to the effect of a juvenile individual. If we consider a semelparous population in which there is no adult survival after one unit of time, then the numbers present in each size category after the elapse of one unit of time are given by

$$
\begin{align}
\text{(a)} \quad & J(t + 1) = nf_1 A_1(t) + nf_2 A_2(t) \\
\text{(b)} \quad & A_1(t + 1) = \pi(1 - \phi(W(t)))J(t) \\
\text{(c)} \quad & A_2(t + 1) = \pi\phi(W(t))J(t).
\end{align} \tag{2.1}
$$

for $t = 1,2,3,\ldots$ Of course we are only interested in non-negative solutions of Eqs. (2.1), and in particular for initial conditions

$$
J(0) \geq 0, \, A_1(0) \geq 0, \, A_2(0) \geq 0. \tag{2.2}
$$

In Eqs. (2.1), π is the probability that a juvenile survives to adulthood and nf_i is the number of offspring produced by an adult of size i during one unit of time. The coefficient n is used here to introduce the inherent net reproductive rate (expected number of offspring per individual per life time at low densities, taking survival into account) explicitly into the model and its analysis. The inherent net reproductive rate is given by $nf_1\pi(1 - \phi(0)) + nf_2\pi\phi(0)$ and will equal n if, without loss in generality, f_1 and f_2 satisfy the normalization

$$f_1\pi(1 - \phi(0)) + f_2\pi\phi(0) = 1. \tag{2.3}$$

These parameters are assumed here to be constants, i.e. not density dependent.

The effects of increased population density on juvenile growth will be assumed to be deleterious, and therefore ϕ is a decreasing function of W. For analytic simplicity only, we will assume here that in the absence of competitive effects, a juvenile will always grow to the larger adult size and that, in the other extreme, at very high (infinitely large) densities all individuals grow to the smaller size. We then have the following condition on the fraction $\phi(W)$:

$$\phi \in C^1(R^+, [0, 1]), \phi(0) = 1, \phi(+\infty) = 0, \phi'(W) < 0.$$

In this case, Eq. (2.3) implies

$$f_2 = 1/\pi.$$

Furthermore, we are interested in the case when larger adults are more fertile so that

$$f_1 < f_2 = 1/\pi.$$

Properties of the Model

Equations (2.1) have the trivial equilibrium $(J, A_1, A_2) = (0, 0, 0)$. It is shown in the Appendix (Theorem 1) that if $n < 1$ then $(0, 0, 0)$ is globally attracting. This makes biological sense in that if an individual cannot at least replace itself at low population densities then the population will go to extinction.

On the other hand, for $n > n_0 = f_2/f_1 = 1/f_1\pi > 1$ it follows from Theorem 2 of the Appendix that the population will increase without bound. This is an artifact of the simplifying assumptions made here. Namely, if the smaller adult class can sustain the population even at arbitrarily large population densities then the population will grow without bound because no controlling density effects on fertility or survival are present in this simple model.

Consequently, we focus our attention on values of the inherent net reproductive rate between 1 and $n_0 > 1$. From the equilibrium equations

$$J = nf_1A_1 + nf_2A_2$$
$$A_1 = \pi(1 - \phi(W))J$$
$$A_2 = \pi\phi(W)J$$

where $W = J + \beta_1A_1 + \beta_2A_2$ follows

$$J = nf_1\pi(1 - \phi(W))J + nf_2\pi\phi(W)J.$$

If $J = 0$, then clearly $A_1 = A_2 = 0$, i.e. the only equilibrium with no juveniles present is the trivial equilibrium. A nontrivial equilibrium must have $J > 0$, in which case we obtain

$$nN(W) = 1, N(W) = f_1\pi(1 - \phi(W)) + f_2\pi\phi(W),$$

an equation which states the biologically obvious fact that at any nontrivial equilibrium the (density dependent) net reproductive rate $nN(W)$ must be 1. We can now derive a formula for the equilibrium states by noting that:

$$N(0) = 1, N(+\infty) = 1/n_0, N'(W) < 0.$$

Thus for $n\epsilon(1, n_0)$ the equilibrium weighted population size is given by:

$$W = W(n) = N^{-1}(1/n).$$

From this and the equilibrium equations above we obtain the equilibria

$$J = \frac{W(n)}{1 + \beta_1\pi + (\beta_2 - \beta_1)\pi\phi(W(n))}$$
$$A_1 = \pi(1 - \phi(W(n)))J, A_2 = \pi\phi(W(n))J. \qquad (3.1)$$

We have arrived at the result that there exists a unique, positive equilibrium for all values of the inherent net reproductive rate n between 1 and n_0. By a positive equilibrium is meant an equilibrium in which all three classes are present.

Note that the branch of positive equilibrium given by (3.1) bifurcates from the trivial equilibrium at the critical value $n = 1$ and becomes unbounded as n approaches n_0. (See Cushing 1988 for a discussion of the generality of such global continuum branches of equilibria in discrete population models.) We will study the stability of these equilibria below, but first we note that the Jacobian of Eq. (2) at the trivial solution

$$
\begin{pmatrix}
0 & nf_1 & nf_2 \\
0 & 0 & 0 \\
\pi & 0 & 0
\end{pmatrix}
\tag{3.2}
$$

has an eigenvalue $\lambda = +1$ when $n = 1$, as is to be expected from the bifurcation of the nontrivial equilibria. Note, however, that when $n = 1$ this Jacobian also has an eigenvalue $\lambda = -1$. This indicates that perhaps there is also present a bifurcating branch of 2-cycles. This borne out by Theorem 3 of the Appendix from which we find that *for* $1 < n < n_0$ *there exists, in addition to a unique positive equilibrium, a unique synchronous 2-cycle.* By a synchronous 2-cycle is meant a period two solution of (2.1) in which the juveniles are synchronized to appear all together at alternating time periods, i.e. a 2-cycle in which $J(t)$ alternates between 0 and a positive value. Specifically, the bifurcating 2-cycles are given by

$$
\begin{aligned}
&(J(t), A_1(t), A_2(t)) \\
&= \begin{cases}
(W(n), 0, 0), t = 0, 2, 4, \ldots \\
(0, \pi(1 - \phi(W(n)))W(n), \pi\phi(W(n))W(n)) \quad t = 1, 3, 5, \ldots
\end{cases}
\end{aligned}
\tag{3.3}
$$

for $1 < n < n_0$.

Of interest now are the stability properties of the equilibria and 2-cycles. Analytically, we can determine local asymptotic stability for n near 1.

First, consider the stability of the trivial equilibrium $(0, 0, 0)$. We have seen that this equilibrium is stable for $n < 1$. For $n > 1$, the eigenvalues of the Jacobian (3.2), which are $\lambda = 0, \pm\sqrt{n}$, show that $(0, 0, 0)$ is unstable. Thus, for $n > 1$ we expect a viable population, although it is not yet clear what the asymptotic dynamics are.

To determine the local stability of the equilibria (3.1) we need to determine the eigenvalues λ of the Jacobian

$$
M = \begin{pmatrix}
0 & nf_1 & nf_2 \\
\pi(1 - \phi(W)) - \pi\phi'(W)J & -\pi\phi'(W)\beta_1 J & -\pi\phi'(W)\beta_2 J \\
\pi\phi(W) + \pi\phi'(W)J & \pi\phi'(W)\beta_1 J & \pi\phi'(W)\beta_2 J
\end{pmatrix}
\tag{3.4}
$$

of equations (2.1) evaluated at the equilibrium (3.1). If $M = M(n)$ is treated as a function of n, then near $n = 1$ we can write $M = M(1) + M'(1)(n - 1) + 0((n - 1)^2)$ and

$$\lambda = \lambda(n) = \lambda(1) + \lambda'(1)(n - 1) + 0((n - 1)^2).$$

If we denote the right and left (row) eigenvectors by

$$v(n) = v(1) + v'(1)(n - 1) + 0((n - 1)^2),$$
$$w(n) = w(1) + w'(1)(n - 1) + 0((n - 1)^2)$$

respectively, then a substitution of these expansions into $Mv^T = \lambda v^T$ yields the formula (see Caswell 1989)

$$\lambda'(1) = w(1)M'(1)v^T(1)/w(1)v^T(1). \tag{3.5}$$

(w^T denotes the transpose of w.) We are interested in the two cases $\lambda(1) = \pm 1$, the eigenvalues of $M(1)$, which is given by (3.2) with $n = 1$. The eigenvectors $v_{\pm}(1)$, $w_{\pm}(1)$ of $M(1)$ corresponding to these two eigenvalues are

$$v_{+}(1) = [1, 0, \pi], \; w_{+}(1) = [\pi, \pi f_1, \pi f_2]$$
$$v_{-}(1) = [1, 0, -\pi], \; w_{-}(1) = [\pi f_2, 0, -\pi]$$

respectively. The derivative $M'(1)$ is straightforwardly calculated from (3.4). These calculations and formula (3.5) lead finally to

$$\lambda_{+} = 1 - \frac{1}{2}(n - 1) + 0((n - 1)^2),$$
$$\lambda_{-} = -1 + \frac{1 - \beta_2 \pi}{2(1 + \beta_2 \pi)}(n - 1) + 0((n - 1)^2).$$

We conclude that for $n > 1$ sufficiently close to 1, the bifurcating positive equilibria (3.1) are stable if $\beta_2 < 1/\pi = f_2$ and unstable if $\beta_2 > 1/\pi = f_2$.

Each of the triplets in the 2-cycle (3.3) is an equilibrium of the first composite of the equations (2.1). The 2-cycle is stable if the eigenvalues of the Jacobian matrix of this first composite evaluated at one of the triplets (say the first one ($W(n)$, 0, 0)) are inside the complex unit circle. This Jacobian turns out, amazingly enough, to be easier to analyze than that of the equilibria (3.1). This is because the first column has 0 as its second and third entries, which means the entry in the upper right hand corner is one eigenvalue while the other two can be found from the lower right hand 2×2

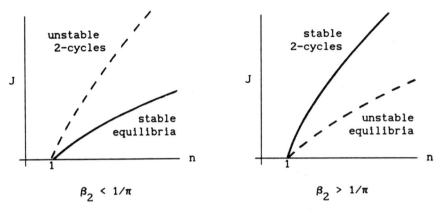

Figure 6.1. Two branches bifurcate from the trivial equilibrium $(J, A_1, A_2 = (0, 0, 0)$ at the critical value $n = 1$, one consisting of equilibria and the other of synchronous 2-cycles. These branches are schematically represented in these graphs of the maximum of the J component of the attractor against n. The solutions on the branches have opposite stability properties, depending upon the magnitude of the competition coefficient β_2.

corner matrix. Skipping the straightforward calculation of the Jacobian, it turns out that one eigenvalue is $\lambda = 0$ and that the other two are

$$\lambda_1 = 1 + n(f_2 - f_1)\pi\phi'(J(n))J(n)$$
$$\lambda_2 = \pi n f_1 + n(f_2 - f_1)\pi\phi(W_1(n))$$

where

$$W_1(n) = nf_1A_1(n) + nf_2A_2(n)$$
$$+ \beta_1\pi(1 - \phi(W(n)))J(n) + \beta_2\pi\phi(W(N))J(n)$$

from which we find

$$\lambda_1 = 1 - (n - 1) + 0((n - 1)^2)$$
$$\lambda_2 = 1 + (1 - \pi\beta_2)(n - 1) + 0((n - 1)^2).$$

Thus for $n > 1$ sufficiently close to 1, the bifurcating synchronous 2-cycles (3.3) have the opposite stability of the equilibria (3.1), namely they are stable if $\beta_2 > 1/\pi$ and unstable if $\beta_2 < 1/\pi$.

The two possible bifurcation diagrams are schematically represented in Fig. 6.1. These results indicate that, at least for populations with small inherent net reproductive rates, strong adult competition affecting juvenile growth and size at maturation is destabilizing. This is implied by the destabilization of the equilibrium and the change from equilibrium to oscillatory dynamics as β_2 is increased.

Whether these stability conclusions concerning the equilibria and the 2-cycles remain intact, and also whether other asymptotic dynamics are possible, for larger values of the inherent net reproductive rate n can be investigated numerically. Fig. 6.2 shows typical bifurcation diagrams, using n as a bifurcation parameter, for the two cases $\beta_2 < 1/\pi$ and $\beta_2 > 1/\pi$. Notable from these diagrams are the facts that an equilibrium branch originally stable ($\beta_2 < 1/\pi$) can lose its stability through a "Hopf" bifurcation to an invariant circle in which the attractors are aperiodic, and then regain it again for larger n. In the case where the bifurcating 2-cycle is stable ($\beta_2 > 1/\pi$), the 2-cycle ultimately loses its stability, sometimes to a restabilized equilibrium and sometimes to a period doubled 4-cycle (not shown). It can also happen that for certain n values there exist both a stable equilibrium and a stable 2-cycle. Thus, equations (2.1) can exhibit exotic dynamics.

Fig. 6.3 shows bifurcation diagrams using the competition coefficient β_2 as a parameter. Typically, with increasing β_2 one sees the loss of equilibrium stability to a stable 2-cycle for either n near 1 (as proved above) or for very large n. For intermediate values of n (where one can find "chaotic" aperiodic dynamics; see Fig. 6.1), one typically observes an opposite scenario, where the dynamics are unstable and "chaotic" for smaller values of β_2, but are stabilized to equilibrium dynamics for large values of β_2. In the latter case, increased intraspecific competition can be said to be stabilizing.

Concluding Remarks

Model equations (2.1) describe in a relatively simple way the dynamics of a size-structured population whose juveniles mature to either a small or a large adult size depending upon the intensity of competition due to population density, and in particular due to competition from the larger sized adults as measured by the competition coefficients β_1 and β_2. Direct density effects on adult fertility have been ignored, as have density effects on juvenile survival. By investigating the asymptotic dynamics of this model, both numerically and analytically, we have seen that strong intraspecific competition from adults of this kind can result in either equilibrium, oscillatory, or aperiodic "chaotic" dynamics, depending upon relative paramter values in the model. The results suggest that for populations with either very low or very high inherent net reproductive rates, this kind of intraspecific competition is destabilizing. In these cases the dynamics of the model change with increased competition coefficient from equilibration to periodic 2-cycles in which juveniles and adults appear in alternate time intervals. That is to

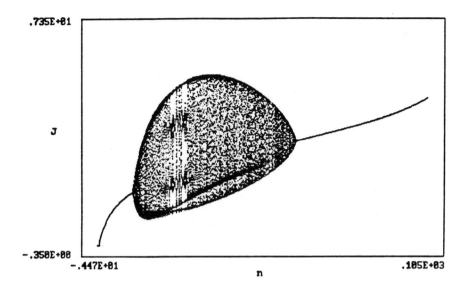

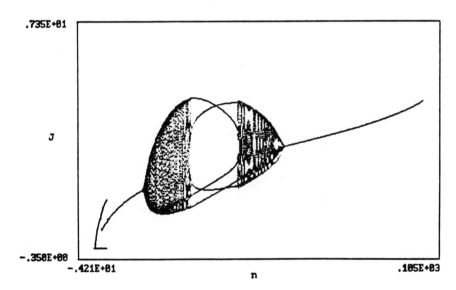

Figure 6.2. The juvenile component of the attractor with $\phi(W) = \exp(-W)$ in (2.1) is plotted against the inherent net reproductive rate n. $\pi = 0.9$, $f_1 = 0.01$, $f_2 = 1.11$, $\beta_1 = 0.5$, $\beta_2 = 0.9$ (top), 1.5 (bottom)

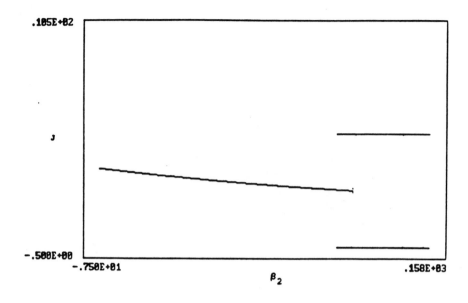

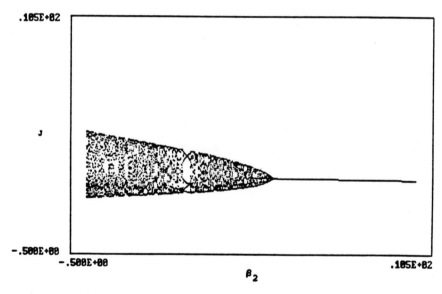

Figure 6.3. The juvenile component of the attractor with $\phi(W) = \exp(-W)$ in (2.1) is plotted against the competitive coefficient β_2. $\pi = 0.9$, $f_1 = 0.01$, $f_2 = 1.11$, $\beta_1 = 0.5$, $n = 50$ (top), 70 (bottom).

say, increased competition results in the populations adjusting so as to avoid juvenile and adult competition. On the other hand, the model also implies that for populations with "intermediate" values of the inherent net reproductive rate, strong intraspecific competition between adults and juveniles is stabilizing in that increases in the competition coefficient cause the dynamics to equilibrate from otherwise chaotic aperiodic motion.

It would be interesting to see how robust these conclusions would be if some of the simplifying assumptions were dropped. In particular, it would be of interest to assume that juvenile survival is density dependent. This is done in Ebenman (1988b) under the assumption that juvenile survival and individual growth are inversely related to competition through a concept of "plasticity." Fertility in general is also density dependent. Model (2.1) assumes that adults reproduce only once (semelparity) and die. How are these conclusions affected by possible adult survival and repeated reproduction (iteroparity)? We plan to study extensions of model equations (2.1) that include these phenomena in future research.

Appendix

THEOREM 1. If $n < 1$ then any solution of (2.1)–(2.2) tends geometrically to $(J, A_1, A_2) = (0, 0, 0)$.

Proof: From (2.1b, c) follows $A_1(t) + A_2(t) = \pi J(t)$ and from (2.1a)

$$
\begin{aligned}
J(t + 1) &= nf_1\pi J(t) + n(f_2 - f_1)A_2(t) \\
&= nf_1\pi J(t) + n(f_2 - f_1)\pi\phi(W(t - s))J(t - 1).
\end{aligned}
$$

Thus

$$0 \le J(t + 1) \le n\pi f_1 J(t) + n(1 - \pi f_1)J(t - 1)$$

and consequently $0 \le J(t) \le x(t)$ where $x(t)$ satisfies $x(0) = J(0)$, $x(1) = J(1)$ and

$$x(t + 1) = n\pi f_1 x(t) + n(1 - \pi f_1)x(t - 1).$$

The quadratic characteristic equation associated with this second order linear difference equation can easily be shown to have two real roots whose magnitudes are less than one when $n < 1$. Thus, in this case, $x(t)$ and hence $J(t)$ tend geometrically to zero. Equations (2.1b, c) then imply both $A_1(t)$ also tend geometrically to zero.

THEOREM 2. *If $n > n_0 = f_2/f_1$, then $J(t) \to +\infty$.*

Proof: This follows immediately from the inequality

$$J(t + 1) = nf_1\left[A_1(t) + \frac{f_2}{f_1}A_2(t)\right] \geq nf_1[A_1(t) + A_2(t)] = \frac{n}{n_0}J(t - 1).$$

THEOREM 3. For $n\epsilon(1, n_0)$ there exists a unique synchronous 2-cycle given by (3.3).

Proof: The first composites of equations (2.1) are

$$J(t + 2) = nf_1\pi(1 - \phi(W(t)))J(t) + nf_2\pi\phi(W(t))J(t)$$
$$A_1(t + 2) = \pi(1 - \phi(W_1(t)))(nf_1A_1(t) + nf_2A_2(t))$$
$$A_2(t + 2) = \pi\phi(W_1(t))(nf_1A_1(t) + nf_2A_2(t))$$

where

$$W_1(t) = nf_1A_1(t) + nf_2A_2(t) + \beta_1\pi(1 - \phi(W(t))J(t) + \beta_2\pi\phi(W(t))J(t)$$

whose equilibrium equations are

$$J = nf_1\pi(1 - \phi(W)J + nf_2\pi\phi(W)J$$
$$A_1 = \pi(1 - \phi(W_1))(nf_1A_1 + nf_2A_2)$$
$$A_2 = \pi\phi(W_1)(nf_1A_1 + nf_2A_2).$$

It is easy to see that $(J, A_1, A_2) = (W(n), 0, 0)$ solves these equations and that this is the only solution in which $J \neq 0$ and $A_1 = 0$.

Acknowledgements

J. M. Cushings's research supported by the divisions of Applied Mathematics and Population Biology/Ecology Divisions of the National Science Foundation under grant No. DMS-8902508. Jia Li's research supported by the Department of Energy under contracts W-7405-ENG-36 and KC-07-01-01

Literature Cited

Bellows, T. S. 1982. Analytical models for laboratory populations of Callosobruchus chinensis, and C. maculatus (Colioptera, Bruchidae). *J. Anim. Ecol.* **51**:263–287.

Botsford, L. W. 1981. The effects of increased individual growth rates on depressed population size. *A. Nat.* **117**:38–63.

Caswell, H. 1989. *Matrix Population Models*. Sinauer Associates, Sunderland, MA.

Cushing, J. M. 1988. Nonlinear matrix models and population dynamics. *Nat. Resour. Model.* **2**:539–580.

Cushing, J. M. 1991. Some delay models for juvenile vs. adult competition, In *Proceedings of the Claremont Conference on Differential Equations and Applications to Biology and Population Dynamics*, S. Busenberg and M. Martelli (eds.), Springer-Verlag, Berlin.

Cushing, J. M. and J. Li, 1992. Intra-specific competition and density dependent juvenile growth, *Bull. Math. Biol.* (in press).

Cushing, J. M. and J. Li. 1989. On Ebenman's model for the dynamics of a population with competing juveniles and adults, *Bull. Math. Biol*, **51**:687–713.

Cushing, J. M. and J. Li. 1991. Juvenile versus adult competition, *J. Math. Biol.* **29**:457–473.

Ebenman, B. 1987. Niche differences between age classes and intraspecific competition in age-structured populations, *J. Theor. Biol.* **124**:25–33.

Ebenman, B. 1988a. Competition between age classes and population dynamics. *J. Theor. Biol.* **131**:389–400.

Ebenman, B. 1988b. Dynamics of age- and size-structured populations: intraspecific competition. In *Size-Structured Populations: Ecology and Evolution*, B. Ebenman and L. Persson, (eds.), pp. 127–139. Springer-Verlag, Berlin.

Ebenman, B. and L. Persson. 1988. *Size-Structured Population: Ecology and Evolution*. Springer-Verlag, Berlin.

May, R. M., G. R. Conway, M. P. Hassell, and T. R. E. Southwood. 1974. Time delays, density-dependence and single-species oscillations. *J. Anim. Ecol.* **43**:747–770.

Prout, T. and F. McChesney. 1985. Competition among immatures affects their adult fertility: Population dynamics, *Am. Nat.* **126**:521–558.

Tschumy, W. O. 1982. Competition between juveniles and adults in age-structured populations. *Theor. Popul. Biol.* **21**:255–268.

Werner, E. E. and J. F. Gilliam. 1984. The ontogenetic niche and species interactions in size-structured populations, *Ann. Rev. Ecol. Syst.* **15**:393–425.

Wilbur, H. M. 1980. Complex life cycles, *Ann. Rev. Ecol. Syst.* **11**:67–98.

7

Parallel Computers and Individual-based Models: An Overview

James W. Haefner

ABSTRACT. **In this expository overview, I briefly review the basics of computer architecture as they relate to parallel computers. Distributed memory, multiprocessor systems are emhasized. I cover methods to parallelize some fundamental types of ecological simulation models: foodweb models, individual-based population models, population models based on partial differential equations, and individual movement models. Recent developments in parallel operating systems and programming tools on multiprocessors are reviewed. Because of complex relationships between parallel computer architecture and efficient algorithms, I conclude that ecological modelers will need to become more acquainted with hardware than previously.**

Introduction

The purpose of this paper is to briefly describe parallel computers and their potential applications to individual-based and individual-oriented models. I begin with a basic description of computer architecture that includes parallel, multiple processor designs. The second part concerns a few key concepts related to programming distributed memory multiprocessor machines. I introduce the tactics for parallelizing an algorithm, the ideas of processor mapping and load balancing, communicating sequential processes and message-passing approaches, and finally, processor synchronicity and deadlock. Part three describes ecological simulation model applications to parallel computers. This includes a description of an application to foodweb simulations based on ordinary differential equations. I propose some approaches to other types of simulation models of direct relevance to individual-based models. A final section outlines a few tools that are available to facilitate programming on multiprocessor machines. The discussion is intended primarily for ecologists unfamiliar with these ideas, and references are limited to the standard texts and survey articles, and not the primary literature.

Hardware

To limit the discussion and focus on more meaningful aspects of parallel computing, I will ignore three aspects of "low-level" parallelism (Duncan 1990). Low-level parallelism is small-scale parallel characteristics of computing systems that most computers (including microcomputers) contain. Examples of low-level parallelism are: (1) buffered input/output (I/O, e.g., to and from serial ports), (2) functional units to perform arithmetic, logical, and floating point operations, and (3) CPU functional units that decompose elementary operations into "sub-instructions" (e.g., decode instructions, calculate addresses, load operands, etc.). Because of the ubiquity of these aspects of parallel processing, I will ignore them in the following discussion.

Within this narrowed focus, the term "parallel processing" encompasses a wide range of computer architectures and applies to many, disparate machines. It incorporates machines with very tightly coupled multiple processors as well as those with a single CPU but having the ability to perform simultaneous operations on elements of vectors in pipelines. Because of this diversity, computer scientists have developed several taxonomic schemes and descriptive languages to order and inter-relate different architectures. Flynn (1972) decomposed computer structure into two components or "streams": data and instructions. For each category there are two possibilities: single or multiple streams. Thus, all computer designs were described as 1 of 4 possibilities from a 2×2 table (Fig. 7.1): one instruction applied to one piece of data (single-instruction, single-data: SISD), one instruction applied to several pieces of data (single-instruction, multiple-data: SIMD), multiple instructions applied to multiple data (MIMD), and multiple instructions applied to a single piece of data (MISD). The classical computer as described by the mathematician John von Neumann is SISD: a single CPU sequentially performs operations on a series of single pieces of data. SIMD computers are those with an array of tightly coupled processors (array processors) under the control of a single master CPU. A pipelined, vector computer is also considered to be SIMD by some authors (Hockney and Jesshope 1988). Computers with multiple processors are able to simultaneously compute different algorithms on different data sets, and thus are MIMD. The last category (MISD), implies a computer that applies multiple algorithms to identical data. Although some authors associate this design with pipelined vector machines (e.g., Desrouchers 1987), Flynn and others believe there are no examples.

Although Flynn's scheme has great heuristic value, it is a very coarse categorization and does not distinguish processor arrays from pipelines (Hockney and Jesshope 1988). Consequently, several authors have attempted more elaborate categorizations. Hockney and Jesshope (1988) and Dasgupta (1990) developed notational schemes to describe all possible de-

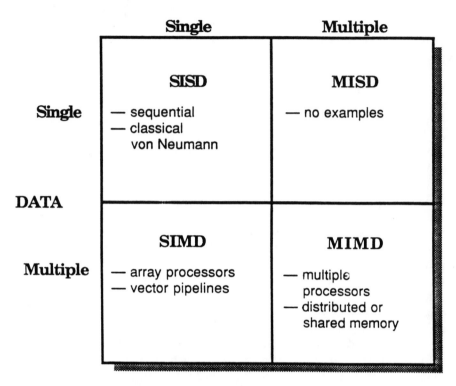

Figure 7.1. Flynn taxonomy of hardware architecture. See text for details.

signs. Although useful for characterizing individual designs, these do not provide a single portrait of the inter-relationships between designs that Flynn's original method did (Fig. 7.1). Recently, Duncan (1990, Fig. 7.2) provided a taxonomy that usefully differentiates most of the known designs. Duncan distinguishes synchronous (all processors in lockstep), MIMD (classical multiple processors), and MIMD paradigm (a heterogeneous collection of hybrid designs that use multiple processors).

Space does not permit a description of each of these designs, but Hwang and Briggs (1984), Hockney and Jesshope (1988), and Duncan (1990) give an overview and entries to the literature for each. Many of the designs were constructed for a special purpose or are only available at a few research sites. I will briefly describe the most important and more commonly available types of machines.

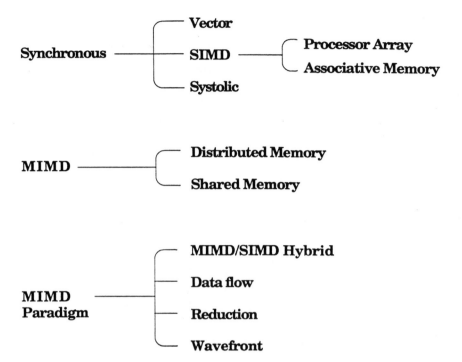

Figure 7.2. Duncan taxonomy of hardware architecture. (Copyright 1990 IEEE; redrawn from Duncan (1990) with permission.) See text for details.

Vector Pipeline Designs

Pipelines are structures in CPUs that decompose a larger operation (instruction) into a series of smaller operations in which the output of a previous operation is the input for the next one. Thus, an initial input datum is sequentially transformed by the pipeline until the desired final output is achieved. This design is said to be parallel since new initial data can enter the pipeline befor the final output of the previous datum is produced (Fig. 7.3). Pipelined machines can differ by the scale of the initial operation that is decomposed. Figure 7.3 illustrates pipelining a low-level operation (integer addition). At larger operational scales, instructions for floating point multiplication and division can be pipelined in order to speed the necessary alignments of exponents and the addition of mantissas. In any case, if each sub-instruction (or "stage") is performed by separate hardware, those components will be operating in parallel.

Filling the pipe requires time that provides little benefit and the proportion of a program occupied in this way should be minimized. This can be achieved

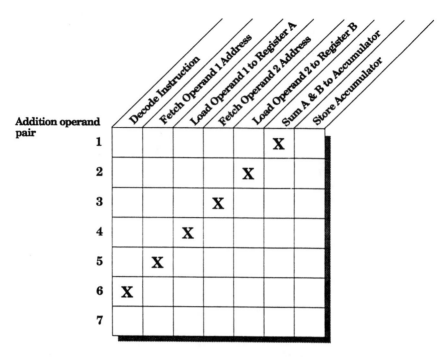

Addition operand pair

	Decode Instruction	Fetch Operand 1 Address	Load Operand 1 to Register A	Fetch Operand 2 Address	Load Operand 2 to Register B	Sum A & B to Accumulator	Store Accumulator
1						X	
2					X		
3				X			
4			X				
5		X					
6	X						
7							

Figure 7.3. Instruction level pipelining. Columns represent hardware implementations of steps in computing the sum of two operands. 'X' indicates the position of the operand pair (rows) in the pipeline. In this example, seven pairs of numbers can be added in parallel.

if one is performing large numbers of identical arithmetic operations sequentially, e.g., adding two long vectors. This situation produced designs that incorporated both pipelines and special vector registers. Vector pipeline machines are those that move as many numbers as possible into relatively long CPU registers and sequentially feed individual numbers from several registers into a multi-stage pipeline (Fig. 7.4). The Cray-1 was an early example of this approach.

SIMD

SIMD designs are synchronous systems with many processing elements (PEs) connected to a few neighbors, most often as a 2D grid or a hypercube (defined below). The PEs are controlled by a single control unit that issues sequential instructions to each PE. Additional hardware feeds data to the PEs from memory so that all PEs are executing the same instruction, but on different input data.

Vector Register A

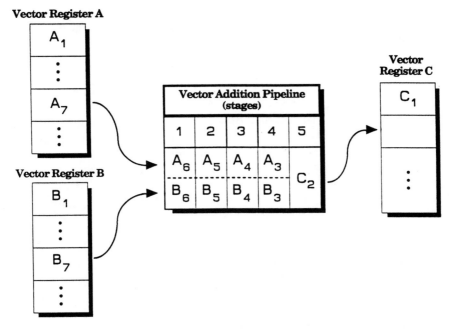

Figure 7.4. Vector pipelining. Elements of two vectors are fed from vector registers into a pipeline in which columns represent stages required to multiply two floating point numbers. Results are stored in a third vector register. (Copyright 1990 IEEE; redrawn from Duncan (1990) with permission.)

SIMD PEs can be either bit-parallel (reading the instruction or data streams as collections of bits, e.g., bytes) or bit-serial (reading a single bit at a time). The ILLIAC IV was a bit-parallel design, while the Connection Machine (CM-2, see Hillis 1985) uses many, relatively simple, bit-serial PEs. The CM-2 consists of 65,536 PEs in a $12th$ order hypercube (see below) with 2,048 floating point units. Processor arrays are especially useful for problems in which the data are distributed in a regular geometry that can be mapped onto the processor topology (e.g., a discrete solution grid for two-dimensional PDEs). The fact that a CM-2 won the 1989 Gordon Bell Prize (Dongarra et al. 1990) demonstrates clearly that this architecture can achieve supercomputer levels.

Two additional SIMD designs are somewhat specialized: systolic arrays and associative memory. A systolic design (Kung 1980, Fountain 1987, Hockney and Jesshope 1988) is SIMD in which the PEs are arranged as a pipeline (but which may be 2D). Associative memory machines operate on vertical slices of memory (e.g., bit 3 from all 8 bit memory words). This design is commonly used for parallel searching and sorting databases (Soucek and Soucek 1988, Chisvin and Duckworth 1989).

MIMD

MIMD designs are those most commonly associated with parallel computers. They are typically composed of an intermediate number of relatively complex PEs. Two major classes of MIMD machines are distinguished by the relation of the PEs to memory. Distributed memory machines provide each PE with its own local memory that cannot be directly addressed by other PEs. Shared memory MIMD machines provide a pool of memory which can be accessed by any PE. In many shared memory designs, individual PEs have a small amount of local memory used primarily as a cache to reduce memory access bottlenecks (see Dubois and Thakkar 1990 and associated articles).

Both designs are characterized by relatively little common supervised control, compared to SIMD array processors. As a consequence, independent action by PEs can produce two different problems in the two designs. Distributed memory machines must communicate with each other in order to pass information or results as required by the algorithm. This is typically accomplished with a message-passing procedure (see section Communicating Sequential Processes). In addition to the time costs of exchanging data, deadlock can occur when two or more PEs attempt to read from each other simultaneously. Special care must be used when programming to avoid this situation, but operating systems and programming environments are being developed to simplify this.

PEs of shared memory machines can communicate indirectly using shared memory, and do not incur the overhead or programming problems of interprocessor communication. To preserve data integrity, however, the values contained in shared memory must not be accessed by more than one PE at a time. Consequently, code must be provided to prevent such an occurrence. A number of techniques have been developed: monitors, semaphores, programmable hardware interrupts (Lusk et al. 1987, Perrott 1987, Strenstrom 1990). These techniques are not immune from deadlock (see below).

MIMD Paradigm

"MIMD paradigm" is Duncan's term for a miscellaneous group that includes MIMD-SIMD hybrids, dataflow designs, and a collection of other special purpose machines. Hwang and Briggs (1984), Hockney and Jesshope (1988), and Duncan (1990) describe these architectures. Dataflow machines may have applications in individual-based models which simulate a collection of individuals in different states and/or having different input data requirements that are produced by different ecological processes executing in parallel (e.g., weather, numbers of individuals of a given size, etc.). The PEs of dataflow computers use processes that begin execution only when all data requirements of a process have arrived at the PE (Perrott 1987). Program execution is controlled by the flow of data. This contrasts with SISD machines in which a single program counter initiates instruction execution and requests the necessary data (control flow machines).

Multiple PE Topology

The specification of the topology or interconnections of PEs in a network of processors is well-studied (Lipovski and Malek 1987, Hockney and Jesshope 1988). Topology may be fixed (static) or dynamic (using switching network hardware) during program execution. In general, the objective is to minimize communication time (in distributed memory MIMD) or synchronization and memory addressing complexity (in shared memory MIMD). Common topologies are: hierarchical trees, 2D or 3D grid (mesh), pipeline (unidirectional or bidirectional), ring, and hypercube. The latter connects nodes so that the number of neighbors is equal to the dimensionality (or "order") of the cube and the binary representation of a node and its neighbors differ by 1 bit. Thus, a 2D hypercube is a square with 4 nodes and each node has 2 neighbors. The binary representation of the nodes is (clockwise): (00, 01, 11, 10). A 3D hypercube is a cube with 8 nodes each of which has 3 neighbors. In a 3D hypercube, there are 2 faces labeled (000, 001, 011, 010) and (100, 101, 111, 110) connected so that each neighbor differs by 1 bit. Higher order hypercubes are defined in the same way, and the number of connections among PEs increases as $\log_2 N$, where N is the order of the hypercube. Multiple connections between PEs has been a serious design problem for large numbers of PEs. A typical solution is to cluster PEs into "supernodes." For example, the CM-2 clusters 16 PEs together into supernodes that are connected as a 12th order hypercube. This design gives 4096 (2^{12}) supernodes of 16 PEs each for a total of 65,536 PEs.

Parallel Architecture Characteristics

The different taxonomies mentioned above emphasize morphology and form. An alternative, graphical method based more on function than form was developed by Desrouchers (1987). He plots different computer designs as a region in a three-dimensional space whose axes are: granularity, control, and topology (see also Fox and Furmanski 1989). Granularity refers to the power of each PE and ranges from very complex CPUs [coarse grain (strong)] to bit-serial, single bit elements of an SIMD processor array [fine grain (weak)]. Control refers to the degree ("loose" to "tight") to which the elements producing parallel operation are synchronized and controlled to act in lockstep

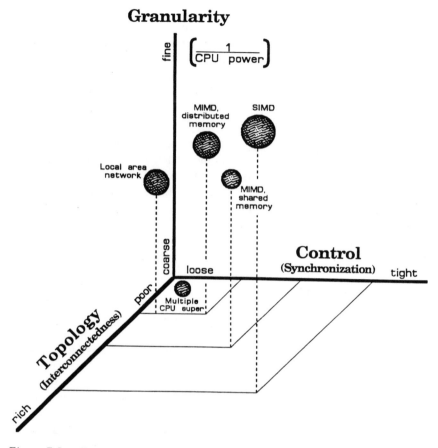

Figure 7.5. General 3D view of computer architecture. Spheres represent classes of computer designs. See text for explanation.

concert. Topology describes the degree ("rich" to "light") to which PEs are interconnected. Topology is also correlated with the number of processors used.

Figure 7.5 shows the approximate placement of the major designs as they are typically implemented. With only three dimensions, it does not provide the detail of Hockney's taxonomy (Hockney and Jesshope 1988), but permits broad comparisons of a variety of machines in a single figure. This scheme also permits the separation of radically different machines (vector *vs* multiprocessor) while at the same time indicates the similarities of certain designs (e.g., vector and processor array). A classical von Neumann (SISD) machine is positioned at the origin. This has coarse granularity since the single CPU is usually very powerful, and has mimimal interconnections and

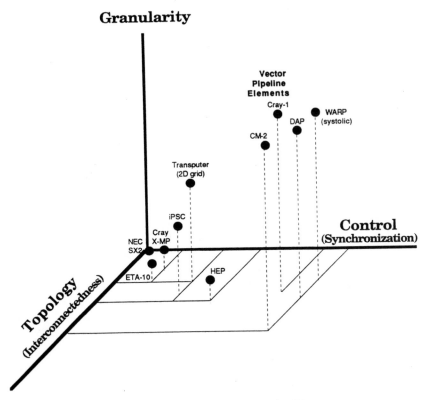

Figure 7.6. Relative positions of specific machines in 3D space.

control synchronization since there is only a single PE. SIMD designs (including vector pipelines) occupy the other extremes in this space being relatively weak CPUs heavily interconnected under strict control of another processor. Pipelines are included here because the focus is on the individual stage which is interpreted as one of the parallel PEs. Distributed and shared memory MIMD designs are assigned slightly different positions on the assumption that PE "communication" by shared memory locations provides greater inter-connectedness of individual PEs, and memory locking implements greater control and synchronization compared to message-passing between nodes of a distributed memory MIMD.

Figure 7.6 shows the location in this space of several important machines. Since a particular machine can incorporate several parallel designs (e.g., multiple processors and vector pipelines) its location depends on the scale of the machine being plotted. If the scale of interest is the loosely connected multiprocessors, the machine will be plotted near the origin. If the scale is the hardware elements constituting a pipeline, the machine will be near SIMD designs. Many more machines have been built or designed than are shown. The placement is meant to be accurate only in a relative sense; the purpose here is to provide a conceptual framework in which to compare machines.

Individual machines within one of Flynn's categories can be positioned by qualitative ranks on the dimensions. For example, different classes of topological connections of MIMD multiprocessors have quantitatively different numbers of neighbors per PE. A pipeline uses 2 neighbors/PE, a 2D grid pattern uses 4 neighbors/PE, a 3D cube has 8 neighbors/PE, and so on. Topology also incorporates the total numbers of PEs in a machine, so a machine with many PEs connected in a pipeline can have greater interconnection richness than one with a few PEs in a more complex inter-connection scheme.

The details of architecture and performance of the machines shown are available in more general texts (Hwang and Briggs 1984, Kowalik 1985, Hockney and Jesshope 1988, Houstis et al. 1988), but I will briefly describe some that are shown. The NEC-SX2 is a single processor vector pipeline supercomputer; I placed it at the origin to emphasize the single CPU. The Cray-XMP is also a vector pipeline, but can be configured with up to 8 of the powerful CPUs. The ETA-10 (Control Data Corporation) was a system of 8 vector pipeline supercomputers using shared memory. The iPSC (Intel Personal Super-Computer) is a distributed memory MIMD usually configured as a hypercube of between 128 to 1024 processors. In such a configuration it has slightly greater interconnectedness than a network of Inmos transputers connected in a 2D grid. (Transputers can be connected in other arrangements). The PEs of both of these last machines are relatively less powerful than the HEP (Heterogeneous Element Processor, Kowalik 1985), thus, they are assigned a position higher on the granularity axis. The HEP has shared memory and multiple processing element modules (PEMs) which comprise several PEs, as well as several instruction pipelines (Jordan 1985). The CM-2 is a massively parallel SIMD machine, and occupies the extremes of topology and control. The ICL DAP is similar, but possesses fewer PEs. I placed the Cray-1 near these as an illustration of a vector pipeline.

This plot indicates some broad regions of possible designs that have not been fully explored. For example, large numbers (rich topology) of medium-grained, but loosely controlled PEs have only recently been constructed. Two machines based on the T800 transputer have been assembled: a 400 node "Computing Surface" by Meiko, Ltd for the University of Edinburgh, UK and a 400 node "SuperCluster" manufactured by ParaCom, Ltd for Shell Oil Corporation. As very large scale integration (VLSI) continues to dominate CPU/microprocessor design (May 1989), and the cost per PE declines, more of these machines will be built (e.g., Rettberg et al. 1990).

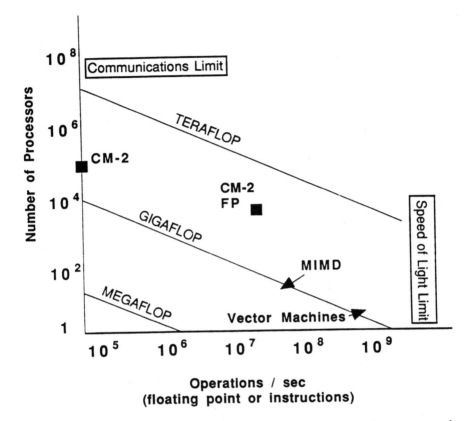

Figure 7.7. Limits to computing speed. Diagonal lines are lines of constant speed equal to the product of processor speed and number of processors. Various classes of machines are indicated, CM-2 = Connection Machine, FP = floating point unit attached to CM-2. (Copyright 1990 American Institute of Physics; redrawn from Boghosian (1990) with permission.)

Performance

Two separate issues limit parallel computer speed. First, there is a limit to the speed of an individual processor and, to date, there has been a design trade-off between processor speed and numbers of processors (Fig. 7.7, from Boghosian 1990). Hecht (1989) discusses some fundamental issues in the physical limits of digital computation in a single processor. Assuming the speed of a complete system to be the product of numbers of processors and processor speed, different designs are positioned on the contours of system

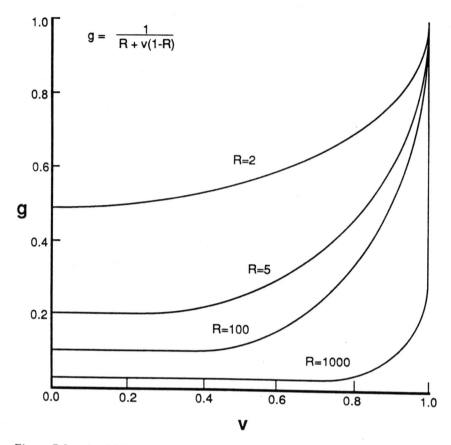

Figure 7.8. Amdahl's law. g = fraction of theoretical maximum speed-up, v = fraction of program that is parallel, R = number of processors or ratio of vector speed to scalar speed. (Copyright 1988 IOP Publishers, Ltd; redrawn from Hockney and Jesshope (1988) with permission.)

performance (Fig. 7.7). The objective for future designs is clearly to move most rapidly up the gradient to teraflop performance.

The second limit is Amdahl's Law (Fig. 7.8) which describes speed-up (time on 1 processor ÷ time on P processors) as a function of the proportion of time the algorithm is performing calculations in parallel. It was originally formulated for vector pipeline machines, but it also seems to apply to multiprocessors (Hockney and Jesshope 1988):

$$g = \frac{1}{R + v(1 - R)}$$

where g is the fraction of the theoretical maximum speed-up, R is the maximum ratio of parallel speed to sequential speed (using 1 processor for multiprocessors), and v is the fraction of time (or number of calculations) that a particular program spends performing in parallel. Figure 7.8 makes two points. First, for any R value, as the amount of parallel code (v) increases, the fraction of maximum performance increases monotonically. Second, as the number of processors (proportional to R) increases, greater amounts of parallel code are required to achieve equivalent levels of speed-up. Note that g is standardized to the maximum value for a given R.

Because at large R the curve rises sharply at relatively large values of fraction of parallel code, Amdahl's Law was initially thought to seriously limit the utility of parallel computers. Subsequent practical experience has shown, however, that for a variety of parallel designs and algorithms it is relatively easy to develop programs that achieve greater than 0.95 fraction time in parallel computations (Denning and Tichy 1990). Moreover, the classical measure of speed-up has recently been challenged (Gustafson 1988). He argues that speed-up should incorporate the increased size (or computational load, Flatt and Kennedy 1989) that can be accommodated on multiple processors since this is an important application on scientific problems. The scaled speed-up is defined as: $S_k = [kT(1,1)/T(p,k)]$, where k is the problem size ratio on 1 and p processors, $T(1,1)$ is the time required for the largest problem that will fit on 1 processor run on a single processor, and $T(p,k)$ is the time required for the problem increased by a factor of k and run on p processors (Karp and Flatt 1990). Using this metric, Gustafson et al. (1988) observed scaled speed-ups of about 1020 for 3 different problems run on a 1024 element distributed memory MIMD. Using the classical definition, the speed-ups ranged from 639 to 351. Starting with Amdahl's Law, Karp and Flatt (1990) derived a new performance metric: fraction of an algorithm that is executed in serial mode (f) They found that LINPACK (a common benchmark) had 0.97 parallel fractions ($1 - f$) and the tests in Gustafson et al. (1988) achieved 0.999 parallel fractions. Thus, Amdahl's Law does not appear to be a fatal problem, but it is a function of the size of the problem and (for MIMD) the number of processors.

Programming Concepts for MIMD Architectures

There are a few basic concepts that apply to many programming problems on MIMD machines: (1) the types of programming structures that can be made parallel, (2) the resolution or granularity at which the programming structure is parallelized, (3) the problem of mapping a parallel structure onto a particular MIMD topology to maintain processor balancing, (4) communicating sequential processes (CSP), and (5) synchronicity and deadlock of CSPs. Below, I briefly review these concepts and provide some references.

Parallelizable Program Structures

Data and code are the two fundamental structures that constitute a computer program, and these structures can form the basis for parallelizing an algorithm. Data parallelism (Jamieson 1987, also "structure parallelism," Hockney and Jesshope 1988, or "domain decomposition," Fox and Furmanski 1989) occurs when one splits the data upon which a program operates over the processors. In the pure case, all processors perform the same operation and consequently it is especially suited for SIMD architectures. MIMD designs can also use this approach. The best applications are problems with a dense, geometric distribution (e.g., 2D PDEs for heat diffusion). By a dense, geometric distribution I mean a problem for which the output or variable of interest is continuous over the geometric space. These are basically those that can be discretized using a lattice so that each processor computes similar functions on different regions of space. The application of this need not be limited to Cartesian space. Many operations on dense matrices (e.g., inner products, Gelernter 1987, Akl 1989) can be parallelized by assigning different combinations of rows and columns to different processors.

Other problems, for example particle-based movement of sparsely spaced individuals in space, also have a geometric distribution, but unless large numbers of individuals are modeled, distributing space across processors will be inefficient because much of space is empty and those processors will be idle.

The alternative to data parallelism is function [or "process" (Hockney and Jesshope 1988) or "algorithmic" (Hey 1989)] parallelism in which the overall algorithm is decomposed into sub-tasks, some of which can be performed in parallel on different processors. For example, the function $y = x^2 + wz - \exp(x)$ comprises three additive terms that can each be computed in parallel. Vector pipeline machines, at the level of the hardware that performs the different stages of the pipeline, use function parallelism.

The hybrid approach divides both data and functions across processors. MIMD architecture does this best. An advantage of MIMD is that it can implement all three approaches. Depending on the problem, however, the performance of a non-MIMD machine may exceed that of a MIMD implementing the approach for which the other hardware was specifically designed. There is, consequently, a trade-off between obtaining higher performance on a smaller set of problems using SIMD and sacrificing speed for greater generality on MIMD. It may be that the set of problems for which SIMD is efficient is so large and ubiquitous that complete generality is not an important trait.

A key consideration in choosing between data and function parallel approaches is the degree to which the code may be scaled-up to larger numbers of processors. In any algorithm, there is a limit to the number of arithmetic operations that can be distributed on processors because (a) there are only a finite number of operations in an algorithm, and (b) data dependencies limit parallelization. Function parallel problems will not benefit from additional processors in this case. Data parallel problems can use more processors effectively if the limitation is the amount of data acted upon. Examples of this case are spatially explicit problems and individual-based problems where there is the need to increase the space or numbers of individuals simulated. This approach was part of the success of the Sandia National Laboratory in scaling computational fluid dynamics problems to 1000 processors (Gustafson et al. 1988, Jenkins 1989).

Granularity

Granularity is the resolution or scale at which a problem is parallelized. It applies to both data and function parallelism (Kung 1980). Fine-grained parallelism in data parallel approaches is parallelism over "small" pieces of data (e.g., a single number); coarse-grained parallelism distributes large data structures across processors. For example, while searching a database in parallel, sequentially distributing subsets of single records to processors is fine-grained in contrast with a coarse-grained approach that distributes subsets of the entire database across processors. Applied to function parallel approaches, vector pipelines are fine-grained since they parallelize at the level of individual arithmetic operations. Coarse-grained approaches are those where each processor performs many computations. As noted below, granularity interacts with load balancing and synchronicity: too large a grain size for a problem may cause some processors to be inefficiently idle, too small of a grain size may reduce performance on distributed memory MIMD machines due to excessive time spent communicating intermediate results. Kruatrachue and Lewis (1988) presented a method for optimally determining grain size in MIMD machines.

Processor Mapping and Load Balancing

The mapping problem is that of assigning a set of instructions to a particular set of processors that either share memory and/or are interconnected with one another according to some fixed scheme (Berman 1987). This is an optimality problem since a poor mapping will degrade performance, as noted above. The most common source of performance degradation is improper load balancing which arises when some processors are idle waiting for needed information to be computed on other processors.

Load balancing may be static or dynamic. Normal procedure requires that the programmer assign processor mapping and load balancing by hand. Some automatic tools are available (Berman 1987, Kruatrachue and Lewis 1988). Fox and Furmanski (1989) and Udiavar and Stiles (1989) used a simulated annealing model to adjust load dynamically. When the granularity of parallelism is on the level of individual user jobs (tasks) in a multiprogramming operating system, Hubermann and Hogg (1988) have proposed a "computational ecology," where tasks compete for computer resources (e.g., CPUs, memory blocks, etc.) and the "best" competitors are allowed to execute. Other approaches to load balancing in distributed operating systems were the focus of the May 1990 issue of *Computer*.

Communicating Sequential Processes

A central concept for programming MIMD machines is that of a communicating sequential process as formalized by Hoare (1985). While there are several interpretations of 'process' (compare Lusk et al. 1987 and Hoare 1985), for the present purposes, a process will be a segment of computer code currently operating on input data to produce an output. Processes and programs are different. A program is a set of binary symbols (possibly stored in a file) that can be interpreted as data and machine instructions; it exists in the absence of the computer. A process exists only when it is executing (or waiting to execute) on an actual processor. A process may be any amount of code, varying from a complete, large program to a portion of a loop within a program.

Multiple processes, in the context of MIMD, can be acting simultaneously. If the processes are being executed by a single processor, then "simultaneity" is achieved by a time slicing method. If the processes are on different processors, then they may be physically running in parallel. Processes may be created by another process and assigned to a particular processor. The assignment of processes to processors is part of the mapping and load balancing problem.

In shared memory MIMD machines, processes communicate via global memory. In general, there are no direct connections between processors. In distributed memory machines, processes typically communicate by passing messages. This is similar to asynchronous communication over a serial data line between two computers connected by a modem. A hardware protocol exists to indicate that a connection is available for reading or writing. Data may be buffered or not depending on the processor hardware. A processor attempting to communicate with another processor that is busy may block and wait for the other processor or may time-out after a number of tries to perform I/O. Processors that must communicate with others to which there is not a direct connection, must use software and/or switching hardware to route the message to the intended processor.

The Inmos transputer (tm) illustrates the fundamental importance of CSPs. The transputer is a distributed memory MIMD CPU specifically designed for parallel processing and based on CSPs (Homewood et al. 1987, Soucek and Soucek 1988). The transputer's special language "occam" has a key word (par) that automatically creates and executes parallel CSPs that are defined by code blocks that follow. Other MIMD computers and compilers provide similar facilities in subroutine libraries.

Synchronicity and Deadlock

There are very few problems that can be solved on multiprocessor computers in which the PEs do not need to exchange data. One case which minimizes the frequency of exchange is a Monte Carlo simulation when each replicate executes completely on a single PE. By definition, the replicates are independent and only when the replicate run is complete is it necessary to move data between PEs to communicate results to a peripheral device for output. In almost all other applications, PEs must interact due to the data requirements of the algorithm being performed. The output of process *A* may be required by *B* before *B* can execute. In this situation, *A* and *B* must be synchronized so that data may be exchanged. In shared memory MIMD machines, two processes can not be allowed to simultaneously read or write the same memory location. The processes must be synchronized to the extent that process *A* can prevent process *B* from accessing a resource (memory location) that *A* is using. Dinning (1989) and Graunke and Takkar (1990) review available methods in shared memory MIMDs. Distributed memory MIMD processors exchanging data by message-passing are usually synchronized by the read/write process. One implementation is for inter-processor reading and writing to block or wait until both processors are available to communicate (Dinning 1989).

Deadlock occurs when two processes each attempt to access resources (e.g., data) held by the other (Singhal 1989). In shared memory systems, this can occur when the process must read or write memory locations locked by the other (e.g., deleting two adjacent members of a double linked list, Lusk et al. 1988, Brawer 1989). In message-passing systems, deadlock occurs when both processes attempt to read or write from the other simultaneously. Shumway (1989) and Singhal (1989) review methods for deadlock detection and prevention.

Ecological Examples

Relatively few applications of parallel computers to simulations of the type performed by ecological modelers have been implemented. Vector machines are being used (e.g., Costanza et al. 1990, Stockwell and Green 1990), although explicit discussion of vectorizing strategies for algorithms unique to ecological problems is rare (Casey and Jameson 1988). Excluding PDE solvers, there has been very little use of MIMD parallel computers in continuous system simulation. Fountain (1987) presented some of the problems of mapping a non-linear model of a national economy onto a MIMD machine, but gave no results. Hamblen (1987) compared Euler and Runge-Kutta methods for solving a system of 2 linear ODEs on 2 transputers. Haefner (1991) compared communication strategies and costs for Euler and Runge-Kutta integration of 96 non-linear ODEs on 2 transputers. Pearce et al. (1986) reviewed parallel hardware architecture on which continuous system simulations have been performed. Lei et al. (1986) and Wang et al. (1986) gave some techniques for partitioning ODEs on to a network of processors. Most of these tools are restricted to linear systems. Finally, Dekker (1984) reviewed parallel continuous system simulation from the perspective of general systems theory. More applications of multiprocessor computers to discrete event simulation (especially time-warp simulation) have been made (Concepcion 1985; Zeigler 1985, 1990; Reed and Fujimoto 1987; Wilson 1987; Zeigler and Zhang 1990).

The infrequent use of parallel computers in ecological modeling may be due to the relatively small problems that ecologists have previously attacked and the relative inaccessibility of MIMD machines to the working ecologist. Both these conditions will change as ecologists adopt an individual-based approach and as the technology becomes cheaper, more accessible, and easier to program. Because of the lack of real examples, the following discussion will primarily be hypothetical but will illustrate the potential. In some cases (e.g., PDEs), the ecological problems follow standard numerical methods used in other disciplines (e.g., physics) where the parallel implementations have been developed. In these situations, the benefits of parallel computation are documented.

Food-web Models as ODEs

Food-web models are often formulated as systems of ordinary differential equations (ODEs) or finite difference equations. These may be elaborated with explicit, time-dependent driving variables or with a coarse-grained spatial component that models physical flows between a relatively small number of spatial cells. Many models are composed of components (e.g., state variables) whose basic structure is repeated within the system. This can occur when (1) a trophic level or functional group is resolved into a set of subgroups (e.g., species), and (2) when the system is described with a coarse-grained spatial distribution. In case (1), identical forms of the derivatives or equations of rates of change are used for all members of the trophic level. For example, the primary producers in an aquatic model may be composed of a set of phytoplankton species each of whom absorbs nutrients according to a Michaelis-Menton relation, but with different parameter values and different connections to herbivores. In case (2), differential equations are replicated over all the state variables in each of the cells that constitute the spatial structure of the system. In writing code for MIMD machines, it is

```
typedef struct {
    double val,workval;              /* state */
    int numconn;                     /* num connections to other st.vars.*/
    int conn[MAXCONN][MAXSV];        /* id of connecting st. vars.*/
    double parms[MAXCONN][MAXPARM];  /* params of derivatives*/
    double (*df[MAXCONN])();         /*addresses of derivative functions*/
} SV;
```

Figure 7.9. C code for structure of a state variable in a simulation model. MAX-
CONN is the maximum number of state variables connected to a given
state variable, MAXSV is the maximum number of state variables in
the system, and MAXPARM is the maximum number of parameters
that are used in any component of the derivative. val is the value of
the state variable after the last integration step; workval is a temporary value of the state variable used in the integration routines.
numconn and conn [ ] [ ] are the number and identities (respectively) of state variables involved in the derivative. df [ ] is an array
of functions using parameters parms [ ] [ ].

advantageous, to generalize the structure as much as possible in order to
experiment with mapping and load balancing. Swartzman and Kaluzny (1987)
recently described a very clean and general code structure for nonlinear models
that addresses many of these problems. I have extended and generalized their
approach to provide greater flexibility in defining system topology and applications to multiple processors (Haefner 1990).

A first-order ODE can be described by the derivative of a state variable
equated to a summation of terms interpreted as flows into and out of the
variable. For example, $dx/dt = A + D - B - C$ describes the rate of change
of a state variable x with two flows (A, D) into the compartment represented
by x and two flows out (B, C). Usually, the components A, B, C, and D
are non-linear functions of other state variables and external driving variables. The flexibility and generality that we need for load balancing and
programming ease on multiprocessors can be embodied in a suitable data
structure of state variables and the components of their derivatives. Figure 7.9
shows the C code for a structure defining a state variable that contains all the
information that a state variable needs to compute its derivative. The equations needed to represent the components of the total derivative of the state
variable are specified in an array of addresses of functions returning a double
precision real number: df[ ]. The two-dimensional array parms[ ] [ ]
contains the parameters needed by each of the derivative component functions. Each component of a differential equation (e.g., A, B, C, D, above)
could be represented by a different function and its address stored as an
element of df[ ]. For example, A might be the logistic equation, D an immigration constant, C Lotka-Volterra predation, and B an emigration con-

stant. A system is an array of these structures: SV System[MAXSV];. If derivative component A is the logistic equation for the ith species, then its structure is initialized with: System[i].df[0] = logistic, where logistic is a function to compute density-dependent growth rate.

This structure, which has some features of classes in object-oriented programming languages (Wiener and Pinson 1988), is especially useful in a distributed memory MIMD context because it becomes easy to create integration routines that are identical for all processors. This provides tools for creating code that is easier to debug.

A simulation loop using this structure on a single processor is:

```
while (t < MaxT) {
Integrate(t,Num_State_Vars,Delta_T,System[ ])
}
```

where t is the current time, Num_State_Vars is the number of state variables simulated, Delta_T is the solution time step, and System[ ] is the array of state variable structures.

The function Integrate( ) loops through the array System[ ] summing the rate of change of each component of the derivative of each state variable by invoking the functions contained in each structure System[i].df[j]( ). Figure 7.10 illustrates this for Euler integration. This function does not change if different differential equations are used, but it must be altered for different integration methods (e.g., Runge-Kutta).

There are two basic approaches to parallelize code to solve systems of ordinary differential equations. The first, a fine-scale (function parallel) approach, parallelizes the integration routine over the processors and, effectively, steps through the state variables serially using the parallel version of the integration method. Kumar and Lord (1985) review some parallelizing methods for a variety of integration techniques applied to missile flight dynamics. The second, more coarse-grained (data parallel) approach uses a sequential version of an integration method, but distributes the state-variables across processors.

To illustrate these concepts, consider a routine to perform Runge-Kutta integration on a system of ODEs in which different ODEs (i.e., state variables) are distributed across processing elements. The central problem is to ensure that all processes use the current values of state variables for their integrations. Since the current values are contained on remote processors, we require a method to synchronize the communication of intermediate results. The CSP model works well in this application. Most compilers for multiprocessing CPUs provide functions to create and run processes in parallel. In the following, I briefly sketch the method and skip many details.

Assume a 4-processor, distributed memory MIMD such as Fig. 7.11. To exchange data between nodes and the root, we create and run in parallel

```
void Integrate(t,numsv,h,System)
int numsv;          /* number of state variables in system */
double t,           /* current time */
       h;           /* integration interval */
SV System[ ];       /* all system objects */
{
      int i,j;
      double dyt[MAXSV];
      int *np;


      /* init all workvals */
      for(i=0;i<numsv;i++)
        System[i].workval = System[i].val;
      for(i=0;i<numsv;i++) {
        dyt[i]=0.0;
              /* compute rates */
        for(j=0; j<System[i].numconn; j++)
          dyt[i] += (*System[i].df[j])(t,i,j,System);
      }
              /* update states */
      for(i=0;i<numsv;i++)
        System[i].val += h*dyt[i];
}
```

Figure 7.10. C code for ODE integration on single processor using a general data structure for state variables. System[i].df[j] refers to the address of the jth component function of the derivative for the ith state variable. (*System[i].df[j])() signifies the contents of the address: the function. += adds the right-hand-side to the left-hand-side of the assignment.

processes to read from and write to the root the current values of all the variables required on the node (Fig. 7.12). While these occur in the root, each node performs a sequential operation to integrate its equations which may involve one (Euler) or more (four for Runge-Kutta) sequences of reading from the root, performing part of the integration, and writing to the root. Even though processes are acting independently, integration on nodes proceeds only with current, correct data because I/O operations performed on a node wait for each root process to complete its own I/O operation. This synchronizes the processes.

To implement this approach, all processors (root and nodes) execute identical code (Fig. 7.13). The advantage of an explicit approach to simulation model structure using data structures that contain addresses of derivative functions (Fig. 7.9) in simplifying code is evident in this figure. Each node appears to perform the same computation (Fig. 7.13), but the integration function (Fig. 7.10) uses different derivatives, because the array of function

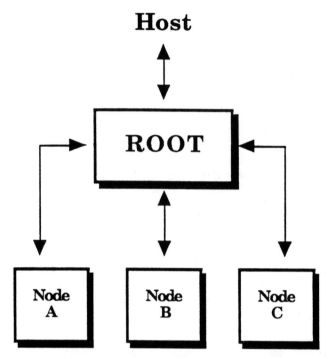

Figure 7.11. Hierarchical network of distributed memory MIMD processors.

addresses differs across PEs. This scheme is easily extended to permit the root to perform some of the integrations or to permit the nodes *A, B, C* to pass intermediate results directly among themselves without involving the root. Moreover, this approach should work well on other message-passing, distributed memory MIMD machines such as the NCube and the Intel iPSC. Significantly more complex message-passing code is required if more nodes are used and are not directly connected to the root. This problem requires appropriate operating system software, programming environments, or library functions (see below).

To test the speed-up on an ecological problem, I (Haefner 1991) compared times for a three-trophic level foodweb with 96 species split across two transputers. I observed a 1.9 speed-up comparing a simulation on 1 and 2 nodes. I also estimated communication costs when all state variables were passed between processors compared to passing only those needed by each processor. The differences in costs were about 5–9%, indicating for these types of simulations that computation times are more important than communication times.

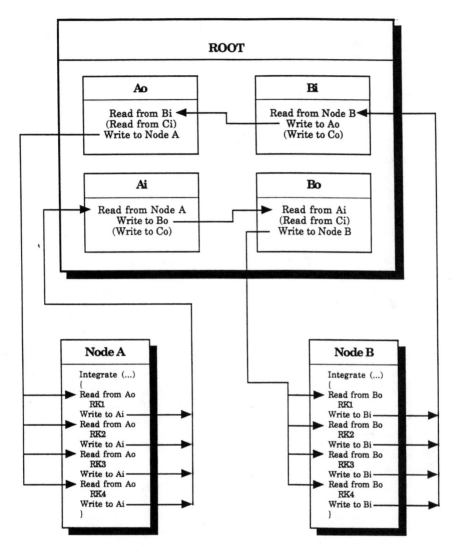

Figure 7.12.   Box diagram for CSP synchronization control of Runge-Kutta integration. Arrows represent communication channels. Node *C* is omitted for clarity. RKi (i = 1..4) represents the calculation of the four intermediate derivatives in fourth-order Runge-Kutta integration.

```
if(ROOT) {        /* this processor is the root */
  :Write initial conditions to Nodes A, B, C
  :Create processes Ai, Ao, Bi, Bo, Ci, Co to run in parallel
  :Start processes in parallel (time slicing on Root)
}
else {            /* this processor is a node */
  for(t=0;t<MaxT;t++)
     Integrate(...);       /* From Fig. 10 */
}
```

Figure 7.13. C code fragment for CSP control of integration synchronization on distributed memory processors. ROOT is true if the processor executing the code is the root; it is false otherwise.

Individual-based (oriented) Population Simulation

Following Metz and de Roos (this volume), I distinguish individual-oriented models, where individuals are separately modeled, as a subset of individual-based models which may include other approaches (e.g., PDEs). While Huston et al. (1988) cite many examples of individual-oriented models, for definiteness I will focus on one (DeAngelis et al. 1991, see Fig. 7.14) to illustrate some applications to parallel computers. Most individual-based models are based on complex finite difference equations which describe the dynamics of a few characteristics of individuals. DeAngelis et al. (1991) track individual size (condition), numbers of offspring produced, habitat position (water column or benthos), and probability of dying. Periodically, selected states of individuals are summed over all individuals to produce population states. Typically, individual state dynamics are driven by stochastic environmental variables (e.g., prey availability, and cover and nest site availability, Fig. 7.14).

A powerful metaphor for coding individual-oriented models is the finite-state automaton ("FSA," Hopcroft and Ullman 1969). Markov processes, which are frequently applied in ecology, are a simple, stochastic example of FSAs. In a FSA, an individual is represented as being in one of a finite number of discrete states (e.g., age classes). As the individual or its environment change, the state of the individual can change. The possible dynamics of a FSA can be represented as a state transition graph (Fig. 7.15) in which circles are states and solid lines are transitions between states.

The basic concept of a FSA must be modified, however, to accommodate the flexibility required by realistic ecological individual-oriented models. In addition to the finite states, a given model may need to describe individuals by continuous variables or "properties" (e.g., size or condition in DeAngelis et al. 1991). Also, FSAs of individual-oriented models use transition functions that are non-linear functions of individual states and properties and

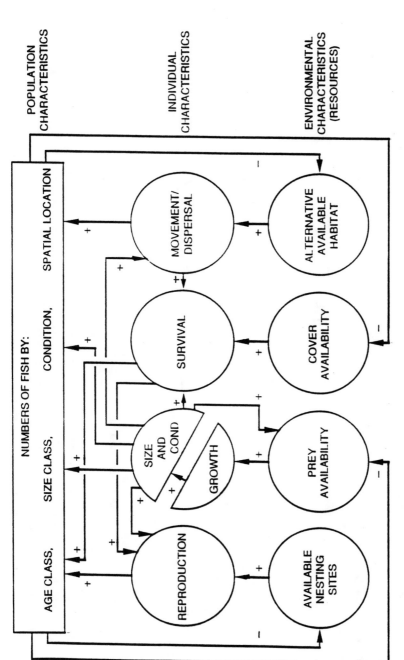

Figure 7.14. Conceptual framework of an individual-oriented model for a fish population. Solid arrows indicate influences with '+' and '−' indicating positive and negative effects. (Copyright 1991 Elsevier Press; reproduced from DeAngelis et al. (1991) with permission.)

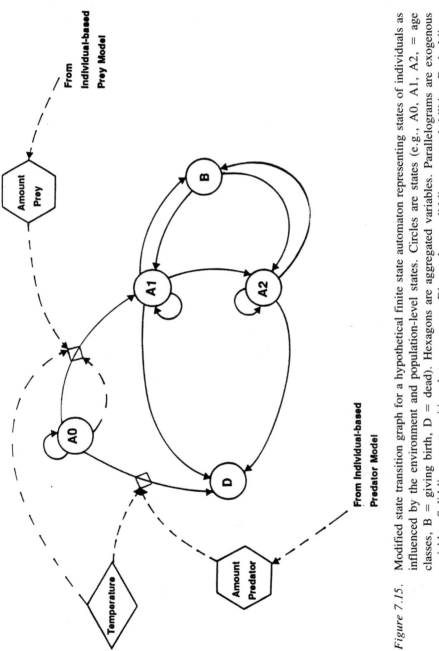

Figure 7.15. Modified state transition graph for a hypothetical finite state automaton representing states of individuals as influenced by the environment and population-level states. Circles are states (e.g., A0, A1, A2, = age classes, B = giving birth, D = dead). Hexagons are aggregated variables. Parallelograms are exogenous variables. Solid lines are transitions between states. Diamonds on solid lines are probabilities. Dashed lines indicate functional dependence of other variables or individual state properties (e.g., size of A1). Only a few dependencies are shown.

environmental conditions. Using the concepts of Forrester diagrams for ODEs (Forrester 1961), complicated FSAs can be diagrammed (Fig. 7.15).

This diagram conceals the fact that realistic individual-oriented models use several discrete classifications to characterize individual states. For example, DeAngelis et al. (1991) distinguish 3 age classes (adult, "egg-to-swim-up," and "swim-up-to-1-year-old"), 2 sexes, 2 habitat positions (benthos, water column), and 2 positions relative to the nest (near, dispersed). Moreover, not all combinations are meaningful (e.g., adults and 1-year-olds are not classified by nest proximity). Therefore, in addition to diagramming tools, a data structure that simplifies parallel programming is also needed.

A *C* construct well-suited to this problem is a data structure that defines individuals' states using "bitfields." Bitfields associate groups of individual bits within an unsigned integer. For example, two bits of the integer can be used to code the age state, the third bit can be used for sex, the fourth bit for habitat position, and so on. The state of each individual is one of these data structures, and an individual's state is altered by changing one or more of the bits. By interpreting the composite of all the bitfields as an integer, an array can be constructed such that each element contains the address of a transition function for each state. This array, in conjunction with an array of structures that record individual properties, permits a simple loop to simulate individuals analogous to the loop to calculate rate of changes in ODEs (Fig. 7.10). For example:

```
for(i=0;i<Num_Individuals;i++)
(*transitions[state_structure[i]](... arguments ...)
```

where state-structure[i] is the state of the *i*th individual, transitions[ ] is an array of function addresses for each possible state (*transitions[ ] is the function with arguments ... arguments ...).

Once the array of state transition function addresses is initialized, this loop is all that is needed. As the states of individuals change during a simulation (by action of (*transitions[ ])), the correct function is automatically selected from the array in the above loop. Undefined combinations of bitfields are assigned a function address that returns an error message.

PDE Individual-based Population Models

Most PDE single population models are not so large as to require parallel processing. Multi-species models may require it, but the computational requirements will always be much smaller than individual-oriented models. In any case, parallel versions of PDE solvers have received much attention. Ortega and Voigt (1985) emphasize vector methods and Reed and Fujimoto (1987), Rice (1987), and Akl (1989) describe multiprocessor methods. As an example, Jong and Stiles (1989) compared several implementations of the flux corrected transport (FCT) algorithm on different shared- and distributed-memory MIMD machines. On a network of 8 transputers hosted by an IBM-AT compatible microcomputer, they obtained a 7.7 speed-up for a 32 × 32 points 2-D FCT. The speeds were about twice those for a VAX-8650 and a 16 node NCUBE. These numerical methods will apply directly to population models (Metz et al. 1988), which is an advantage of this approach.

Movement Models

An area in great need of parallel models is individual movement in two or three dimensions. These models are similar in structure to *N*-body simulations that are being used with increasing frequency in astrophysics, plasma physics, molecular dynamics. Greengard (1990) gives a general overview and an entry bibliography. The technical literature is reviewed by Hockney and Eastwood (1981) and in a variety of symposia (e.g., Brackbill and Cohen 1985, Houstis et al. 1988). Although some parallel concepts from physics carry over to biological simulations, the method of calculating "forces" that affect movement are obviously tied to theory in the underlying discipline. The physical field equations are well-developed and therefore effort in developing parallel algorithms has general application beyond a particular simulation. In behavioral ecology, at this point, there is less agreement on theory where some of the great diversity of simulations stems from differences in the sensory modalities used by different organisms. Thus, models developed for fish (e.g., Paul Jacobsen, personal communication) or krill (Simon Levin, personal communication) rely on visual or pressure receptors to generate the "force field" driving individual movement. Some models use chemical cues (ants: Haefner and Crist, in prep.; bacteria: Jackson 1987). Differences in the physical nature of light, pressure waves, and chemical diffusion and advection produce large differences in the structure of models and algorithms. Until more experience on implementing these models on parallel machines is available, few generalizations or recommendations can be offered. Data parallel approaches in which the individuals are distributed on processors (as opposed to dividing space over processors) will probably be most efficient. Programming complexity can be reduced, particularly when movement occurs in complicated regions (e.g., repelling borders, vegetation, etc.), if a FSA or object-oriented approach is used.

Limitations of Vector Machines

Three common features of individual-oriented models limit the usefulness of vector pipeline computers. First, depending on the complexities of the behavioral states modeled and of the geometry within which movement occurs, individual-based model code may contain large numbers of conditional statements. Typically, vectorizing compilers compute all branches, then use the correct branch depending on conditionals. Many if-thens will degrade vector performance below that of a scalar version that computes only the correct branch. Second, random creation and removal of objects combined with spatial position and movement that is independent of an object's location in the array containing all objects (i.e., in machine memory) means that for many computations within an individual-based model memory accessing will often be random. This eliminates data locality and reduces the efficiency of vectorization. Third, most vectorizing compilers will not vectorize a loop containing a call to a subroutine (unless the subroutine has been vectorized). In particular, loops using random numbers must remove calls to random number generators, perhaps by storing random numbers in separate arrays outside the main loop. This is an easy solution but emphasizes one of the basic alterations in programming style that can require large code re-organization in order to use vector pipelines effectively. If PDEs are used, then vector machines have been demonstrated to be effective. Many of the above problems can be addressed by explicit, hand-coded vectorization (e.g., using gather and scatter vector subroutines, Casey and Jameson 1988). Nevertheless, coding to minimize execution time on vector machines will probably always be time consuming and require significant expertise.

Programming Developments

Major effort has been devoted to simplifying the use and programming of parallel computers. The vendors of large vector pipeline, multiprocessor machines (e.g., Cray, Alliant, Stardent, etc.) supply FORTRAN and C optimizing compilers that provide some level of automatic vectorization of loops. They also supply extensions to the basic languages to permit synchronization for shared memory MIMD designs. Typically, these are implemented as special comments to provide some level of portability of the code to single processor machines without the extensions. Triolet (1987) and Polychronopoulos (1988) review some of the basic methods and problems of vectorizing loops. Perrott (1987) and Hockney and Jesshope (1988) give examples for several machines and compilers. Other languages with parallel constructs have been developed. Parallel versions of the object-oriented language C++ exist (Qi and Wagner 1989, Beck 1990). Also, ADA and Modula-2 have some facilities for concurrent processes or "co-routines" (Perrott 1987). Occam, as mentioned, was explicitly designed for message passing CSP on transputers. The primary language on the Connection Machine is a parallel form of Lisp (Hillis 1985), but other languages are now available. Other parallel Lisp implementations are discussed in Goldman and Gabriel (1989) and Zorn et al. (1989). A parallel version of the Continuous System Simulation Language (CSSL) has been developed (Hay 1986). A number of other specialized languages exist (e.g., Baldwin 1989, Tick 1989).

New operating systems (including additions to extant systems) and programming environments have also been created to facilitate programming multiple, distributed processors. Among new operating systems are Trollius (Braner 1989) and GENESYS (Hudson and Bradshaw 1990) for transputers, Helios (UNIX-like for transputers, Noble, et al. 1990), SEDOS (distributed memory MIMD from the ESPRIT project, Diaz and Vissers 1989), Mach (UNIX derivative, Black 1990), and Amoeba (has a UNIX emulation facility, Mullender et al. 1990).

Environments with graphical interfaces as adjuncts to existing operating systems and programming tools for MIMD machines are available (Browne et al. 1989, Guarna et al. 1989). Hypertool is a non-graphical environment that automates the conversion of a sequential program into one that will run on a generic, message-passing multiprocessor machine (Wu and Gajski 1990). Zhang and Wagner (1989) developed an environment for load balancing and process mapping based on C++ for transputers. Debuggers for concurrent programming exists (Lehr et al. 1989, McDowell and Helmbold 1989). General and portable *C* code for sophisticated message-passing primitives is described in Lusk et al. (1987) and Gabber (1990). Similar tools for FORTRAN exist (Dongarra and Sorensen 1987, Peir et al. 1987, Allen et al. 1988, Appelbe et al. 1989). Express (Parasoft Inc.) and ParaScope (Computer System Architects, Inc.) are programming and debugging environments for transputers. Linda (Gelernter 1985, 1989), a framework for a memory model to facilitate communication between distributed memory processors, has been implemented on large arrays of transputers at the University of Edinburgh (MacDonald 1989) and Cogent Research, Inc. (Leler 1990) and as an environment for expert systems development (Busalacchi 1989). Overviews of additional programming environments and performance tools have been collated by Harrison (1990) and Nichols (1990). Major advances in all these areas, especially in interactive debuggers, are occurring very rapidly. Slower progress in porting standard numerical libraries to MIMD parallel networks is occurring, but efforts are underway (Anonymous 1989). Overall, great strides are being made to simplify parallel coding which will ultimately make these machines accessible to practising ecologists.

Conclusions

This overview has emphasized the applications of multiprocessor architecture to individual-based modeling. These machines permit the study of larger and more complex models by a data parallel strategy. This is not necessarily a good thing, but is sometimes unavoidable, especially when small spatial and temporal scales are required. Future developments in individual-based models that would benefit from advanced computing are: multiple species (including competitors and predator-prey relationships), models of greater numbers of individuals within a population, greater realism in the behavioral and physiological mechanisms of movement, and additional individual states (e.g., genetic differences). Although vector pipeline designs will continue to be applicable to individual-based models using PDEs, individual-oriented models will benefit most from multiprocessors. It is uncertain now whether SIMD or MIMD (shared or distributed memory) machines will prove the most useful to future models. It is possible that no single design will suffice and ecological modelers will need to choose the hardware according to the nature of the problem. Recent improvements in continental networks have made this a reality.

Whatever the architecture used, the characteristics of parallelism imply that new programming paradigms and approaches must be acquired by ecological modelers. Distributed memory MIMD machines, which offer one of the most cost effective avenues to parallel computing, embody many of these challenges. In particular, the concepts of communicating sequential processes, process synchronicity, and deadlock will be important. Second, experience indicates that program development and debugging is relatively easy in data parallel problems, if differences in the code compiled for each processor is minimized. As illustrated above, using data structures with embedded functions to compute derivatives in foodweb simulations or transitions in FSAs of individual-oriented population dynamics models can reduce the differences between processors to that of input data determining the appropriate function addresses. Since these data structures are similar to classes in object oriented languages, ecological modelers using multiprocessor machines will be aided by such language constructs. We can expect better compilers for parallel versions of C++ in the near future. Finally, modelers must develop and acquire a new set of tools to assist in mapping processes to processors, passing messages among PEs in different topologies, and assessing network performance. All of this implies that in individual-based ecological modeling, as in other areas requiring high-performance computing, future modelers will probably need to know more, not less, about the details of hardware design and its relation to the numerical solution.

Acknowledgments

This work was primarily supported by grant SB-1182 from the Utah State University Office for Research. Additional support was provided by NSF grant BSR 88-17358 to J. A. MacMahon. I thank M. Braner (Cornell University Theory Center) and G. Stiles (USU) for helpful discussions of transputers.

Literature Cited

Akl, S. G. 1989. *The Design and Analysis of Parallel Algorithms*. Prentice-Hall, Englewood Cliffs, NJ.

Allen, F., M. Burke, P. Charles, R. Cytron, and J. Ferrante. 1988. An overview of the PTRAN analysis system for multiprocessing. In: *Supercomputing*. E. N. Houstis, T. S. Papatheodorou, and C. D. Polychronopoulos (eds.), pp. 194–211. Springer-Verlag, Berlin.

Anonymous. 1989. Project Survey: Software research in ESPRIT's second phase. *IEEE Software* **6(6)**:54–57.

Appelbe, B., K. Smith, and C. McDowell. 1989. Start/Pat: A parallel-programming toolkit. *IEEE Software* **6(4)**:29–38.

Baldwin, D. 1989. Consul: A parallel constraint language. *IEEE Software* **6(4)**:62–69.

Beck, B. 1990. Shared-memory parallel programming in C++. *IEEE Software* **7(4)**:38–48.

Berman, F. 1987. Experience with an automatic solution to the mapping problem. In *The Characteristics of Parallel Algorithms*. L. H. Jamieson, D. Gannon, and R. J. Douglas. pp. 307–334. MIT Press, Cambridge, MA.

Black, D. L. 1990. Scheduling support for concurrency and parallelism in the Mach operating system. *Computer* **23(5)**:35–43.

Boghosian, B. M. 1990. Computational physics on the Connection Machine. *Comput. Phys.* **4(1)**:14–32.

Brackbill, J. U. and B. I. Cohen (eds.). 1985. *Multiple Time Scales*. Academic Press, New York.

Braner, M. 1989. Trollius: a software solution for Transputers and other multicomputers. In *NATUG 2, Proceedings of the First Conference of the North American Transputer Users Group*. G. S. Stiles (ed.), pp. 1–5. NATUG, Utah State University, Logan, UT.

Brawer, D. 1989. *Introduction to Parallel Programming*. Academic Press, Boston.

Browne, J. C., M. Azam, and S. Sobek. 1989. CODE: A unified approach to parallel programming. *IEEE Software* **6(4)**:10–18.

Busalacchi, P. J. 1989. Linda on a transputer-based personal computer. In *NATUG 2, Proceedings of the Second Conference of the North American Transputer Users Group*. J. A. Board, Jr. (ed.), pp. 223–228. NATUG, Utah State University, Logan, UT.

Casey, R. M. and D. A. Jameson. 1988. Parallel and vector processing in landscape dynamics. *Appl. Math. Comput.* **27**:3–22.

Chisvin, L. and R. J. Duckworth. 1989. Content-addressable and associative memory: Alternatives to the ubiquitous RAM. *Computer* **22(7)**:51–64.

Concepcion, A. I. 1985. Mapping distributed simulators onto the hierarchical multibus multiprocessor architecture. In *Proceedings 1985 Conference on Distributed Simulation*. pp. 8–13. Simulation Councils, Inc., San Diego, CA.

Constanza, R., and T. Maxwell. 1991. Spatial ecosystem models using parallel processers. *Ecol. Modell.* **58**:159–183.

Costanza, R., F. H. Sklar, and M. L. White. 1990. Modeling coastal landscape dynamics. *Bioscience* **40**:91–107.

Dasgupta, S. 1990. A hierarchical taxonomic system for computer architectures. *Computer* **23(3)**:64–74.

DeAngelis, D. L., L. Godbout, and B. J. Shuter. 1991. An individual-based approach to predicting density-dependent dynamics in smallmouth bass populations. *Ecol. Model.* **57**:91–115.

Dekker, L. 1984. Concepts for an advanced parallel simulation architecture. In *Simulation and Model-based Methodologies: An Integrative View*. T. I. Oren, B. P. Zeigler, and M. S. Elzas (eds.). NATO ASI Series, Vol. F10, Springer-Verlag, Berlin.

Denning, P. J. and W. F. Tichy. 1990. Highly parallel computations. *Science* **250**:1217–1222.

Desrouchers, G. R. 1987. *Principles of Parallel and Multiprocessing*. Intertext Publications, Inc., McGraw-Hill, New York.

Diaz, M. and C. Vissers. 1989. SEDOS: Designing open distributed systems. *IEEE Software* **6(6)**:25–32.

Dinning, A. 1989. A survey of synchronization methods for parallel computers. *Computer* **22(7)**:66–77.

Dongarra, J., A. H. Karp, K. Kennedy, and D. Kuck. 1990. Special Report: 1989 Gordon Bell Prize. *IEEE Software* **7(3)**:100–104, 110.

Dongarra, J. J. and D. C. Sorensen. 1987. SCHEDULE: Tools for developing and analyzing parallel Fortran programs. In *The Characteristics of Parallel Algorithms*. L. H. Jamieson, D. Gannon, and R. J. Douglas. (eds.) pp. 363–394. MIT Press, Cambridge, MA.

Dubois, M. and S. Thakkar. 1990. Cache architectures in tightly coupled multiprocessors. *Computer* **23(6)**:9–11.

Duncan, R. 1990. A survey of parallel computer architectures. *Computer* **23(2)**:5–16.

Flatt, H. P. and K. Kennedy. 1989. Performance of parallel processors. *Parallel Comput.* **12**:1–20.

Flynn, M. J. 1972. Some computer organizations and their effectiveness. *IEEE Trans. Comput.* **C-21**:948–960.

Forrester, J. W. 1961. *Industrial Dynamics*. MIT Press, Cambridge, MA.

Fountain, T. 1987. *Processor Arrays: Architectures and Applications*. Academic Press, London.

Fox, G. C. and W. Furmanski. 1989. The physical structure of concurrent problems and concurrent computers. In *Scientific Applications of Multiprocessors*. R. Elliott and C. A. R. Hoare (eds.), pp. 55–88. Prentice-Hall, New York.

Gabber, E. 1990. VMMP: A practical tool for the development of portable and efficient programs for multiprocessors. *IEEE Trans. Parallel Distrib. Syst.* **1**:304–317.

Gelernter, D. 1985. Generative communication in Linda. *ACM Trans. Program. Lang. Syst.* **7(1)**:80–112.

Gelernter, D. 1987. Programming for advanced computing. *Sci. Am.* **257(4):**90–98.

Gelernter, D. 1989. The metamorphosis of information management. *Sci. Am.* **259(8):**66–73.

Goldman, R. and R. P. Gabriel. 1989. Qlisp: Parallel processing in Lisp. *IEEE Software* **6(4):**51–59.

Graunke, G. and S. Takkar. 1990. Synchronization algorithms for shared-memory processors. *Computer* **23(6):**60–69.

Greengard, L. 1990. The numerical solution of the N-body problem. *Comput. Phys.* **4(2):**142–152.

Guarna, V. A., D. Gannon, D. Jablonowski, A. D. Malony, and Y. Gaur. 1989. Faust: An integrated environment for parallel programming. *IEEE Software* **6(4):**20–27.

Gustafson, J. L. 1988. Reevaluating Amdahl's law. *Commun. ACM* **31:**532–533.

Gustafson, J. L., G. R. Montry, and R. E. Brenner. 1988. Development of parallel methods for a 1024-processor hypercube. *SIAM J. Sci. Stat. Comput.* **9:**609–638.

Haefner, J. W. 1990. Communication costs of continuous system simulation on parallel computers: An ecological example. *Proceedings UKSC 90 Simulation Conference,* pp. 86–89. Brighton, UK.

Haefner, J. W. 1991. Foodweb simulation on parallel computers: Inter-processor communication benchmarks. *Ecol. Model.,* **54:**73–79.

Hamblen, J. O. 1987. Parallel continuous system simulation using the Transputer. *Simulation* **49:**249–253.

Harrison, W. 1990. Tools for multiple-CPU environments. *IEEE Software* **7(3):**45–52.

Hay, J. L. 1986. ESL—advanced simulation language for parallel processors. In *Parallel Processing Techniques for Simulation,* M. G. Singh, A. Y. Allidina, and B. K. Daniels (eds.), pp. 171–182. Plenum Press, New York.

Hecht, J. 1989. Physical limits of computing. *Comput. Phys.* **3(4):**34–40.

Helios. 1990. *The Helios Parallel Programming Tutorial.* Distributed Software, Ltd., Bristol, UK.

Hey, A. J. G. 1989. Reconfigurable transputer networks: practical concurrent computation. In *Scientific Applications of Multiprocessors,* R. Elliot and C. A. R. Hoare (eds.), pp. 39–54. Prentice-Hall, Englewood Cliffs, NJ.

Hillis, W. D. 1985. *The Connection Machine, 2nd ed.* MIT Press, Boston.

Hoare, C. A. R. 1985. *Communicating Sequential Processes.* Prentice-Hall International, London.

Hockney, R. W. and J. W. Eastwood. 1981. *Computer simulation using particles.* McGraw-Hill, New York.

Hockney, R. W., and C. R. Jesshope. 1988. *Parallel Computer: 2 Architecture, Programming, and Algorithms, 2nd ed.* IOP Publishing Ltd., Bristol, UK.

Homewood, M., D. May, D. Shepherd, and R. Shepherd. 1987. The IMS T800 transputer. *IEEE Micro* October, 10–26.

Hopcroft, J. E. and J. D. Ullman. 1969. *Formal Languages and Their Relation to Automata.* Addison-Wesley, Reading, MA.

Houstis, E. N., T. S. Papatheodorou, and C. D. Polychronopoulos (eds.). 1988. *Supercomputing.* Springer-Verlag, Berlin.

Huberman, B. A. and T. Hogg. 1988. The behavior of computational ecologies. In *The Ecology of Computation,* B. A. Huberman (ed.), pp. 77–115. Elsevier B. V., North-Holland.

Hudson, P. and S. Bradshaw. 1990. *GENESYS: An operating system for parallel computers.* Transtech Technical Note, N. 2. Transtech Devices Limited, Buckinghamshire, UK.

Huston, M., D. DeAngelis, and W. Post. 1988. New computer models unify ecological theory. *BioScience* **38**:682–691.

Hwang, K. and F. A. Briggs. 1984. *Computer Architecture and Parallel Processing.* McGraw-Hill, New York.

Jackson, G. A. 1987. Simulating chemosensory responses of marine microorganisms. *Limnol. Oceanogr.* 32:1253–1266.

Jamieson, L. H. 1987. Characterizing parallel algorithms. In *The Characteristics of Parallel Algorithms,* L. H. Jamieson, D. Gannon, and R. J. Douglas (eds.), pp. 65–100. MIT Press, Cambridge, MA.

Jamieson, L. H., D. Gannon, and R. J. Douglas (eds.). 1987. *The Characteristics of Parallel Algorithms.* MIT Press, Cambridge, MA.

Jenkins, R. A. 1989. New approaches in parallel computing. *Comput. Phys.* 3(1):24–32.

Jong, J-M. and G. S. Stiles. 1989. A comparison of parallel implementations of flux corrected transport codes. In *Developing Transputer Applications,* J. Wexler (ed.), pp. 113–128. IOS Publishers, Amsterdam.

Jordan, H. 1985. HEP architecture, programming and performance. In *Parallel MIMD Computation: The HEP Supercomputer and Its Applications.* J. S. Kowalik (ed.), pp. 1–40. MIT Press, Cambridge, MA.

Karp, A. H. and H. P. Flatt. 1990. Measuring parallel processor performance. *Commun. ACM.* **33**:539–543.

Kowalik, J. S. (ed.). 1985. *Parallel MIMD computation: The HEP Supercomputer and its Applications.* MIT Press, Cambridge, MA.

Kruatrachue, B. and T. Lewis. 1988. Grain size determination for parallel processing. *IEEE Software* 5(1):23–32.

Kumar, S. P. and R. E. Lord. 1985. Solving ordinary differential equations on the HEP computer. In *Parallel MIMD Computation: The HEP Supercomputer and Its Applications,* J. S. Kowalik (ed.), pp. 231–273. MIT Press, Cambridge, MA.

Kung, H. T. 1980. The structure of parallel algorithms. *Adv. Comput.* **19**:65–112.

Lehr, T., Z. Segall, D. F. Vrsalovic, E. Caplan, A. L. Chung, and C. E. Fineman. 1989. Visualizing performance debugging. *Computer* **22(10)**:38–51.

Lei, S., A. Y. Allidina, and K. Malinowski. 1986. Clustering technique for rearranging ODE systems. In *Parallel Processing Techniques for Simulation*, M. G. Singh, A. Y. Allidina, and B. K. Daniels (eds.), pp. 31–43. Plenum Press, NY.

Leler, W. 1990. Linda meets Unix. *Computer* 23(2):43–54.

Lipovski, G. J. and M. Malek. 1987. *Parallel Computing Theory and Comparisons*. Wiley, New York.

Lusk, E., R. Overbeek, J. Boyle, R. Butler, T. Disz, B. Glickfeld, J. Patterson, and R. Stevens. 1987. *Portable Programs for Parallel Processors*. Holt, Rinehart and Winston, New York.

MacDonald, N. 1989. Focus 2: Linda—A model of generative communication. *Edinburgh Concurrent Supercomputer Newsletter* 9:14–17.

May, D. 1989. The influence of VLSI technology on computer architecture. In *Scientific Applications of Multiprocessors*, R. Elliott and C. A. R. Hoare (eds.), pp. 21–37. Prentice-Hall. New York.

McDowell, C. E. and D. P. Helmbold. 1989. Debugging concurrent programs. *ACM Comput. Surv.* **21**:593–622.

Metz, J. A. J., A. M. DeRoos, and F. van den Bosch. 1988. Population models incorporating physiological structure: A quick survey of the basic concepts and an application to size-structured population dynamics in waterfleas. In *Size-structured Populations: Ecology and Evolution*, B. Ebenman and L. Persson (eds.), pp. 106–126. Springer-Verlag, Berlin.

Mullender, S. J., van Rossum, A. S. Tanenbaum, R. van Renesse, and H. van Staveren. 1990. Amoeba: a distributed operating system for the 1990s. *Computer* **23(5)**:44–53.

Nichols, K. M. 1990. Performance tools. *IEEE Software* **7(3)**:21–30.

Ortega, J. and R. Voigt. 1985. *Solution of PDE's on Vector and Parallel Computers*. ICASE Report 85–1, Langley Research Center,

Pearce, J. G., P. Holliday, and J. O. Gray. 1986. Survey of parallel processing in simulation. In *Parallel Processing Techniques for Simulation*, M. G. Singh, A. Y. Allidina, and B. K. Daniels (eds.), pp. 183–202. Penum Press, New York.

Peir, J-K, D. D. Gajski, and M-Y Wu. 1987. Programming environments for multiprocessors. In *Supercomputing: State-of-the-art*, A. Lichnewsky, C. Saquez (eds) pp. 73–93. Elsevier B.V., Netherlands.

Perrott, R. H., 1987. *Parallel Programming*. Addison-Wesley, MA.

Polychronopoulos, C. D. 1988. Advanced loop optimizations for parallel computers. In *Supercomputing*, E. N. Houstis, T. S. Papatheodorou, and C. D. Polychronopoulos (eds.), pp. 255–277. Springer-Verlag, Berlin.

Qi, R. and A. Wagner. 1989. Programming for data parallelism using C++. In *NATUG 2, Proceedings of the Second Conference of the North American Transputer Users Group*, J. A. Board, Jr. (ed.), pp. 307–323. NATUG, Utah State University, Logan, UT.

Reed, D. A. and R. M. Fujimoto. 1987. *Multicomputer Networks: Message-Based Parallel Processing*. MIT Press, Cambridge, MA.

Rettberg, R. D., W. R. Crowther, P. P. Carvey, and R. S. Tomlinson. 1990. The Monarch parallel processor hardware design. *Computer* **23(4)**:18–30.

Rice, J. R. 1987. Parallel methods for partial differential equations. In The Characteristics of Parallel Algorithms. L. H. Jamieson, D. Gannon, and R. J. Douglas (eds.). MIT Press, Cambridge, MA.

Shumway, M. 1989. Deadlock-free packet networks. In *NATUG 2, Proceedings of the Second Conference of the North American Transputer Users Group*, J. A. Board, Jr. (ed.), pp. 139–177. NATUG, Utah State University, Logan, UT.

Singhal, M. 1989. Deadlock detection in distributed systems. *Computer* **22(11)**:37–48.

Soucek, B. and M. Soucek. 1988. *Neural and Massively Parallel Computers: The Sixth Generation*. Wiley, New York.

Stockwell, D. R. B., and D. G. Green. 1990. Parallel computing in ecological simulation. *Math. Comp. Simul.* **32**:249–254.

Strenstrom, P. 1990. A survey of cache coherence schemes for multiprocessors. *Computer* **23(6)**:12–24.

Swartzman, G. L. and S. P. Kaluzny. 1987. *Ecological Simulation Primer*. Macmillan, New York.

Tick, E. 1989. Comparing two parallel logic-programming architectures. *IEEE Software* **6(4)**:71–80.

Triolet. R. 1987. Programming environments for parallel machines. In *Supercomputing: State-of-the-art*, A. Lichnewsky and C. Saguez (eds.), pp. 131–152. Elsevier B. V., Netherlands.

Wang, L., A. Y. Allidina, K. Malinowski, and M. G. Singh. 1986. Multi-level hierarchical structures for the solution of large sets of ordinary differential equations. In *Parallel Processing Techniques for Simulation*, M. G. Singh, A. Y. Allidina, and B. K. Daniels (eds.), pp. 45–63. Plenum Press, New York.

Wiener, R. S. and L. J. Pinson. 1988. *An Introduction to Object-Oriented Programming and C++*. Addison-Wesley, Reading, MA.

Wilson, A. 1987. Parallelization of an event driven simulator for computer systems simulation. *Simulation* **49**:72–78.

Wu, M-Y. and D. D. Gajski. 1990. Hypertool: A programming aid for message-passing systems. *IEEE Trans. Parallel Distrib. Syst.* **1**:330–343.

Udiavar, N. and G. S. Stiles. 1989. A simple but flexible model for determining optimal task allocation and configuration on a network of transputers. In. *NATUG1: Proceedings of the First Conference of the North American Transputer Users Group*, G. S. Stiles (ed.), pp. 31–40. NATUG, Utah State University, Logan. UT.

Zeigler, B. P. 1985. Discrete event formalism for model based distributed simulation. In *Proceedings 1985 Conference on Distributed Simulation*. pp. 3–7, Simulation Councils, Inc., San Diego, CA.

Zeigler, B. P. 1990. *Object-oriented Simulation with Hierarchical, Modular Models: Intelligent Agents and Endomorphic Systems*. Academic Press, Boston.

Zeigler, B. P. and G. Zhang. 1990. Mapping hierarchical discrete event models to multiprocessor systems: concepts, algorithm, and simulation. *J. Parallel Distrib. Comput.* **9**:271–281.

Zhang, Y. and A. Wagner. 1989. The design and implementation of a topological programming environment for parallel C++ on transputers. In *Proceedings of the Second Conference of the North American Transputer Users Group.* J. A. Board, Jr. (ed.), pp. 349–367. NATUG, Utah State University, Logan. UT.

Zorn, B., H. Kinson, J. Larus, L. Semenzato, and P. Hilfinger. 1989. Multiprocessing extensions in Spur Lisp. *IEEE Software* **6(4)**:41–49.

8

From Local to Global: The Lesson of Cellular Automata

Michel J. Phipps

ABSTRACT. This contribution examines a class of simple models, the Cellular Automata (CA), and discusses their relevance to particular aspects of ecological theory. The emphasis is put on spatial differentiation, and more precisely, on how ecological processes can engender and also maintain patchiness. The Neighborhood Coherence Principle (NCP) is presented as a general rule of spatial differentiation and pattern formation. Simulations using the SISPAQ model demonstrate how NCP's basic tenets generate dynamical processes. An application of NCP to the problem of tissue homeostasis—a problem of "cell ecology"—suggests how CA may provide new insights in the understanding of a phenomenon resulting from otherwise complex mechanisms. The advantages of CA are finally discussed in the light of four particular aspects: individuality, spatial realism, simplicity and stochasticity.

Introduction

A variety of dynamic models today offer a wide range of solutions to those adept at simulation. Some make use of state variables to describe overall behaviors in groups of individuals, while others incorporate individualistic features that are believed to drive global behaviors (Huston et al. 1988). Some unfold in continuous time whereas others use a discrete time frame. Models also range from simple to highly complex ones. Compared to their more sophisticated counterparts, some may look extremely straightforward, not to say simplistic. However, the dynamic behavior that they display goes far beyond the level of complexity normally expected. Such a remark, by no means new (May 1976, Couclelis 1988), should arouse our curiosity and suggest a basic question: would it be wise to use such simple models and find out what we may learn from them about the real world, before invoking more complex ones?

The present contribution examines a particular class of simple models, namely, the Cellular Automata, which are well suited to mimic systems with strong local interaction. After reviewing the nature of these models, we shall discuss their implications in ecological theory and their relevance with regard to the prerequisites that are deemed necessary in ecological modeling. This will include considerations of individuality in ecological systems, and the spatial realism which controls effective interaction.

Cellular Automata Models

Although the first cellular automaton, invented by von Neumann, goes back to 1948 (von Neumann 1963), this class of model long remained an oddity confined to the mathematical game section of scientific magazines (Gardner 1971, Mazoyer 1989). Despite notable exceptions, few obvious applications were in sight. By the end of the seventies, however, this state of affairs took a new turn. The prospect of building parallel processing computers with a cellular architecture (Toffoli and Margolus 1987) and the large number of applications then appearing (Peterson 1987) gave this topic a new impetus. The literature on the subject is relatively recent (Wolfram 1983, 1984a,b, 1985, 1986, Langton 1984, Demongeot et al. 1985, Gutowitz 1990a). Mathematics and computing science are obvious home disciplines to Cellular Automata, but a large array of applications also appear in physics (Gerhardt et al. 1990, Fredkin 1990), biochemistry (Langton 1986), biology (Atlan 1985, Inghe 1989, Victor 1990, Sieburg et al. 1990, Phipps et al. 1990), and geography (Couclelis 1988, Phipps 1989), among other disciplines.

In Wolfram's own words, Cellular Automata (CA) "are systems of cells interacting in a simple way but displaying complex overall behavior" (Wolfram 1985). They may be generically characterized by a few salient features: i) the system substrate is made up of a 1-, 2-, or 3-dimensional cell network; each cell interacts with its neighborhood (a limited number of cells chosen according to a design proper to each CA), ii) each cell can adopt any one of m possible discrete states, iii) the system follows a discrete time dynamic; at each time unit, each cell updates its state according to a *transition rule* taking into account the state of neighboring cells; an initial configuration is given at the beginning of a simulation run; then the system is allowed to evolve under the same conditions, over a finite number of time units. As outlined by Jen (1990), CA form a class of spatio-dynamical models where space, time and states are discrete.

Theoretical investigations of CA generally consist in taking a given rule and discovering its mathematical properties. Gutowitz (1990b) then speaks of a forward problem. Most natural science applications take an inverse approach. It consists in discovering a rule whose properties mimic those of the process known—or believed—to govern the studied phenomenon. The nature of these rules (deterministic or probabilistic) offers a convenient framework for the following discussion.

Deterministic Cellular Automata

In most CA, the transition rule is deterministic; that is, the new state of a cell is determined rigidly on the basis of its actual state and states present in neighboring cells. Let us consider the following example: in a 2-dimensional CA with a regular hexagonal cell arrangement and a possibility of 3 states, a deterministic rule can be as follows: if the distribution of states in the neighborhood of a cell is as biased as 3/2/1 (or more biased, e.g. 4/1/1 or 4/2/0, etc.), then this cell adopts the dominant state; otherwise (including the case 3/3/0), it keeps its present state. Another example is given by the famous "Game of Life" devised by John Conway in 1970. It is a square cell network with two possible states: live (L) or dead (D), with the following rule: any L cell dies unless it has 2 or 3 L in its neighborhood of 8; any D cell can become L only if it counts exactly 3 L in its neighborhood (see Greussay 1988, Bays 1988 and Bak et al. 1989 for further considerations on the properties of the "Game of Life").

In each example, the rule is simple. It clearly suggests the assumption made on the basic character of the phenomenon to be simulated. In the first example, the assumption is that of a relatively stable phenomenon developing a strong neighborhood dependence. The rule of the "Game of Life" obviously assumes that simulated organisms thrive in intermediate conditions, thus avoiding isolation and overcrowding as well. In each case, the rule is also deterministic: it leaves no room for uncertainty and the fate of a cell, at any time in the dynamical process, rigidly depends on the distribution of states in that particular cell and its neighborhood. The types of dynamical behaviors that deterministic CA produce have been abundantly documented, particularly as far as elementary CA are concerned (linear CA with a neighborhood of 2 and a possibility of 2 states).

Stochastic Cellular Automata

A transition rule may also incorporate some stochastic element. In stochastic (or probabilistic) CA, the selection of a state at each time step, is done subject to a probability function (Lee et al. 1990). Thus, Inghe (1989) has developed a 2-dimensional stochastic automaton to study survivorship in a plant population composed of individuals produced either by sexual reproduction (genets) or by clonal lateral growth (ramets). In this CA, an empty cell can be colonized by a ramet emitted from one of the 12 neighbors (4 neighboring cells are not directly adjacent to the central cell) subject to probabilities which vary with the distance between the central cell and its neighbors. Disturbances are imposed upon the system by randomly chosen areas of mortality with various degree and regime. This latter feature represents a second stochastic element of the model.

A number of probabilistic automata (or stochastic CA) have been developed with various objectives (Atlan 1985, Demongeot 1985, Phipps 1989). They are meant to purposefully express the variability inherent in natural objects, be this variability intrinsic or just conventional. This, of course, introduces a departure from the strictly "automatic" nature of CA and some theoretical properties thereof may not entirely apply to this subclass of CA. Conversely, stochastic CA may offer interesting features. Lee et al. (1990) have investigated the case where the probabilistic transition function includes some reinforcement element whereby parameters of the function are modified so as to provide "reward" or "punishment." The result is a CA with cells being able to learn adaptive behaviors.

Dynamical Behaviors of Cellular Automata

As a preliminary step toward a CA general theory, Wolfram (1984b) undertook a systematic study of behaviors yielded under specific rules and neighborhood conditions. In his now classical paper, he discussed four classes of behaviors that can be produced by elementary CA. These are qualitatively characterized as follows (Wolfram's own words in italics):

Class I: *Evolution leads to a homogeneous state;* after a certain number of time units, all cells have a similar state irrespective of the initial configuration.

Class II: *Evolution leads to a set of separated simple stable or periodic structures;* in this case, a spatial organization appears with distinct spatial domains in which two types of temporal patterns may exist: homogeneous and stable state or periodical change of states;

Class III: *Evolution leads to a chaotic pattern;* irrespective of the initial conditions, CA yield aperiodic patterns indistinguishable from the initial pattern;

Class IV: *Evolution leads to complex localized structures, sometimes long-lived;* in this class, some very complex spatial patterns may arise and reproduce over long periods of time; these patterns may also exhibit intriguing spatial propagation despite a perfect conservation of their shape.

These results suggest a rich potential of dynamic modes and spatial forms. In a simulation of the spatio-temporal dynamics of a rodent population, using a basic cellular automaton model, Couclelis (1988) has shown that three of Wolfram's classes were observed (I, II and III). The conclusion reached by Couclelis is that complex unexpected spatial dynamics could merely result from simple spatial interactions. As far as spatial patterns are concerned, class II is of particular interest since it points to the emergence of a spatial differentiation resulting exclusively from neighborhood interactions. Probabilistic CA are much less known as to their dynamical properties. However, classes I and II were shown to occur depending on initial probability conditions (Phipps 1989). Both dynamical modes could be interpreted respectively as self-maintaining and self-organizing capacities.

Short Range Interactions in an Unrestricted Space

The most interesting features of CA pertain to the way in which space and spatial interactions are considered. Although space in a CA is subdivided into discrete units or cells, it is clearly unrestricted; that is, no further a priori divisions prevent the circulation of information. Rohlf and Schnell (1971) and Comins (1982) have used the analogy of stepping-stones to describe short range interactions in such a spatial system. Whereas this analogy evokes the idea of an animal able to jump from any stone to the next one, it might not be entirely appropriate. Each cell, in turn, plays the role of emitter (source) and receiver (sink) of spatial interaction and the strict neighborhood constitutes the unique important set for every single cell. Interaction must therefore be seen as a non-source, non-directional and short-range process (Phipps 1989) able to produce complexity through propagation, and amplification by local feed-back among individual cells (Couclelis 1988). If such an interaction may conceivably propagate or expand across the space from cell to cell like falling dominos, its effect is actually reshuffled in each cell given the state configuration in the local neighborhood. This produces a wealth of complex, often unexpected behaviors that defies verbal description. Structures can emerge in some location, expand or propagate, or shrink and disappear. In other instances, distinct structures originating from different locations can expand and face edge to edge, thus ending up in a stable boundary that divides the space. There is no doubt that most intriguing properties that have been found in CA result from this complex interplay between the local and global levels.

Cellular Automata and Ecological Theory

Although only few recent papers explicitly refer to CA theory (Inghe 1989, for instance), there exists a growing body of ecological literature dealing with problems amenable to CA applications. Strikingly, they all relate in some way to the question of spatial heterogeneity or patchiness, to put it in plain terms. Since MacArthur and Wilson (1967), the influence of spatial structures on ecological processes at various levels of organization has gained increasing attention. May and Southwood (1990), for instance, have recently reviewed various results showing how environmental patchiness affects ecological processes such as foraging behavior, life history, population dynamics and survival, population interaction, and overall biotic diversity. All studies mentioned in that paper address the forward question of how the biological processes respond to a pre-existing environmental heterogeneity (see also Sutherland 1990). There is however, the inverse question: how biological processes can create, enhance, or maintain ecological patchiness. From this standpoint, the origin of heterogeneity is the very question at stake. The non-trivial question of patchiness emerging in an initially homogeneous environment is of prime importance.

The works by Rohlf and Schnell (1971) and Turner et al. (1982) provide a remarkable illustration of this inverse question. Both are inspired from the *isolation by distance* model. They predict the heterogeneous spatial distribution of the frequency of alleles in a heterozygous plant population reproducing under the rule of nearest-neighbor pollination and limited seed dispersal. In both examples a patchy gene distribution is predicted and resembles available empirical observations. These examples are interesting on two accounts: i) the type of question addressed in these works is clearly akin to probabilistic CA but without mentioning explicitly this class of models, and ii) more importantly, they show how the genetical systems under scrutiny can differentiate, genetically and spatially, solely by virtue of short range interaction in a homogeneous unrestricted space. This phenomenon may be understood as self-organization whereby a new differentiation emerges, thus giving way to a metapopulation (Levins, cited in Gilpin 1990) made up of distinct genetical sub-populations, spatially segregated. The fact that this phenomenon occurs despite the lack of any previous difference in space and therefore does not imply any environmental adaptive selection constitutes, of course, the thrust of this research.

The Neighborhood Coherence Principle (NCP)

A large number of patterns that can be observed in natural as well as human environments seem to result from processes which operate along the line of those described above. The patchy metapopulations evoked by Rohlf and Schnell (1971) and Turner et al. (1982) are but one example. Landscape patches of distinct communities, patchy plant communities, or urban ethno-cultural concentrations do not necessarily result from heterogeneity of their substrate. Along with many other examples, they suggest that, beyond differences in nature of causative processes and scale, the emergence of these patterns follows a similar underlying rule (Phipps 1989). In this latter paper, such a rule has been spelled out, purposefully in ubiquitous terms, as the neighborhood coherence principle (NCP). In plain terms, it reads: a state which exists in any particular cell (or site) tends to impose itself upon neighboring cells (or sites), thus leading to local coherences. In order to test the theoretical properties of this rule, a probabilistic CA and a computer pro-

Table 1. Simplified flow chart of the simulation program SISPAQ.

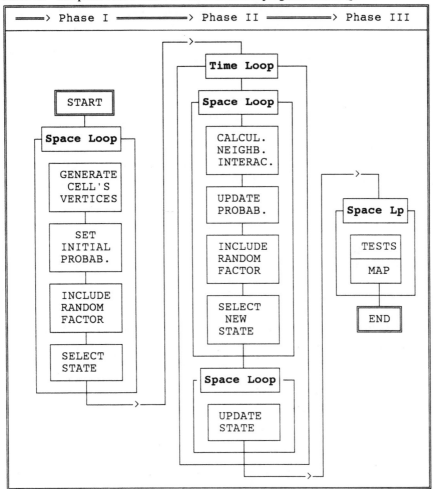

gram (SISPAQ) have been developed to implement NCP. A simplified flow chart of this program is shown in Table 8.1. It consists of a 2-dimensional hexagonal cell system in which each cell *u* may assume one of *m* discrete states, subject to a set of probabilities p_j ($j = 1, m$). At each time unit *t*, for a given cell *u*, probabilities are updated in a way which takes into account the states present in cells that belong to v_u, the neighborhood of *u*. Accordingly, each cell of v_u sends a message to the effect of promoting the occurrence of its own state in all its neighbors (including cell *u*), thus implementing the basic tenet of NCP. The mathematical structure of SISPAQ has been fully developed in Phipps (1989).

The two key elements in the model are: i) the transition rule and ii) the procedure of state selection subject to probabilities p_j. The transition rule is given by the following equation:

$$\Delta p_{ujt} = [r\, a_{ujt}\, p_{ujt}(k_{ujt} - p_{ujt})] + \epsilon_{ujt}$$

where Δp_{ujt} is the change of probability of cell u being in state j between time t and time $t + 1$. This is a finite difference equation of the logistic function with a growth rate a_{ujt} representing the neighborhood interaction and a limit k_{ujt} for a given state j. All terms p, a and k are bounded to [0, 1]. The need to consider m competitive states necessitates the use of a matrix form of the logistic equation (Phipps 1989). This latter feature warrants that:

$$\sum_{j=1}^{m} p_{ujt} = 1, \quad \forall u, t.$$

At each time unit t, the limit values k_{ujt} of the function are updated and depend on the probability values. This feature introduces a cell memory and plays the role of a reinforcement element, thus making SISPAQ similar to an adaptive CA (Lee et al. 1990). Following the updating procedure, the new state of a cell is selected using a random generation function subject to the new probability values. Factors r and ϵ_{ujt} can be omitted. They respectively represent a scalar ($0 \leq r \leq 1$) and a random factor; the condition $r = 0$ entails the relaxation of the neighborhood constraint and lets the cell system evolve under purely random variation. This feature is of interest since it allows simulation of a situation whereby an initial period of normal functioning ($r = 1$) is followed by a disruption of the cell-cell communication ($r = 0$). When this disruption occurs in a limited portion of the system, the comparison between the normal portion and the affected one is highly demonstrative of the role of NCP in the system homeostasis (Phipps et al. 1990).

SISPAQ is clearly reminiscent of several models of interacting particle systems including spin glass (Mezard et al. 1987), neural networks (Hopfield 1982) and, to a lesser extent, percolation transition models (Peterson 1987). It shares with them the discreteness of states taken by individual entities and the short range interaction. Note that the latter holds only for the Edwards-Anderson model of spin glass (Edwards and Anderson 1975), but not the infinite range SK model. However, the originality of SISPAQ lies in its adaptive behavior, a common trait with neural networks. In SISPAQ, cells locally "learn" from each other a consensual behavior which becomes strongly self-maintaining, as will be seen in the next section.

Some Rules Governing Pattern Formation

SISPAQ simulations have demonstrated several remarkable properties of NCP, the two most salient of which are its capacities of self-organization and self-maintenance (Phipps 1989, Phipps et al. 1990). The respective expression of both capacities depends on initial probabilities. By the sole virtue of cell communication, a system subject to relatively even initial probabilities builds up a highly organized pattern with spatially segregated homogeneous states. This patchiness clearly illustrates Wolfram's class II behavior. Conversely, when initial probabilities significantly depart from evenness, the system evolves toward a homogeneous state, thus following Wolfram's class I behavior. The end-point of this dynamical behavior represents a self-maintaining capacity. Homogeneity is indeed maintained in spite of individual variability by cooperative cell networking which represses the occurrence of variant states. It should be noted that once a pattern has emerged from self-organization, the self-maintaining capacity also operates within each patch, thus allowing for state stability and, ultimately, for pattern homeostasis.

It was further conjectured, and demonstrated in the case of a two-state system, that the departure between both dynamical behaviors was controlled by the critical probability interval $0.42 \leq p_1 \leq 0.58$. Within this critical interval, both states will persist in distinct spatial domains. Outside of this interval, the non-dominant state would neither be able to establish nor build up enough neighborhood coherence, nor maintain itself against the NCP effect. Furthermore, both capacities were shown to be dependent on neighborhood size. For instance, if the range of interaction increases, the neighborhood ratio (i.e. the number of cells which belong to the neighborhood of a single cell) will also increase (e.g. from 6:1, 12:1 or 18:1) and the homeostatic effect will become easier. Conversely, the self-organizing capacity will be lowered. This shows that the emergence of patchiness is eased by short range interaction.

In three-state systems, some of these rules can be visualized by diagrams of the phase-space. At each time unit t, a cell u can be characterized by its probabilities p_{ujt} of being in state j. Similarly, the whole system can be characterized by f_{jt}, the relative frequencies of states $j = 1, m$ at time t. Since:

$$\sum_{j=1}^{m} p_{ujt} = 1$$

and:

$$\sum_{j=1}^{m} f_{jt} = 1$$

a single cell as well as the whole system can be visualized as a point in the phase-space represented by an equilateral triangle. It is then possible to follow a single cell's kinetics by its trajectory in phase-space over the time span of simulation (Figure 8.1) or the distribution of all cells at a specific time period (Figure 8.2). Similarly, the evolution of the whole system can be depicted by its trajectory in the phase-space (Figure 8.3) as well as the distribution of end-point for a set of systems (Figure 8.4)

The Interplay of the Local and Global Levels

Figure 8.5 shows the configuration of a three-state system at various stages of its dynamical process: time 0 (A), 5 (B), 15 (C) and 30 (D). Initial conditions assume a homogeneous space and state equiprobability (maximum entropy). Spatial differentiation develops throughout the simulated time span: at time 5, some kernels around which patches will eventually grow are already visible. From time 15 to 30, changes in configuration are less conspicuous since patches have acquired a relative stability. The NCP homeostatic effect tends to maintain the structure.

Over the time span, each cell individually follows a kinetic trajectory in the phase-space as shown in Figure 8.1 for 4 selected cells. All trajectories start from the centre of the phase-space (equiprobability), and then, tend toward one of the vertices of the triangle. A trajectory may appear relatively straightforward (B and D) or unpredictable (A and C), thus reflecting respectively a greater or lesser instability in the cell's neighborhood. Figure 2 displays the end-points of all the 441 cells of the system at various stages. Globally, the process mimics a centrifugal force whereby cell progressively become more committed to a specific state.

A similar trajectory may visualize the whole system's history. Figure 8.3 shows such trajectories for 4 selected systems with a homogeneous space, but assuming distinct initial conditions (the more central end of each trajectory). Again, the process mimics a centrifugal force. End-points, however, are not always peripheral. In Figure 8.3A, a single state ultimately remains; the system is homogeneous and illustrates Wolfram's type I behavior. The 3 other systems show a type II behavior whereby 2 states (B and D), or 3 states (C) are still extant at the end of the process. In the latter case, the non-dominant states resist eradication owing to the size of their patches, and to the strength of their own NCP effect. This figure strongly suggests the relationships between the origin of a trajectory and its endpoint; that is, between initial and final state probabilities. This is shown in Figure 8.4 by the trajectory end-points of 520 simulated systems. Initial conditions were evenly distributed over the whole phase-space. End-points are concentrated into 7 zones (vertices, mid-sides and centre) which re-

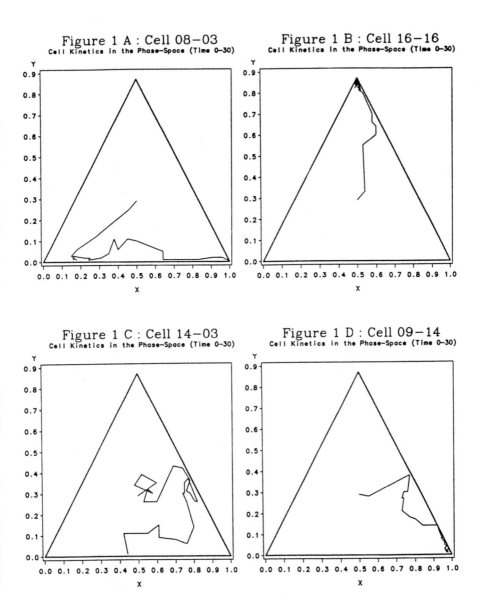

Figure 8.1. Kinetic trajectories of 4 selected cells in the phase-space (from time 0 to time 30)

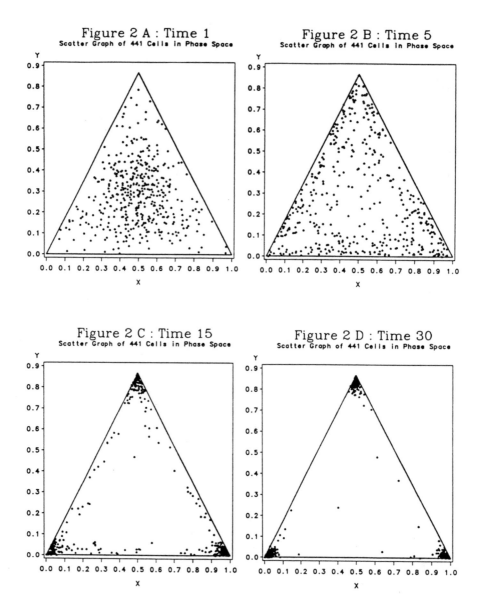

Figure 8.2. Positions of 441 cells in the phase-space at time 1 (A), 5 (B), 15 (C) and 30 (D).

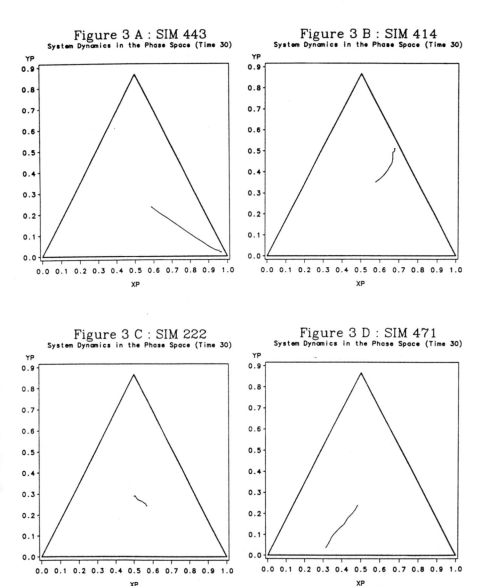

Figure 8.3. Dynamical trajectories of 4 selected systems in the phase-space over the time span 0 to 30 (the trajectory's origin is the most central end). At each time unit, the overall system is represented by a single point subject to the relative frequency of the three states.

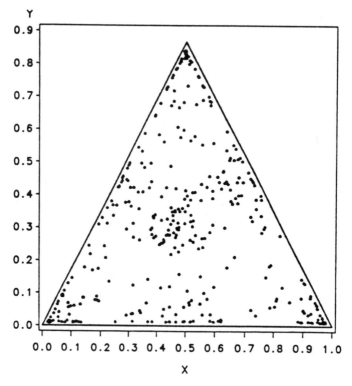

Figure 8.4. End-points of the dynamical trajectories for 520 simulated systems. All systems assume a homogeneous space. Initial conditions are evenly distributed over the whole phase space.

spectively represent single state, 2-state, and 3-state attractors. Other locations are far less stable: a system in such locations, be it in an initial or a transient position, would likely be pulled toward one of the attractors. The fact that a few systems can nevertheless end up in these unstable zones reflects, as mentioned above, that non-dominant states may resist if they form patches of a sufficient size to generate a strong NCP effect.

These elements, as well as other considerations pertaining to the configuration's features (see for instance the skipping-step test in Phipps 1989) may certainly cast some light on the intriguing scaling problem: how can we understand the transformation of local interaction into a global dynamical behavior.

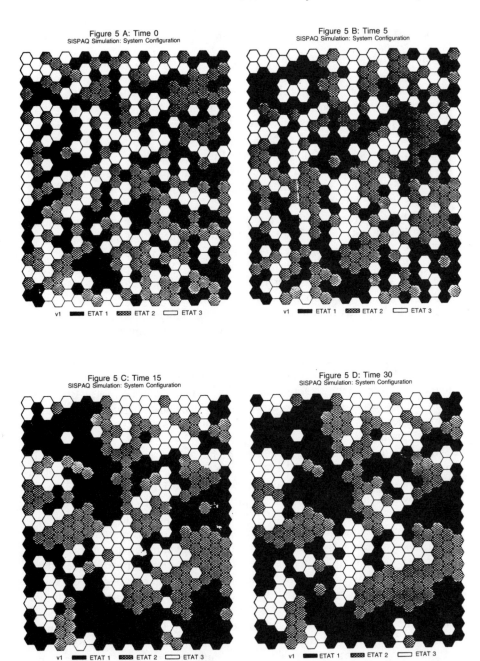

Figure 5 A: Time 0
SISPAQ Simulation: System Configuration

Figure 5 B: Time 5
SISPAQ Simulation: System Configuration

Figure 5 C: Time 15
SISPAQ Simulation: System Configuration

Figure 5 D: Time 30
SISPAQ Simulation: System Configuration

v1 ETAT 1 ETAT 2 ETAT 3

Figure 8.5. System's configuration at 4 stages of the dynamical process. The system assumes a homogeneous space and an initial equiprobability of the 3 states.

Neighborhood Coherence and Cell Ecology

Although sensu stricto ecology does not include the tissue level of organization, there are many common features between tissues and populations. Indeed, the study of cell populations cultured in vitro has often been done using methods borrowed from population ecology (see Skehan 1986). Moreover, in vivo tissues display cells spatially arranged in relatively stable networks. As an example, connective tissues form a 3-dimensional arrangement, but monolayer epithelia display a 2-dimensional arrangement. A tissue may thus be regarded as a system whose individual components interact locally as the components of a population actually do.

Cells can be observed in various states. They may stay in a quiescent, differentiated state and perform specific functions during a span of time after which they die. Other cells in a potentially proliferative state may divide and provide for the replacement of dead cells. The daughter cells will eventually acquire the quiescent differentiated state. Cells have also been shown to exhibit short circuit communication among neighbors through gap-junctions. It was first demonstrated by Loewenstein (1979) that this cell-cell communication was responsible for the establishment of a metabolic co-operation among neighboring cells. But the overall importance and the controls that should be expected to develop from this communication in such processes as embryo development, differentiation and growth are yet to be uncovered. It is indeed, increasingly believed that local interactions play a mandatory role in the development of an animal from a zygote (Boettiger 1989). This would oppose the widely held view that the crux of organismic organization rests on global field mechanisms such as reaction-diffusion systems (Hunding et al. 1990).

In order to explore hypothetical properties which could result from this biological phenomenon, the NCP and the SISPAQ model were applied to simple tissue models in which the neighborhood effect was supposedly mediated by gap-junction communication. On the basis of simulations, Phipps et al. (1990) suggested that an NCP-like phenomenon could well be evoked in biological systems in the developmental process of differentiation as well as in tissue homeostasis. Differentiation would be akin to pattern formation, whereas homeostasis would rest on the self-maintaining capacity. According to this hypothesis, a NCP-like process, operating through cell-cell communication, would be sufficient to account for the maintenance of a tissue in its normal quiescent state. As a consequence, the extinction of the NCP effect, due to any disruption or impairment of cell-cell communication, would determine the loss of homeostasis, and the possibility for single or groups of normally dividing cells to escape the quiescence constraint exerted by surrounding cells. Accordingly, the onset of uncontrolled division (hyperplasia) and tumorigenesis could be understood as a disease of cell-cell communication.

As a corollary to the rule of neighborhood size mentioned in the preceding section, cell systems such as monolayer epithelia with a 6:1 contact ratio should be more prone to deregulation than connective tissues with a 12:1 ratio. This would entail a greater occurrence of cancer in 2-dimensional cell arrangements. This prediction is amply verified by empirical evidence; it is indeed known that 90% of cancers are initiated in monolayer epithelia. These hypotheses are currently undergoing empirical investigations.

Discussion

Two questions should be addressed in this discussion: (1) what NCP can teach us about patchiness, and (2) how it can provide new insights into a wider range of problems.

As far as the emergence of patchiness in homogeneous environments is concerned, NCP and SISPAQ simulations highlight four results that deserve further empirical investigations: i) the existence of self-organizing and self-maintaining capacities giving way to distinct dynamical behaviors; ii) the critical probabilities for which the system switches from one dynamical behavior to the other; iii) the relation between range of interaction, size of the neighborhood and the propensity of the system to build up patchiness; and iv) the strength of the capacities of self-organization and self-maintenance, particularly as far as pattern homeostasis is concerned. Most of these points coincide with the findings of Rohlf and Schnell (1971) and Turner et al. (1982) who based their investigation on the isolation by distance model and used simulation models very similar to SISPAQ. Starting from an even gene distribution, they found a time increase of homozygosity and a spatial segregation of homozygous types. They also found a high stability of segregation patterns and a tendency of small neighborhoods (strict nearest neighbor pollination) to increase patchiness. All these results converge with those obtained with the SISPAQ model which generalize the rules of pattern emergence to all phenomena where the basic generative process may be understood as neighborhood coherence.

Indeed, NCP and SISPAQ demonstrate the way in which a system develops a global spatial differentiation under the exclusive control of strictly local and individual driving forces. Be it through molecular messengers in cell systems, ramet expansion in plant communities, nearest-neighbor pollination in some plant populations, energy and biotic fluxes among adjacents communities in landscape, or through visible cultural signaling in urban communities, the principle of neighborhood coherence accounts for the way in which information is exchanged locally with the ultimate effect of creating local consensus and, furthermore, spatial heterogeneity.

The application of NCP to cell systems certainly represents an interesting and promising avenue. It shows that tissue homeostasis could be conceivably regarded as a cell community phenomenon whereby mutual interaction controls individualistic variability, thus safeguarding tissues from erratic cell behaviors. This obviously challenges the accepted view of a cell's intrinsic stability as well as that of an overall predetermined organismic control. Homeostasis would thus appear as a phenomenon of cell social behavior, a matter of cell ecology in its own right. In this regard, cell biology would learn a lesson from ecology. This line of investigation reminds us of the old concern that biologists had for a positional information and strikingly echoes the new trend toward a topobiology advocated by Edelman (1988).

Conclusion

This contribution pleads for the use of Cellular Automata in ecological modeling. Should these models be considered an alternative to other more familiar models? Until they have proven to give a better understanding of the real world, the answer to this question is no. However, it would certainly be of interest to test them concurrently with well known models. Such a comparison should certainly serve the purpose of scientific investigation. But in so doing, we have to keep in mind two key issues: for what purposes should we have recourse to CA, and on what conditions? Answering these questions touches on various aspects: individuality, spatial reality, simplicity and stochasticity.

To what extent do CA satisfy the requirements which are deemed necessary in ecological modeling, particularly with regard to individual variability (Huston et al. 1988)? It has been clear in this paper that basic units in CA are primarily of a spatial nature. In a sense, we may consider this feature as an extension of the notion of the individual. Moreover, in all applications of CA in ecology, each cell is further considered a site for one biological individual, be it sessile or mobile. While it is not clear how a deterministic CA could incorporate individual variability, this is perfectly conceivable with a probabilistic rule. For instance, differences in the propensity to adopt one particular state (i.e. variable initial probabilities) may well be introduced among individuals, in a random manner or according to some spatial pattern. In the SISPAQ flow chart represented in Table 8.1, the inclusion of a random factor (in phase I) just implements this feature. In this regard, CA fit the objective of individual-based models.

In CA, space clearly outweighs all other considerations. Be it at the local level (e.g. the spatial realism which controls effective interaction), or at the global (e.g. the emergence of patchiness in a homogeneous system), space is in the forefront. This primacy might be the single most conspicuous advantage of using CA, particularly at the present time when the distinctiveness of locations takes on increasing importance (Huston et al. 1988). Individuals in the real world do interact in so far as they are in such relative locations that their contact is likely to occur. The way in which the neighborhood is defined in CA, with a wide range of options, warrants that a large number of real situations could be accounted for. Again, the objectives of CA match those of individual-based models.

Whether dynamical models ought to be simple or complex may seem to be a matter of philosophical discussion. This question, however, is somehow bound to that of individuality. Transition rules which mimic individual interaction, albeit simple, have proven to be able to produce a great deal of complexity. Perhaps there is limited value in building up models that produce complex behavior by adding and combining ever more complicated specifications. Such models would barely have a chance to produce anything different than what they were meant for. Not only would their results be tainted with triviality, but many required assumptions would appear unreasonable. A simple rule producing complexity beyond expectation is of much greater heuristic significance, thus hinting that "if it might be assumed to play a role in the real world, it could also be sufficient to account for the observed complexity."

Whether or not dynamical models ought to incorporate stochasticity may appear as another philosophical question. Indeed, the role assigned to randomness is at stake here. In many models, randomness may just represent an error factor, a final touch which nicely blurs the neatness of mathematical formalism. It may also represent the effect of hidden variables, and statisticians have derived convenient tools to make provision for the temporary imperfection of our knowledge. A different line of thought sees randomness as inherent in the very nature of things (Thom 1980 1981, Morin 1980, Atlan 1981, Phipps 1987). Rooted in the paradigm of self-organization, and popularized by such aphorisms as "order from noise" or "chance and necessity," this view could be the finishing stroke to Laplace's tidy world. Be it through fluctuation in dissipative structures (Prigogine and Stengers 1979, Prigogine 1980), or through haphazard processes (Morin 1980), randomness is viewed as playing an essential role in the emergence of order in nature. Far from settling the debate, the emerging paradigm of deterministic chaos raises new questions. Since random-like behaviors can result from deterministic systems and turn spontaneously into order under critical conditions (Langton 1986, Waldrop 1990), is there still any legitimacy for intrinsically random processes? In the context of this paper, this question refers back to the deterministic or stochastic nature of the rule, thus suggesting that the choice of either of them is not a mere matter of convenience.

If, at any rate, CA should contribute to the progress of ecological modeling, it could also be in fostering a *minimalist* approach combining simplicity and stochasticity. This is not intended to value simple approaches as such. But it is increasingly realized that individual interactions, inherently simple and local, may yield a great deal of structural complexity at higher hierarchical levels of organization. Such an approach would therefore seek to account for the observed behavior in the simplest necessary though sufficient way, all other conditions being as undetermined as possible. In other words, this approach would assume the largest entropy beyond the smallest constraint.

As far as conditions of application are concerned, most available examples, strikingly, make use of probabilistic rules. As noted by Gutowitz (1990b), in applying CA to biological systems "one is tempted to relax the strict definition of cellular automata in order to match as well as possible the design of the system under study." The relaxation of the rigidly deterministic rules illustrates well the point. The "Game of Life" may represent an intriguing mathematical device, however, it teaches us little about real life. Probabilistic transition rules certainly offer a better analogy with natural systems. The extent to which other CA features could be relaxed is, of course, a matter of trade-off. On the one hand, relaxing more features would perhaps provide a more realistic view and a better prediction of the state of the system under study. On the other hand, this gain could be at the expense of the heuristic value of the minimalist approach. As shown by the application of NCP to the problem of tissue homeostasis, a simple model may provide an alternative view and understanding of a phenomenon resulting from very complex mechanisms. In that sense, CA could help us to disentangle the threads of scientific explanation. As Couclelis (1988) noted: ". . . their function is to stimulate thought and to produce insights." This leaves us with the still unanswered question: Does science endeavour to understand or to replicate the real world?

Literature Cited

Atlan, H. 1981. Postulats métaphysiques et méthodes de recherche. *Le Débat* **14**:83–89.

Atlan, H. 1985. Two instances of self-organization in probabilist Automata networks: epigenesis of cellular networks and self-generated criteria for pattern discrimination. In *Dynamical Systems and Cellular Automata*, J. Demongeot, E. Golès and M. Tchuente, (eds.), pp. 171–186. Academic Press, London.

Bak, P., K. Chen, and M. Creutz. 1989. Self-organized criticality in the 'Game of Life'. *Nature* **342**:780–782.

Bays, C. 1988. A Note on the Discovery of a New Game of Three-dimensional Life. *Complex Systems* **2**:255–258.

Boettiger, D. 1989. Interaction of oncogenes with differentiation programs. *Curr. Top. Microbiol. Immuno.* **187**:31–68.

Comins, H. N. 1982. Evolutionarily stable strategies for localized dispersal in two dimensions. *J. Theor. Biol.* **94**:579–606.

Couclelis, H. 1988. Of mice and men: What rodent populations can teach us about complex spatial dynamics. *Environ. and Planning A* **20**:99–109.

Demongeot, J. 1985. Random Automata and random field. In *Dynamical Systems and Cellular Automata*, J. Demongeot, E. Golès and M. Tchuente, (eds.), pp. 99–110. Academic Press, London.

Demongeot, J., E. Golès and M. Tchuente. 1985. Introduction: Dynamic behaviour of Automata. In *Dynamical Systems and Cellular Automata*. J. Demongeot, E. Golès and M. Tchuente, (eds.), pp. 1–14. Academic Press, London.

Edelman, G. M. 1988. *Topobiology: An Introduction to Molecular Embryology*. Basic Books, New York.

Edwards, S. F. and P. W. Anderson. 1975. Theory of spin glass. *J. Phys. F Met. Phys.* **5**:965–974.

Fredkin, E. 1990. Digital mechanics: An informational process based on reversible universal cellular Automata. In *Cellular Automata: Theory and Experiment*, H. Gutowitz, (ed.) pp. 254–270, Physica D **45**. Elsevier, North-Holland.

Gardner, M. 1971. On cellular automata, self-reproduction, the Garden of Eden and the game of "life." *Sci. Am.* **224**:112–117.

Gerhardt, M., H. Schuster, and J. J. Tyson. 1990. A Cellular Automaton model of excitable media including curvature and dispersion. *Science* **247**:1563–1566.

Gilpin, M. E. 1990. Extinction of finite metapopulations in correlated environments. In *Living in a Patchy Environment*, B. Shorrocks and I. R. Swingland (eds.), pp. 177–186. Oxford Science Publications, Oxford, England.

Greussay, P. 1988. L'ordinateur cellulaire. *La Recherche* **19**:1320–1330.

Gutowitz, H. (ed.). 1990a. *Cellular Automata: Theory and Experiment*, Physica D **45**. Elsevier, North-Holland.

Gutowitz, H. 1990b. Introduction. In *Cellular Automata: Theory and Experiment*, H. Gutowitz (eds.), pp. vii–xiv, Physica D **45**. Elsevier, North-Holland.

Hopfield, J. J. 1982. Neural networks and physical systems with emergent collective computational abilities. *Proc. Nat. Acad. Sci. USA* **79**:2554–2558.

Hunding, A., S. A. Kauffman, and B. C. Goodwin. 1990. Drosophila segmentation, supercomputer simulation of prepattern hierarchy. *J. Theor. Biol.* **145**:369–384.

Huston, M., D. DeAngelis, and W. Post. 1988. New computer models unify ecological theory. *BioScience* **38**:682–691.

Inghe, O. 1989. Genet and ramet survivorship under different mortality regimes— A Cellular Automata model. *J. Theor. Biol.* **138**:257–270.

Jen, E. 1990. Aperiodicity in one-dimensional Cellular Automata. In *Cellular Automata: Theory and Experiment*, H. Gutowitz (ed.), pp. 3–18, Physica D **45**. Elsevier, North-Holland.

Langton, C. G. 1984. Self-reproduction in Cellular Automata. *Physica D* **10**:135–166.

Langton, C. G. 1986. Studying artificial life with Cellular Automata. *Physica D* **22D**:120–149.

Lee, Y. C., S. Qian, R. D. Jones, C. W. Barnes, G. W. Flake, M. K. O'Rourke, K. Lee, H. H. Chen, G. Z. Sun, Y. Q. Zhang, D. Chen, and C. L. Giles. 1990. Adaptive stochastic Cellular Automata: theory. In *Cellular Automata: Theory and Experiment*, H. Gutowitz (ed.) pp. 159–180, Physica D **45**. Elsevier, North-Holland.

Loewenstein, W. R. 1979. Junctional intercellular communication and the control of growth. *Biochim. Biophys. Acta* **560**:1–65.

MacArthur, R. H. and E. O. Wilson. 1967. *The Theory of Island Biogeography*. Princeton University Press, Princeton, NJ.

May, R. M. 1976. Simple mathematical models with very complicated dynamics. *Nature* **261**:459–467.

May, R. M. and T. R. E. Southwood. 1990. Introduction. In *Living in a Patchy Environment*, B. Shorrocks and I. R. Swingland, (eds.) pp. 1–22. Oxford University Press, Oxford, England.

Mazoyer, J. 1989. Le problème des fusiliers. *La Recherche* **20**:263–266.

Mezard, M., G. Parisi, and M. A. Virasoro. 1987. *Spin Glass Theory and Beyond*. World Scientific, Singapore.

Morin, E. 1980. Au delà du déterminisme: le dialogue de l'ordre et du désordre. *Le Débat* **6**:105–122.

Peterson, I. 1987. Forest fires, barnacles and trickling oil. *Science News* **132**:220–223.

Phipps, M. 1987. Un mauvais adversaire pour un bon combat: Un commentaire sur "l'analyse des systèmes en Géographie humaine." *Géogr. Can.* **31**:47–49.

Phipps, M. 1989. Dynamical behavior of Cellular Automata under the constraint of neighborhood coherence. *Geogr. Anal.* **21**:197–215.

Phipps, M., J. Phipps, J. F. Whitfield, A. Ally, R. L. Somorjai, and S. A. Narang. 1990. Carcinogenic implications of the Neighborhood Coherence Principle (NCP). *Med. Hypotheses* **31**:289–301.

Prigogine, I. and I. Stengers. 1979. *La nouvelle alliance: Métamorphose de la science*. Gallimard, Paris.

Prigogine, I. 1980. Loi, histoire et . . . désertion. *Le Débat* **6**:122–130.

Rohlf, F. J. and G. D. Schnell. 1971. An investigation of the isolation-by-distance model. *Am. Nat.* **105**:295–324.

Sieburg, H. B., J. A. McCutchan, O. K. Clay, L. Cabalerro, and J. J. Ostlund. 1990. Simulation of HIV infection in artificial immune system. In *Cellular Automata: Theory and Experiment*, H. Gutowitz (ed.) pp. 208–227, Physica D **45**. Elsevier, North-Holland.

Skehan, P. 1986. On the normality of growth dynamics of neoplasms in vivo: a data base analysis. *Growth* **50**:496–520.

Sutherland, W. J. 1990. The response of plants to patchy environments. In *Living in a Patchy Environment*, B. Shorrocks and I. R. Swingland (eds.) pp. 45–61. Oxford University Press, New York.

Thom, R. 1980. Halte au hasard, silence au bruit! *Le Débat* **3**:119–132.

Thom, R. 1981. En guise de conclusion. *Le Débat* **15**:115–123.

Toffoli, T. and N. Margolus. 1987. *Cellular Automata Machines*. MIT Press, Cambridge.

Turner, M. E., J. C. Stephens, and W. W. Anderson. 1982. Homozygosity and patch structure in plant population as a result of nearest-neighbor pollination. *Proc. Natl. Acad. Sci.* USA **79**:203–207.

Victor, J. D. 1990. What can automaton theory tell us about the brain? In *Cellular Automata: Theory and Experiment*, H. Gutowitz (ed.) pp. 205–207, Physica D **45**. Elsevier Science Publisher, North-Holland.

von Neumann, J. 1963. The general and logical theory of Automata. In *Collected Work 5*, A. H. Taub, (ed.), pp. 288–328, Pergamon Press, Oxford.

Waldrop, M. M. 1990. Spontaneous order, evolution, and life. *Science* **247**:1543–1545.

Wolfram, S. 1983. Statistical mechanics of Cellular Automata. *Rev. Mod. Phys.* **55**:601–644.

Wolfram, S. 1984a. Cellular Automata as models of complexity. *Nature* **311**:419–424.

Wolfram, S. 1984b. Universality and complexity in Cellular Automata. *Physica D* **10**:1–35.

Wolfram, S. 1985. Some recent results and questions about cellular Automata. In *Dynamical Systems and Cellular Automata*, J. Demongeot, E. Golès and M. Tchuente, (eds.) pp. 153–167. Academic Press, London.

Wolfram, S. 1986. *Theory and Applications of Cellular Automata*. World Scientific, Singapore.

9

Hierarchical and Concurrent Individual Based Modeling

John B. Palmer

ABSTRACT.   **In nature, events are concurrent and the influence of one event upon another propagates through yet other events. Thus, individual-based or individual-like models, which embody the concept of the interaction of model entities, are more true to nature than classical serial models controlled by high level data structures. However, in nature events and processes at one scale constrain or enable events and processes at other scales. The uniform time step and simple local interactions of models which simulate concurrent behavior are special cases of more complex and rich possibilities of organization. A model medium is presented which allows crafting of simulations embodying the principles of concurrent interacting entities which are also hierarchically ordered. The entities of such models advance the virtual time of the simulation. Actions may operate at different scales as the virtual occasions are composed by the model entities. This hierarchical and concurrent individual based model (HICBM) medium is presented in an object oriented idiom. Code for class methods essential to the concept of the medium is given in the programming language Smalltalk. The medium is created by hierarchical restrictions on the order and scope of message-passing among model entities in the HCIBM. The HCIBM scheme should provide a framework for the design of parallel computers to run concurrent ecological models. It is suggested that a better theoretical understanding of concurrency could refresh our understanding of nature. The notion of individual action appears inextricably involved in the concept of concurrency.**

Introduction

Individual-based models (Huston et al. 1988) share a strong resemblance with models of cellular automata (E. F. Codd 1968, Toffoli and Margolus 1988). They are also very similar to models derived from the field of Artificial Intelligence that simulate biological or ecological processes (Taylor et al. 1989, Hogeweg 1989, Lugowski 1989). These two classes of models have demonstrated that a construct of locally interacting model entities can produce quite interesting and sometimes surprising overall global behavior. These models operate on a common time step during which model entities are updated (change their state as a function of the states of neighboring entities). The change of state of model entities proceeds in parallel, and in virtual model time the model entities are acting concurrently. It should be noted that the models of Taylor et al. and Hogeweg have been applied with apparent initial success to ecological problems. Thus, single step concurrency can produce useful results. On the other hand, it is almost an ecological truism that the time and space scale of interactions needs to be taken into account in simulating nature (e.g., Kershaw 1964, Allen and Starr 1982). Events and processes at one scale constrain or enable events and processes at other scales. A uniform time step and simple local interactions are special cases of more complex possibilities of organization. Progress in giving expression to more complex relations is presented in this paper.

The motive for the present work was to develop a medium for object-oriented, individual-based models that would serve as a template for a multitasking computer paradigm. This paradigm could serve as the design principle of computer architectures for running individual-based ecological simulations (Butler et al. 1987). Langton (1989) pointed out that models which simulate concurrency might be effectively implemented on parallel computers, and that to be more true to nature, such models should be free of the global control of high level data structures typical of modern programming languages on serial computers. Thus, a concurrent computing medium should also seek to replace data structure control with 'ecological' control.

The object of the work reported on here is to build a modeling media which is explicitly both parallel and hierarchical, and, in principle, overcomes the limitation of operating on a simple time step. This medium should permit the formation of interesting problems on one hand, and the mapping of the medium into a multitasking computational system on the other. The initial formulation of this programming medium is reported in this paper.

Approach

The formation of a template for hierarchical concurrent individual-based modeling (HCIBM) was approached by developing a general programming medium in which concurrent acts of individual organisms could be simulated. The object-oriented programming language Smalltalk (Pinson and Wiener 1988) was used as a tool because its programming paradigm embodies some of the aspects of a HCIBM. It was felt that an extension of this language could be developed to allow for the expression of the ecological concurrency of acting organisms.

An object-oriented program is a system of logical entities called objects that can communicate with one another. These communications, called messages, invoke the methods which can be thought of as the object's behavior. Objects respond to messages by doing something, e.g., doing a computation and returning the result, altering their states or those of other objects, or even making new objects of some sort. Thus, a natural system could be represented as the interplay of objects sending and receiving messages and changing some parts of the system in response to the messages. A program could be designed to advance concurrently across an ensemble of objects which change in number, kinds, and state with some independence from each other.

There is no necessary order among programming objects as to the pattern in which messages are sent, acted upon, and new messages sent. However, in practice, because a programming language is implemented upon classical computer architectures, the sequence is serial. In fact, the practical control which one has in programming comes from being able to control the sequence of events. This constraint is implicit in programming practice. In nature, however, events are concurrent, and the influence of one event upon another propagates through other events. However, this influence operates over different temporal and spatial scales. Thus, to simulate the concurrency of events in nature over different scales, the problem was to find a way to make explicit a particular kind of order among the interactions of objects and to allow the objects themselves 'control' over the flow of events. Such a medium holds the promise of creating models in which the diversity of individual actions and the patterns of propagation of influence can be investigated with some ease.

Results

Several concepts which are basic to concurrent modeling have emerged in this study. They will be presented before a programming description is given.

Act Determined Time

The first important concept is that properties implicit in the model entities themselves determine the advance of temporal events. Time causes nothing to happen. In a concurrent model one conceives of the acts of entities as driving time. The interactions of the model entities determine the advance of the simulation. Time is more of a bookkeeping convenience, like predation rate, and not a real thing like eating or being eaten. System time in an ideal HCIBM tracks the sequence of the acts of model entities.

Dynamic Hierarchy

The next important concept is that a concurrent system is partially hierarchical in organization. Hierarchies may operate concurrently, and within hierarchies some of their parts operate concurrently. Model entities may include other model entities, and these entities may operate on different temporal or spatial scales. The hierarchy orders the execution of the functional acts of the model entities. However, the acts of the entities may reorder the hierarchy. This arrangement allows a design balance to be struck between the principle of independent action of an entity and the regulation of that action as part of a system.

Determination of the Model of Acting by Place in Hierarchy

How an entity acts depends upon its place in the hierarchy. The place an entity occupies in the hierarchy determines what it does and how it changes. In a sense, its location determines the parameters which specify its mode of acting. The model entities of a concurrent system are competent to execute complete acts, e.g., capturing prey or surviving a drought, and they function as such only when appropriately situated within the hierarchy. When this action is complete, the model entity refreshes its location in the dynamic hierarchy or obtains a new location. It is now competent to act again, but will not do so until moved to act by the entity of which it is a part.

The entity of the concurrent system which includes a given model entity is the super part of the given entity. The initiation of action begins with the super part moving the given part to act. The first operation of the given part is to determine the kind of super part which moved it to act. This determination establishes the particular mode of action for the given part. A super part corresponds somewhat to a parameter determining how an included part acts. However, such a part is more than a parameter, because it also moves the included part to act, and it may alter the neighborhood of the included part prior to moving it to act. To change its mode of acting, a model entity changes its location in the model hierarchy.

Concurrent Units: Wholes

Concurrent sections of the model may be defined by hierarchies headed by specialized model entities which continue to act until all of their parts have finished acting. Such model entities are regarded as wholes. As long as at least one of the parts of a whole re-enlists the whole as its super part, the occasion of the whole continues and the part is guaranteed repeated opportunities to act. When a whole has no parts left, i.e., when all its parts have established themselves under another super part, the occasion of the whole's act ends. Thus wholes provide for extended durations. If an entity becomes included under another super part which is not a whole, it is not guaranteed to continue acting within its present occasion.

The Inclusive Relations of Model Entities: the Movement to Act.

A model entity may have a super part which moves it to act. Entities which are moved to act constitute part of the functional occasion represented by their super part. Because the relation of being moved to act is an inclusive relation, there may be parts within parts. This relation is intransitive within the structure of an occasion. A given model entity can look down at the model entities which it includes, and which themselves may include parts. It may also look up at its super part and any higher parts which include it. Thus the inclusive relations of a functional occasion map into a simple directed graph which has no loops. The movement to act propagates downward from the top entity of the graph. Within the virtual time of the model this movement defines the initial boundary of the functional occasion. The final boundary of the occasion is determined as the various model entities of the graph complete their acts. This completion is accomplished by the model entities becoming included in another model entity which is not a whole entity of the functional occasion. When one model entity becomes included in another entity, it reverts to a potential phase, and it cannot act unless it is moved to do so by its super part.

The Connections of Model Entities.

Besides the relation of inclusion, model entities may be connected with one another in a structural manner. In particular, model entities which are specialized as wholes may be structurally connected. These structurally connected whole model entities may head separate hierarchies of other model entities, some of which in turn may be wholes. These connections serve for the communication of information within the occasion which the model entity heads. The structural relations of the model entities define the places or static locations of the model. These structural relations can be changed by the action of model entities, but such change is not mandatory as is the dynamic inclusive relationoship of part and super part, which must be re-established at the end of the act of a model entity.

The Potential and Objective Phases of Model Entities

Model entities exist in two phases: objective and potential. When a part elects a super part (or is assigned one in the initialization of a model) it is included as a potential part. It must be advanced to an objective phase before it can be moved to act. Consider a section of a model in which the head entity initiates an act. The first operation of the head is to advance its own potential parts into objective status. This movement is then propagated to its newly objective parts which in turn repeat the operation on their potential parts, advancing them to objective status and so forth until the terminal entities in the section are reached. Since the terminal entities are distinguished by having no parts, the first phase of the new occasion of the section is complete, and all the parts of the section are objective. A new inclusive order has been established for the next occasion. Now, again beginning with the head entity, all of the objective parts are moved to act based upon the fresh objective values of their parts. This movement to act likewise propagates along the branching inclusive pathway from the head to the terminal entities. If there are no whole entities in the pathway, the entire occasion of the pathway would be over in one operation of these two phases. If there are wholes in the pathway some of the model entities may go through a number of cycles of alteration of potential to objective to potential and so on until no whole remains with parts.

The Major Features of an Object Oriented Implementation for HCIBM

The concept of a hierarchical concurrent individual-based model will be illustrated by a simple ecological example expressed in an object-oriented idiom. Figure 9.1 illustrates the structure of this example.

The Ecological Example

The ecological example is of a population of organisms consisting of neonatal, immature, and adult male and female individuals. These are divided into populations in different locations. Within each location there are different functional niches for the neonatal and adult individuals. The individuals compete for resources on a scramble basis. If there is not enough resource to meet minimal needs the immature individuals and adults may migrate to a niche in another location, and the neonatal individuals may seek another niche in their location. The act of an individual organism defines the least occasion in the model. Its duration circumscribes the operations of surviving, obtaining nourishment, growing, and developing. If the individual is an adult female its act may also include reproduction. Neonatal individuals are produced in the location of their female parent. The acts of the individual organisms occur within the occasion of a season at a location. Within each location there is functional niche space. The occupancy of the niches renders them habitat in which individuals interact. The level of nutrient production may vary from season to season and from location to location.

The Object Oriented Model Structure

The model entities represent the organisms, their niches, the locations where their niches exist, pathways between the locations, the season in which the population lives, and the model itself. All of these entities are special-

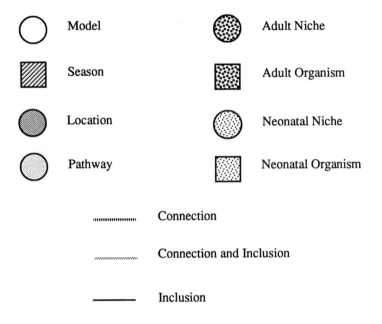

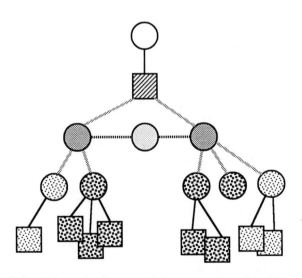

Figure 9.1. Schematic diagram of the types of model objects and their relations in the example for the hierarchical and concurrent individual-based model (HCIBM).

izations of one class of objects. This class is designed to be expressive of dynamic entities which are related through a composition of differing but cohesive space-time scales. With the exception of the organisms and the season, all of these model elements are instances of whole type entities. These whole entities are connected in a static hierarchical configuration which represents the basic model structure. The highest entity is that which represents the model itself. This entity is a logical convenience connecting the model to the modeler. The next model entity in the hierarchy represents the season. The season includes the several locations. The scale of the season includes all that happens at the locations, although in different locations the season may register differently in terms of influencing productivity. The locations are connected through a pathway entity. In this example, the pathway object does not act. It functions only as a link among the locations. This arrangement allows the adult organisms to migrate from location to location within the occasion of a season. Each location includes an adult and several neonatal niches. These represent the functional places that organisms occupy within a location. The two kinds of niches are related in that the levels of production are proportional within a location. Neonatal individuals are produced in the adult niche of their female parent, but move randomly into a neonatal niche in the same location.

Conceptually, the occasion of a season consists of the repeated acts of the individual organisms until all the organisms which were present at run initiation have died or matured and reproduced, or their offspring have done so. During this season the organisms or their offspring may migrate from location to location if densities are high relative to nutrient production.

The Working of the Object Oriented Model

All of the model entities in the HCIBM can respond to the message 'act now' which is sent to them only by the model entity which is their super part. For instance, the season is the super part of the locations. 'Act now' means different things to different model entities. The code which this message invokes is basically what makes different model entities behave differently. The form of the response ensures that the model entities act in the compositional order mapped into the hierarchy, and that the model structure will be reconstituted so that further acts of the model entities may occur. In the example HCIBM, the structure of connections of the season, location and niche entities do not change, but that of the organisms to those structural parts may change. Figure 9.2 shows such changes.

At model initiation, some organisms are included in the niche objects of some locations as potential elements. The season object is moved to act by sending it the 'act now' message. In response, it moves its inclusions, if any, to act. Then it moves its connected locations to act. The locations have connections of adult and neonatal niches which the locations, in turn, move

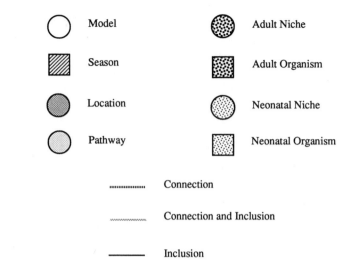

○	Model	●	Adult Niche
▨	Season	▦	Adult Organism
◉	Location	◉	Neonatal Niche
○	Pathway	▢	Neonatal Organism

.... Connection

〜〜〜〜 Connection and Inclusion

───── Inclusion

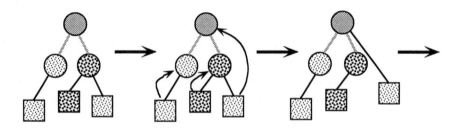

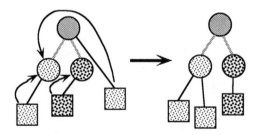

Figure 9.2. Example of changes in inclusive relations during the occasion of the season at a location object. A neonatal organism moves from the adult niche, where it was produced to a neonatal niche in the same location via the medium of the location object. The other organism objects continue to re-establish inclusive relations with their niche objects.

to act by sending to them the 'act now' message. The specific behavior of the location consists in setting the level of productivity for its niches. The niche objects include the organism objects. The specific behavior of the niche objects is to assess the number of interacting organisms and the production available to each. Because it is a whole object, a niche object repeatedly sends 'act now' messages to the organisms included in it until all have transited to a higher model object (the location or the season object). If an organism object transits to a location object, the location object will act again, because the location is an instance of a whole object. The behavior of the organism object will then move it to inclusion in a niche object connected to the location.

The specific behavior of the organisms is the most complex of any of the entities in this example HCIBM. When moved by the 'act now' message, an organism object determines the type of object of its super part. Usually it will be in its niche and will then determine if it survives to obtain nourishment. It does this by sending a message, 'survive duration' to its niche. The value that the message returns informs the organism object if it has survived the duration. If it has not, it reports its state to a data monitor and exits the simulation. If it survives, it obtains its portion of the production from the niche by sending an 'obtain nourishment' message to its niche. If the portion if too meager, the organism ends its act by seeking to be included in another location. This is accomplished by sending a message to its niche's super part, a location object. This object communicates with the pathways object connected to it, which returns to the organism object a randomly selected linkage to one of the location objects.

On the other hand, if the portion is sufficient, the organism object grows and adds one occasion to is age. If now it is either old or big enough, it develops to the next stage. If not, it ends its act by re-including itself with its niche super part. The operation of re-inclusion ensures that it will act again in the current season. A neonatal or immature stage organism may require several such acts of overcoming risk and obtaining nourishment in order to mature to an adult. Because there is an age (as measured in number of these acts) as well as a size threshold for adulthood, the size of newly mature individuals may vary. On the other hand, adult organisms could experience only one operation of risk and nourishment in each seasonal occasion. Although both adult and neonatal model entities coexist in the location object, the time scale is more finely divided for the neonatal individuals. However, an adult individual can move among locations within a season so its world is divided spatially at the scale of locations. Neonatal individuals may move among neonatal niche objects within locations. The scale of the location is thus larger for the neonatal individuals than for the adults. When an adult individual has reproduced, or attempted to, it ends its act by transiting from its niche to the season entity. Since the season is not a whole object, it does not act again in response to the inclusion of organism objects. Individual organism objects included in location or niche objects will cause those objects to continue to move their parts (the organisms) to act. When all have transited to the season entity, the occasion of the season is finally ended. The next season is initiated by the model entity moving the season to act again.

The Major Features of the Example HCIBM in Smalltalk

The major features of the HCIBM as implemented in Smalltalk are presented here. Smalltalk code is presented in the Appendix for the major methods of the system. For those versed in Smalltalk or related languages, this should be sufficient information to develop a system tailored to one's own interests. All HCIBM entities are members of a new Smalltalk class called Part, or members of a subclass of Part. A Part object has the instance variables *inclusions*, *connections*, and *superPart*. The *inclusions* variable is a Value object, also a new class. Value objects store other objects in a potential or objective state. Potential parts are stored by the *include*: message which is sent to a Part entity. A part object advances its potential parts to objective status and quits its current objective parts when it receives the *advanceValue* message. The parts of a Part object are always themselves Part objects.

The *connections* variable is a Set object. It contains parts which may be moved to act or are simply available as linkages to other parts of the model system. All messages among parts must move through the relations (inclusive and connective) of the Part model entities. The *superPart* variable identifies the particular Part object in which a part is included. The value of *superPart* may be nil, but in this case, the model entity must include or be connected to other model entities, because all Part entities must be connected into a whole in order to operate.

The chief methods to which Parts must respond are *actNow*, *actEnd* and *advanceValue*. The *actNow* method has the code for the specific behavior of a model entity. It is always the next received message after a model entity has received the *advanceValue* method. Both methods propagate through the simple branched directed graph structure of the model. When *advanceValue* moves the potential inclusions into objective status, it also establishes a new hierarchical order of inclusive relations and, in effect, erases the old. The propagation of the *actNow* method follows upon the complete establishment of the new inclusive order.

The higher model entities can make objective evaluations of the lower entities within the occasion of their acts. For instance, in the example model, the niches count the number of objectified organisms before the organisms are moved to act and before they obtain nourishment from the niches. Because of this hierarchical relation, the niche objects can properly constrain the organism objects present together in the niche. The *actNow* method must always end with the propagation of the *actEnd*, if a part has inclusions. This method propagates the *actNow* method down the hierarchy.

There is only one major subclass of the Part class. This is the class Whole. A whole entity keeps sending *actNow* messages to its inclusions until it has no more inclusions. An unspecialized Part moves its inclusions to act only once. A Whole object moves its part to act, and then it checks to determine if it has gained any new parts while doing so. A variant on this theme is the method *goOn* which may be sent by a part to its set of connections. This causes the structurally connected parts to continue to act as long as they gain any inclusions by acting.

Discussion

Parallel Computation

It appears that the present HCIBM, or one like it, could serve the original project goals of providing a modeling medium for which a multitasking computer architecture could be developed. The general model for parallel computation is a series of logical processors among which the computational task has been divided (Madisetti et al. 1990). A given processor may get ahead of other processors in the virtual computational time of the simulation. Thus a message may arrive at a processor containing information which it should have used to calculate as far as it did. In this case, the message is said to arrive in the virtual past of the processor, and its calculations must be rolled back to take account of the neglected information. This situation, called time warp, is analogous to some large corporate undertaking in which a large task is broken into subtasks, each assigned to committees which forge ahead on their own in partial independence, only to find later that some of the committees have developed agendas that conflict. Work must then be done to undo what was done wrong in order to keep the entire task on track.

If logical processors are allowed to forge ahead and run the risk of roll back the scheme is referred to as optimistic. The computational scheme is referred to as conservative if it is controlled so that no time warp is allowed, e.g., individuals circulate among the committees to make sure their work is always consistent and in synchrony. There are many detailed schemes for dealing with parallel processing using both of these general approaches. What helps in both cases is that information is available to *look ahead*, and be able to determine it is safe to proceed without risk of being called back. A programming medium such as HCIBM provides information of this nature. For instance, assume two logical processes A1 and A2 are associated with the execution of the *actNow* code for two model entities included under a third entity B. B's processing must start before A's, and end after they are complete. Once they start, both can be run to completion without risk of roll back. When they have terminated, B's process may be resumed, so the processing resources applied to B can be assigned elsewhere until A1 and A2 terminate. The hierarchical modeling medium thus can provide a fair amount of information as to which processes are safe and which may be interrupted. In principle it appears possible to utilize this information to develop computational schemes to optimize the allocation of parallel processing for HCIBMs.

Practical Applications

On a practical level, individual-based models offer a way to bring laboratory data into expression at the population level (Nisbet et al. 1989). One may be able to build up complex models of ecological systems, against which theory could be tested and developed prior to, or as part of, planning and optimizing expensive or long term field experiments. Of course these arguments do not apply only to concurrent models, but a concurrent model allows a direct way to incorporate more realism into a simulation. Such simulations could be interrupted experimentally to alter selected parts which are concurrently acting, e.g., manipulating part of a forest while the rest continues on as the impact of the manipulated part progressively spreads into it. Highly realistic and detailed HCIBMs could be of particular use in the management of populations of endangered species with relatively few individuals. In such situations, the near term consequences of alternate management scenarios could be explored in detail for population level consequences, even down to the level of releasing or removing individual organisms, since the model could be keyed to simulate actual individuals.

Fairly realistic simulation of genetic properties could be built from HCIBMs. Models could be developed to investigate the ontological details of cell lineages and bisexual populations. The hierarchical properties of Part objects are a natural medium for simulation of cell lineages. Organisms in populations would be composed of Part objects themselves and could grow by the addition of new parts. Such enterprises would probably require the power of extensive parallel processing.

Parts, Wholes, and Potential

Restrictions on the order of message passing among model entities in HCIBM create a formal discipline within which a simulation can be designed and interpreted as a composition of occasions. These occasions are interrelated on different spatial and temporal scales. The grain of the spatial scale or the divisibility of the temporal scale depends upon the structure of the model and the functionality of its entities. A complete occasion, i.e., the act of one model entity, can be totally isolated or integrated to varying degrees within other occasions. A flat time-slice is only an option, because occasions or sub-occasions end as a function of several factors, and may thus have different relative durations. The model ideal seeks to embody a co-ordinate system which is completely based on the relations and acts of model entities. For instance, a simulation of the learned pathways of small mammals might represent the finest grain spatiality in a HCIBM. However, the virtual model time is independent of process time, e.g., the most rapid occasions may tie up most computer resources.

Labeling the class of model entities as *parts* reflects the design ideal that model entities must be bound together in order to function. *Wholes* are made subclasses of parts to further emphasize this idea. A whole is generally intended to represent an occasion associated with a situation which is random and instigated by self-acting entities. For example, the niche objects act only if there are organisms present in the location of which the niche is a structural part. It is activated by the organisms and expresses the way they relate to each other in obtaining nourishment.

The idea of a potential state is required to utilize position in the hierarchy to function as the primary parameter for the mode of behavior of an acting part. The duration of the occasion of an act of a model entity is bounded by its potential states (potential–actual–potential). The bindings of model entities through inclusion and connection define the paths along which new potential relations may be established. The major restrictions in the curent formulation are 1) that a model entity can establish only one potential inclusive relation to terminate the end of its act; and 2) any model entity which is directly higher or connected to a higher entity is a possible super part. Restricting or liberalizing these rules would create new classes or HCIBMs. The behavior of different formulations of HCIBMs can be explored through computer simulations without attempting to create a simulation of any particular ecological example. Thus, a theory of hierarchic concurrent processes awaits development.

Ecological Concurrency and Individuality

The idea of concurrency as an ecological topic arises naturally from the endeavor to find a medium to express individual-based simulations. Langton (1989) notes "the fundamentally parallel hardware of nature" in reviewing papers of a symposium on the simulation and synthesis of living systems. He also notes that the general parallel structure of many of the models discussed in that symposium could, in principal, be mapped to parallel computers. These two themes, the parallelism of nature and the implementation of concurrent models on parallel computers focus attention on the conception of computer simulations that mediate between machine and nature. This focus spawns the idea that success in this venture would improve our understanding of nature. Taylor et al. (1989) point out that the use of computers as a modeling medium can have the salubrious influence of forcing theorists to make their theories explicit enough to be implemented and tested "empirically" on computers. This should especially be the case for developing theories about concurrent hierarchical systems. Even without parallel computers for concurrent models, computing the media such as the HCIBM in Smalltalk, could be used to investigate the general properties of such systems on available serial computers.

Ecologists, and most physical scientists, do not conceptualize the systems they study in terms of individuals. Nor do they focus on how parts of the system are both acting concurrently and influencing each other's mode of action. It is customary (and very effective) to look for ways to simplify system properties which arise from the maze of interactions among entities in nature. Thus the fact that individual predators eat individual prey is treated as a predation rate, which may be a function of other variables such as the densities of predator and prey. In this way, the individuality of predator and prey organisms is removed from the ideas which represent the system. However, individuality of the parts of systems is a feature of nature which has always been of interest, or annoyance, and is now easier to address with modern computers. The present symposium, to which this paper is a contribution, demonstrates this point.

Łomnicki (1988) showed that analytic consideration of individual differences could lead to a more complete comprehension of classic population models such as the logistic. Huston et al. (1988) have argued that the realism of individual-based modeling may make some problems tractable that are analytically difficult, and can serve to link functions at different levels within a system, e.g., population dynamics and trophic relations can be built into an individual based model. However, L. Ginzburg (personal communication) argues that individual based models are merely equivalent to more complex formulations of analytical ones, and add little to basic theoretical understanding. For instance, treating size differences in a population as Łomnicki has done is not, in principle, different from analytically handling the distribution of momentum in a gas as is done in thermodynamics. It is merely a matter of pushing the level of analysis down a level. Thus, the unitary individuals in the population are merely distinguished by more variables, e.g., in terms of size in addition to age.

However, differences in individual behavior, expressed by size or other variables, can make real differences in the outcome of population level processes (Shugart 1984; Łomnicki 1988). In dynamics, the existence of classes of processes in which small intial differences can significantly influence the outcome has long been recognized and has been brought to attention by the recent interest in *chaotic* processes (Abraham and Shaw 1984). Could such differences in dynamic systems be introduced by individual differences between parts? Slobodkin (1961) pointed out a case in which a "Napoleonic" prairie dog could influence the population dynamics of a colony for several generations after it had died. These notices are suggestive of a general role for the importance of individual differences, which are usually subsumed in theoretical developments. Consideration of concurrency, i.e., explicitly treating the fact that processes may be going on in a system in parallel and in partial independence, takes the ideal of treating individual differences towards a limit. Consequently, a general formulation of concurrent processes, may be a way in which the role of individuality can be handled theoretically.

The present effort is presented as a step towards getting a handle on the elements of a formalism to conceptualize the essence of the logic of concurrent systems. A principal result, which is clear to me from grappling with the issue of giving concrete expression to concurrency, is that the notion of individual action is inextricably tied up in any concept of the concurrency of nature. In discussions of parallel computer systems, the distinction of acting independently and acting in response to a message appears to be a fundamental analytical distinction (D. Nicol, personal communication). Real existence implies the complementary properties both of acting and responding to actions. The inseparable union of these properties must be at the root of any theoretical concept of a unitary individual that is intended to represent more than the joint realization of values for several variables. But the concept of acting and responding implies relations among individuals which serve to constrain and enable actions, because individuality also immediately implies involvement with other individuals. Thus principles of coordination of action are also logical prerequisites of a theoretical concept of acting individuals.

The possibility of success in such an intellectual enterprise implies that the proper view of nature may be that of a system of independently acting, but mutually constraining, individuals. Thus natural individuals would have those systematic properties which theoretical individuals of a properly formulated theory of concurrent processes would be found to possess. Such a development should eventually prove a touchstone for the scientific ideal of discovering single value transformations to characterize real systems (Ashby, 1963). Conversely, where there appear to be good canonical representations of natural systems (e.g., angular deviation and angular velocity to characterize a pendulum), there must be a corresponding implicit concurrent formulation of the system involving real individuals.

Acknowledgements

I wish to thank Don DeAngelis for inviting me to participate in the now-defunct "Evolution Machine" project of ORNL. Several stimulating discussions with L. Williams and W. L. Zabriskie of the Instrumentation and Controls Division of ORNL stimulated my interest in the subject of concurrency. Special thanks are due to my son-in-law, John Palmer-Kamprath for weeding out many knotty sentences.

Appendix

The Smalltalk code for some of the principle methods of the basic classes of objects used to construct an HCIBM is provided to give more insight into how the concept for the HCIBM is expressed. Comments are given to help get an idea of what the code means. Enough information is given here to develop an HCIBM. Complete code can be obtained from the author.

The Principle Methods of Part Objects

advanceValue "This is one of the two basic methods of the class Part. It advances the potential inclusions to objective status. Inclusions is a Value object. It receives the message *advance* (see Value object methods below). Object names are always written left of messages. The *do:* method iterates through all the members of a Collection object, in this case *parts*. The |*parts*| construction declares a local variable *parts*. The *:=* is the assignment symbol. The special word *self* refers to the object which receives the message. The message *superPart;* establishes the sending object as the super part of the receiving object."

```
|parts|
inclusions advance.
connections := nil.
parts := self inclusions.
parts size > 0 ifTrue:[
   parts do:[:part| part superPart:self.
      part advanceValue]]
```

actNow "This is the other basic method of the HCIBM. It is applied immediately after *advanceValue* has established a new hierarchy.

Specific code for behavior of model entity goes here. See example of Organism object code below."

```
self actEnd
```

actEnd "This method propagates the *actNow* method to all the parts of a Part object. The *actNow* method for any Part object with parts must end with *actEnd*."

```
|parts|
parts := self inclusions.
parts size > 0 ifTrue:[
   parts do:[:part|part actNow]
```

goOnFor:type "Moves connected parts of type to act until each part ceases to generate potential inclusions."

connections do:[:part| (part isKindOf:type ifTrue:[
    [part shouldContinue] while True: [part advanceValue.
                       part actNow]]]

include:part "This method establishes the potential inclusion between two Parts. It is used to build the initial hierarchy and maintain its subsequent inclusive structure."

inclusions potential add: part

The Principal Methods of Whole Objects

actNow "A whole moves moves all its inclusions (parts) to act until they have transited or exited the model. The *shouldContinue* message returns true if the part has any potential inclusions.
Specific code for behavior of whole entity goes here."
    [self shouldContinue] whileTrue:[
        self advanceValue.
        self actEnd]

initialize "Most objects of any class are initialized in some way. All wholes start off with the type of the instance variables *connections* and *inclusions* established."
    connections := Set new.
    inclusions := Value new setPotential:Set new; advance

The Principal Methods of Value Objects

"Value objects are used to manage the potential and objective parts of Part objects. The *inclusions* variable must be a Value object."
advance "This method rolls the potential values to their objective phase and flushes the old objective values."
    objectiveValue := potentialValue copy.
    potentialValue := potentialValue class new

setPotential:object "Generally *object* is a subclass of Collection."
    potentialValue := object

Code for actNow method of an Organism Object.

actNow "This is an example of code for the specific behavior of an Organism object. A model is developed by making specialized subclasses of the part and Whole classes. This method calls a number of Organism object methods for which the code is not given here. However, the names of the methods are descriptive of their function. The carrot, ^, indicates a point of return from the method."

```
|nextSuperPart|
    superPart class = Locale ifTrue:[
        nextSuperPart := superPart oneConnectionOf: nicheClass.
        nextSuperPart = nil ifTrue: [^ self reportStatus].
        ^nextSuperPart include:self].
    superPart class = nicheClass ifTrue:[
        (superPart surviveDuration:survivalFactor) not
        ifTrue:[^self reportStatus].
    nourishment := superPart obtainNourishment:(biomass * maximum-
NourishmentFactor).
    self grow.
    (stage size > 1 & self should Develop)
        ifTrue:[ ^self develop].
    self shouldReproduce
        ifTrue:[ ^self reproduce].
    self should Emmigrate
        ifTrue:[ ^self emmigrate].
    self shouldQuit
        ifTrue:[ ^self transitTo:Season].
    superPart transitTo:Locale
```

Literature Cited

Abraham, R. H. and C. D. Shaw. 1984. *Dynamics—The Geometry of Behavior Part 2, Chaotic Behavior.* Aerial Press, Santa Cruz, CA.

Allen, T. F. H. and T. B. Starr. 1982. *Hierarchy Perspectives for Ecological Complexity.* University of Chicago Press, Chicago.

Ashby, R. A. 1963. *An Introduction to Cybernetics.* Wiley, New York.

Butler, P. L., D. L. DeAngelis, W. B. Dress, M. A. Huston, W. M. Post, and W. L. Zabriskie. 1987. *Feasibility Study for a Specialized Computer: The Evolution Machine (A Collaborative Proposal).* The Environmental Sciences and The Instrumentation and Controls Division, Oak Ridge National Laboratory, Oak Ridge, TN.

Codd, E. F. 1968. *Cellular Automata.* Academic Press, New York.

Hogeweg, P. 1989. MIRROR beyond MIRROR, Puddles of LIFE. In *Artificial Life. The Proceedings of an Interdisciplinary Workshop on the Synthesis and Simulation of Living Systems,* C. Langton (ed.) pp. 297–316. Addison-Wesley, Redwood City, CA.

Huston, M., D. DeAngelis and W. Post. 1988. New computer models unify ecological theory. *BioScience* **38**:682–694.

Kershaw, K. A. 1964. *Quantitative and Dynamic Ecology*. Edward Arnold, London.

Langton, C. 1989. Preface. In *Artifical Life*. *The Proceedings of an Interdisciplinary Workshop on the Synthesis and Simulation of Living Systems*, C. Langton (ed.) pp. xv–xxvi. Addison-Wesley, Redwood City, CA.

Łomnicki, A. 1988. *Population Ecology of Individuals*. Princeton University Press, Princeton, N.J.

Lugowski, M. W. 1989. Computational Metabolism: Toward biological geometries for computing. In *Artificial Life*. *The Proceedings of an Interdisciplinary Workshop on the Synthesis and Simulation of Living Systems*, C. Langton (ed.) pp. 341–368. Addison-Wesley, Redwood City, CA.

Madisetti, V., D. Nicol, and R. Fujimoto. 1990. *Advances in Parallel and Distributed Simulation*. Society for Computer Simulation, San Diego, CA.

Nisbet, R. M., W. S. C. Gurney, W. W. Murdoch, and E. McCauley. 1989. *Structured Population Models: A Tool for Linking Effects at Individual and Population Levels*. Bio. J. Linn. Soc. **37**:79–99.

Pinson, L. J. and R. S. Wiener. 1988. *An Introduction to Object-Oriented Programing and Smalltalk*. Addison-Wesley, Reading, MA.

Shugart, H. H. 1984. *A Theory of Forest Dynamics*. *The Ecological Implications of Forest Succession Models*. Springer-Verlag, New York.

Slobodkin, L. B. 1961. *Growth and Regulation of Animal Populations*. Holt, Reinhart, and Winston, New York.

Taylor, C. E., D. R. Jefferson, S. R. Turner, and S. R. Goldman. 1989. RAM: Artificial Life for the Exploration of Complex Biological Systems. In *Artificial Life*. *The Proceedings of an Interdisciplinary Workshop on the Synthesis and Simulation of Living Systems*, C. Langton (ed.) pp. 275–295. Addison-Wesley, Redwood City, CA.

Toffoli, T. and N. Margolus. 1988. *Cellular Automata Machines*. *A New Environment for Modeling*. MIT Press, Cambridge, MA.

Part III

Models of Animal Populations and Communities

Section Overview

The individual-based modeling approach found application over the past few decades to a number of different types of ecological problems involving animal populations. With one or two important exceptions, the nature of the applications to animals differ considerably from those to terrestrial plant populations and communities discussed in Section IV. Whereas plants are usually static and predominately local in their interactions with other plants, animals usually move and can and have the possibility of interacting with other animals over a large spatial area. One exception is that of sessile animals, such as corals, which have been modeled by individual-based approaches analogous to plant communities (Maguire and Porter 1977). Also, for certain problems territorial animals have sometimes been modeled as individuals having mostly local interactions with other animals. An example is that of the spread of rabies through a spatially-distributed fox population, where, except for occasional long-distance movements, territorial foxes interact mainly with neighbors (David et al. 1982).

Aside from these exceptions, the focus of individual-based models of animal populations has been on different questions than those addressed for plant populations. Plant models have attempted to obtain a more accurate representation of spatially local intra- and interspecific interactions. For animals, the emphasis has often been on movement patterns, although this may change, as we learn the extent to which even highly mobile animals may receive direct influences from only a limited portion of the population. Mathematical representations of hypothesized mechanisms that guide the movements of individual animals have been used, for example, to analyze the observed flocking patterns of birds.

Another question with many theoretical and practical implications is how the internal age and size structure of populations affects the dynamics of the population. This is a matter of particular importance in species such as fish and many aquatic invertebrates, which have plasticity in their growth rates. The distribution of sizes within an animal cohort, not the numbers alone, may be the determinants of population dynamics. While many plants as well have highly plastic growth, this tends to be more closely governed by spatially-local interactions than is the case for animals, whose growth is determined more by a spatial average of interactions between individuals.

The papers of this section definitely reflect the concern with properly representing the internal structure of populations and the mechanisms that affect and are affected by that internal structure. Botsford reviews the range of models that have been developed to study animal populations, especially those that attempt to understand and predict recruitment of important marine fish and invertebrate species. He shows how the elaboration of simple stock-recruitment relations (e.g., Ricker 1954) to include greater detail concerning ages, stages, and sizes within the population have led to greater ability to compare model predictions with data and to reject or accept specific hypothesized mechanisms for population behavior. As he points out, this trend has led from *i*-state distribution models to more detailed *i*-state configuration simulation models [see Metz and Diekmann 1986 or Caswell and John (this volume) for definitions]. The latter type of model, while allowing greater realism of description that may sometimes be essential, also has some difficulties and dangers, including the lack of sufficiently detailed data at the individual level and the problem of checking the reliability of large simulation models.

Crowder et al. discuss specifically the crucial role that size plays in fish interactions and the usefulness of the individual-based approach in describing such interactions. The role of the size structure of populations may be a vital component in understanding the variability of recruitment of young fish to the adult populations. Classical fishery models do not account for this variability, possibly because they represent all individuals as identical. It is quite possible, however, that atypical individuals (for example those that by various circumstances grow faster than others) may constitute the bulk of recruits. The individual-based approach provides a means to test this. Foraging and energetics models for the individual fish have been developed and are being used to explore how size variability develops in populations. In this connection, Crowder et al. discuss their work on the recruitment of a fish species in Lake Michigan and how important size and growth are to the young fish in an environment in which predation mortality can depend strongly on size.

An example of the strong significance of size in population dynamics is exhibited by MacKay's model of sea lamprey interactions with host fish in Lake Michigan. The mortality in fish populations subjected to the lampreys is not a simple function of lamprey numbers, since whether a host fish can recover from an attack depends on the sizes of the host and parasites. MacKay uses an individual-based approach to analyze this problem, coupling a bioenergetics model for the change in sizes of lampreys with a foraging model for their feeding on various size classes of prey. The model is able to make predictions, such as the occurrence of seasonal pulses in host mortality when the sizes of lampreys are large, which models without detailed size structure could not make.

Anderson is also concerned with the individual characteristics of fish and how they integrate into the population-level behavior of survivorship. His emphasis is not on size alone, but the whole suite of factors that affect whether a fish is going to survive another unit of time. This includes both directly lethal accidents, such as catastrophic environmental events, as well as sublethal factors. Sublethal factors, including stress, loss of disease resistance, and slow growth, cumulatively affect the vitality of a fish. These individual factors are usually too numerous to model explicitly, but incorporation of this concept into a simple, abstract mathematical representation of vitality has given Anderson a powerful means to compare survivorship in different populations or the same population at different times. The different patterns of survivorship curves allow the separate influences of accidental mortality and decrease in vitality to be recognized.

Kooijman's paper aptly illustrates some of the reasons for taking an individual-based approach as a starting point for predicting population-level processes. Biomass conversion at the population level, or the conversion of an input of prey biomass into biomass of consumers, including fish or grazing mammals that can be harvested by humans, has traditionally been described by models that include little or no size structure or other types of variation within the population. These approaches may be overly simplistic since biomass conversion is highly dependent on the characteristics of the individuals in the population, such as their sizes and patterns of energy allocation. Kooijman derives a model for population-level biomass conversion based on a model of the process of conversion by an individual. The model is able to make a realistic assessment of how the efficiency of energy conversion depends on the internal structure of the population and how it is harvested.

An example of the use of the individual-based approach to better understand spatial patterns is the paper by Taylor. Its specific objective is to explain the observation that the spatial distribution of animal population density shows systematic dependence on mean density of the population. In particular, the variance of samples taken to estimate population number density increases as a power of the mean population density. Taylor's hypothesis is that behavior of individuals, in particular movements that depend on the location of other individuals in their local neighborhood, can generate the observed power law. He presents an algorithm based on Voronoi tessellations that allow him to generate the spatial distributions resulting from given assumptions about individual behavior in response to neighbors and to reproduce the observed power law patterns in many cases.

While Taylor's paper is the only one here using individual-based modeling to explore animal spatial pattern, the individual-based approach is likely to play as essential a role in understanding spatial distributions as it is becoming in understanding the dynamics of structured populations. Approaches which couple animal movement patterns with spatially-explicit landscapes will also no doubt become more prevalent. See Hyman (1990) for one example in this direction.

Literature Cited

David, J. M., L. Andral, and M. Artois. 1982. Computer simulation model of the epi-enzootic disease of vulpine rabies. Ecol. Modell. **15**:107–125.

Hyman, J. B. 1990. A Landscape Approach to the Study of Herbivory. Ph.D. dissertation. The University of Tennessee, Knoxville, TN, USA.

Maguire, L. A., and J. W. Porter. 1977. A spatial model of growth and competition strategies in coral communities. Ecol. Modell. **3**:249–271.

10

Individual State Structure in Population Models

Louis W. Botsford

ABSTRACT. There is an increasing trend toward the use of population models that include variables such as age, size, stage, and space to describe how individuals differ. These models have for the most part been written in terms of distributions of numbers over these individual state variables, but there has also been increasing use of models that describe each individual in a population explicitly, as being potentially different from all others. I outline reasons for the general use of individual state structure in models used to (1) increase our understanding of population dynamics, and (2) predict population behavior for management. The logical basis underlying both of these requires that models be *realistic*, that is that their component structure be comparable to the population itself, through observation. This realism can be achieved by adequate description of the *state* of individuals in a population and the state of the population. Over the past 15 years of study of recruitment to marine populations, the use of age, size, and spatially structured models has led to advances in our understanding of stability and the effects of random environment not possible with traditional lumped models. The fact that population models with individual state structure connect the population level of integration to the individual level can be used to draw implications for large scale (both spatial and temporal) population behavior from small scale studies at the individual level. For some populations, modelers turn to models which keep track of all individuals explicitly because of convenience or efficiency, but in other cases this approach is required because of the nature of interactions within the population. Although it is not clear just when this approach is absolutely necessary, it seems to be most appropriate for populations that are locally small, with little mixing of individuals in the individual state space. Continued development of a general understanding of population dynamics will require cautious use of this new approach as we move slowly away from the mathematical foundations of distribution-based approaches in increments small enough to allow complete understanding of new aspects of population behavior.

Introduction

In the past decade or so, there has been a rapid increase in the proportion of ecological population models that explicitly account for population structure, be it age, size, spatial or other. Whereas population models of theoretical and applied ecology in the 1960s and 1970s were for the most part written in terms of total numbers of identical individuals or total biomass, the dependent variable is now more commonly a density function of a number of variables over which individual characteristics vary (e.g., n(age, size, latitude, time)). Although the mathematical seeds for such models have long existed (e.g., Euler (1760), Lotka (1907), M'Kendrick (1926), Leslie (1945)), the recent, growing awareness of the biological importance of individual heterogeneity (e.g., Werner and Gilliam (1984)) has motivated a dramatic increase in their development (e.g., Nisbet and Gurney (1982), Metz and Diekmann (1986), Caswell (1989)). Our understanding of population dynamics is increasing rapidly, in large part due to the fact that recent models specifically account for the fact that individuals at different age, size, stage, and location contribute differently to reproductive and mortality rates. The most recent step in this trend is the development of models that consist of computer simulations that keep track of a number of individuals, each of which is potentially unqiue.

In this paper I trace the motivations and advantages that underly this trend. These are primarily aspects that I have found useful in my own work on animal population dynamics, most of which is associated with practical problems in applied population biology. I begin with some philosophical background on why realism is a desirable characteristic of population models and how it can be achieved through incorporation of individual state structure. I then describe examples from the study of recruitment to marine populations, in which models with age structure gave different results and were necessary to study and describe these populations. After mentioning the empirical benefits of individual-based models in field research programs, I then describe potential problems with distribution-based approaches to describing individuals, and their possible solution through an individual simulation approach.

Discussion of the motivation or justification of certain types of modeling requires prior definition of the uses for which the models are constructed. My discussion here is in the context of two of these: (1) the use of models as scientific tools to learn more about populations, and (2) the use of models as practical tools to better manage populations. The former is essentially the process of gaining new knowledge about, or a better understanding of, a population's dynamic behavior. The latter involves using models to predict the population consequences of various management tactics. These two goals form the basis for most current population modeling efforts, one exception being the practice of formulating simple models solely for pedagogical purposes.

Realism

One of the motivating factors that leads some to include individual differences in population models is the idea that a model should be a realistic description of the population being modeled. By the term realistic, I mean that the model provides the opportunity for comparison of its functional basis with the real population, that it be formulated in terms of observable mechanisms and measureable parameters. The term does not necessarily mean that the model correctly contains *the* essential mechanisms that underly population behavior, but rather that it is possible to determine whether it does by comparison of model components to nature. This is distinct from models which may reproduce population-level behavior (i.e., abundance or total biomass versus time), but whose mathematical components do not correspond to real biological objects of mechanisms. The mechanisms underlying population behavior typically act through individual vital rates (e.g., growth, mortality, fecundity, etc.), hence these would be included in a realistic population model. To illustrate this with a counterexample, because the logistic model does not explicitly contain the mechanisms that limit the rate of population increase at higher population levels, it would not be considered realistic. The notion of realism in ecological models has its origins in earlier work by Holling (1964) and Levins (1966) who listed it as one of several desired characteristics, achievable in only some models. It is similar to the term mechanistic as applied to ecological models (e.g., Schoener 1986), but implies more about the modeling process.

The requirements for realism in population models can be seen in the arguments that underlie the two uses of models stated above: the scientific use of models to learn more about populations, and the practical use of models in the management of populations. The basis for the former can be described in terms of a research program of modeling and observation designed to determine the cause(s) underlying a specific population behavior (e.g., irruptions, cycles, protracted decline). For such problems, several possible explanations have typically been proposed, and the object of the program is to determine which is (are) the cause(s). Each is incorporated into a model (which involves other information termed auxilliary assumptions or subhypotheses), and the population dynamic consequences of each is determined. The basic question then asked is, for each proposed explanation, does the model produce the observed behavior of interest? Ecologists differ in the way they interpret the outcome of such a program (as well as whether one can even formulate a program with distinct hypotheses, e.g., Quinn and Dunham (1984)). Some put more emphasis on negative outcomes (following Popper (1935)) while others strive for a positive outcome (e.g., May (1981)). However, these differences are not the focus here. The important point is that without the ability to compare real biological mechanisms to the functional mechanisms in the model, carrying out this program is impossible. For example, whether density-dependent individual growth could be the cause of a protracted decline in abundance cannot be determined if individual growth is not explicit in the model. Investigators need to be able to compare the model to the object being modeled in order to compare not only the mechanism(s) of primary interest, but also other assumptions necessary to formulate the model (i.e., auxilliary assumptions or subhypotheses).

The practical use of models can be most easily supported by a purely inductive argument, in contrast to the hypothetico-deductive scheme that might underlie the scientific use. The object of the practical use of models is to predict future behavior, possibly under a variety of untried management policies. This use of modeling can be justified using an argument by analogy (Salmon 1973); the model resembles the population in certain observable or measurable ways (e.g., growth, reproductive and mortality rates, time series of past abundance or catch), hence it will resemble the population in another, currently unmeasurable way (i.e., future behavior, such as catch in a fishery). The important point to be drawn here is that the strength of an argument by analogy improves as the number of relevant similarities increases in this case as the number of ways in which the model resembles the population increases. These can be increased by increasing the length of the time series over which model behavior matches population behavior, *and* by increasing the realism of the population model by providing similarities at the individual level.

From these brief arguments, we can conclude that realism is a necessary characteristic for either use of models. However, to maintain a balanced view in this brief presentation, it should be pointed out that these two arguments are neither universally used nor universally accepted. Most modeling efforts, of course, include no such explicit philosophical considerations, and realism in models is often argued against. Some applied population modelers argue that models that require fewer assumptions and less data which often means they are less realistic, are better. Ludwig and Walters (1985) have actually shown that one can produce better management policy from time series data by assuming a different population model than the one which created the series (see Botsford 1987, Ludwig 1987).

The Concept of State

Even after deciding that a realistic model is required, there remains the problem of how to construct one. A valuable guide in this process is an understanding of the concept of state from general systems theory, as adapted and introduced to ecology by Caswell et al. (1972). Simply put, the *state* of a system is the information required to specify its behavior in the immediate future, in response to all possible inputs (see Caswell (1989) and this volume for more details). For realistic models, we require that the model be formulated in terms of variables which adequately describe the state of a system. The concept of state can be illustrated by a counterexample: for most populations any model formulated in terms of total number in the population (e.g., any form of the logistic model) will not uniquely specify the future behavior of a population. For example, the number of deer in a population next year is not specified by the fact that there are 100 deer this year, one needs to at least know how many are adult males, adult females and juveniles. Thus, total number, itself, is not an adequate description of state.

The concept of state is valuable, not just in explaining why individual state structure is required in population models, but also in indicating which individual state variables are necessary to adequately describe a specific population. Extending the reasoning from the example above, if reproductive and mortality rates are uniquely determined by specifying an organisms's age, then age is an adequate description of individual state. If these rates are uniquely determined by specifying an individual's size, then size is an adequate description of individual state. If growth rate does not vary with time (i.e. is dependent on neither environment nor density), then either age or size is an adequate individual state variable; but if growth is plastic, then size must be used as the state variable. These considerations for size and age as examples are summarized in Table 10.1.

Thus the selection of model state variable(s) depends on determination of the structural variable(s) (e.g., age, size, stage) on which reproduction and mortality depend, and whether the relationship between various structural variables varies with time. For fish, a group noted for plastic growth, there is a long history of attempts to determine, for example, whether fecundity depends on size or age (see Bagenal 1978, Wooton 1979 for reviews). General, rigorous techniques for determining these dependencies are being developed (Caswell 1989).

After determining the variable(s) appropriate for describing the state of an individual, the same criterion for an adequate description of state must be applied at the population level to select a population model. One could, of course, keep track of all individuals modeled with adequate individual state variables, and that would be an adequate population state variable. However, most modelers to date have chosen to describe populations in terms of the number of individuals in each individual state, that is, in terms of a distribution or density function of individuals. These models have been

Table 10.1 Types of models required to adequately represent the state of a population when reproductive, mortality, and growth rates vary in specific ways.

		GROWTH RATE	
		Fixed	*Time varying*
			(i.e., density-dependent or environmentally forced)
	age dependent	age or size specific	age specific
REPRODUCTION and MORTALITY	*size dependent*	age or size specific	size specific
	age and size dependent	age or size specific	age and size specific

formulated either in terms of partial differential equations (continuous state variables) or matrices (discrete state variables). Metz and Diekmann (1986) have determined conditions under which a population model formulated in terms of a distribution of individuals over individual state variables is an adequate description of population state. For models which are obviously deterministic, these conditions can be briefly paraphrased (see Metz and Diekmann (1986) for details) as stating (1) that all individuals experience the same inputs, and (2) that one be able to calculate the outputs of individuals, including not just how the population affects the outside world, but also how individuals in the population affect other individuals in the population. It seems that the former need not be required for individuals that are distributed over space because the physical environment can vary with space. When possible stochastic effects are considered, two more conditions are required to guarantee that the population appear deterministic: (3) that numbers are large enough to iron out chance events (e.g., demographic stochasticity),[1] and (4) that random inputs are a specified function of time (Metz and Diekmann 1986).

These conditions can be qualitatively summarized by stating that distri-

---

[1]Demographic stochasticity refers to the variability in numbers surviving a certain time period, that arises from the fact that a small number of discrete individuals is present. The result of n individuals surviving with probability p is binomially distributed with mean np and variance $np(1 - p)$. For large n, the standard deviation is small relative to the mean, hence the deterministic result (np) is an adequate portrayal of survival. However, for small n, the standard deviation is large relative to the mean, and the binomial process must be explicitly retained. The resulting random variability is termed demographic stochasticity.

bution models will work if the population is nowhere locally small for a substantial period of time, that is, if (1) there is no location in the individual state space at which the number of individuals is so small that effects such as demographic stochasticity or a similar effect involving binomial trials in inter-individual interactions (other than mortality) are important, or (2) if there is such a location, there is enough "movement" (e.g., migration) that any effects of discrete, random events are averaged out by rapid mixing of individuals. Note that this notion of local smallness implies not just spatial proximity, but also proximity in other individual state variables such as size or age. This implies that populations with small subpopulations, local interactions between individuals, and little migration through space would be poorly described by distribution models. Note that social interactions with a historical characteristic would tend to lead to locally small models because the number of individuals with a specific history quickly becomes small.

Different Behavior: Recruitment in Marine Populations

While philosophical arguments provide greater generality, they are not as immediately convincing as examples of the differences in behavior between realistic models which account for individual state structure and models which do not. Examples given in the past include a discussion of how incorporation of age and size structure led to an understanding of (1) density-dependent recruitment, (2) multiple equilibrium due to plasticity in growth rate, and (3) optimal harvest policy not possible with simple models (Botsford 1981b). More recently, Hastings (1988) has discussed the advantages of including age structure in the analyses of cannibalism and juvenile competition, spatial structure, and adult predation and competition. Also, Nisbet et al (1989) have reviewed the advantages of state-structured models in their analysis of *Daphnia* population dynamics. Here I will briefly describe how progress over the last 10 years in our understanding of marine populations with density-dependent recruitment has depended on consideration of population structure.

An early step in our understanding of the dynamics of marine populations with density-dependent recruitment involved the stock-recruitment model of Ricker (1954). He formulated this model,

$$S_{t+1} = R_t = a \, S_t \, e^{-bS_t}, \tag{1}$$

where S_t is stock abundance in year t and R_t is recruitment in year t, primarily for the management and analysis of semelparous populations. Ricker noted that for semelparous populations, the model was locally unstable when the slope at equilibrium was less than -1.0 (Fig. 10.1). As the slope at equi-

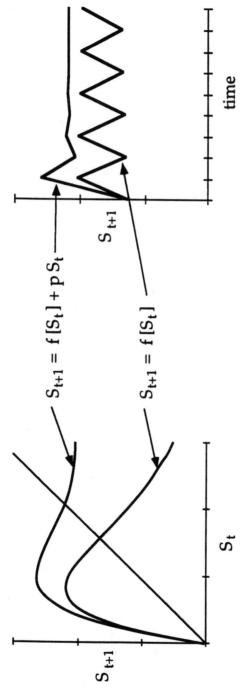

Figure 10.1. The stabilizing effect of adding age structure. On the left, the dependence of stock abundance at time $t + 1$ on stock abundance at time t for a semelparous population and an iteroparous population with the same dependence of recruitment on stock. On the right, the time series resulting from each.

Table 10.2 Normalized slope of the recruitment function, below which a simple
age-structured population model becomes locally unstable and goes to
cyclic behavior. The model assumes a fixed lag T to reproductive age,
and constant survival p thereafter (after Clark 1976).

T	p					
	0.0	0.2	0.4	0.6	0.8	1.0
0	−1.0	−1.5	−2.33	−4.0	−9.0	∞
1	−1.0	−1.2	−1.67	−2.5	−5.0	∞
2	−1.0	−1.13	−1.37	−1.86	−3.33	∞
3	−1.0	−1.08	−1.23	−1.57	−2.65	∞
4	−1.0	−1.05	−1.16	−1.41	−2.23	∞
5	−1.0	−1.04	−1.12	−1.31	−1.97	∞

librium is further decreased below −1.0, the cycles produced undergo successive period doubling and behavior is eventually chaotic (May and Oster
1976). This model, and similar nonlinear expressions, has also been used
as the general, discrete-time version of the logistic equation for iteroparous
as well as semelparous populations. They have had a substantial impact on
theoretical ecology because they can produce chaotic behavior, which is virtually indistinguishable from random behavior (May and Oster 1976).

An alternative, more realistic approach to describing the more general
iteroparous case is to add age structure explicitly by allowing a fraction of
the reproductive stock to survive each year.

$$S_{t+1} = R_t + p\, S_t = g[S_t] + p\, S_t \qquad (2)$$

where $g[S_t]$ is the stock-recruitment relationship from above (or a similar
expression), and p is survival. From this model the effects of adding age
structure can easily be seen to be stabilizing, by noting that the survival
term adds p to the slope of the right hand side of the expression for recruitment (Fig. 10.1). The next step in increased realism is to allow for a
fixed delay from recruitment at age zero to maturity (Clark 1976),

$$S_{t+1} = R_t + p\, S_t = g[S_{t-T}] + p\, S_t \qquad (3)$$

If we modify our definition of slope by normalizing numerator and denominator to equilibrium values, we see that a more negative slope is required
for local instability for all values of adult survival p and maturation lag T
(Table 10.2). Although detailed discussion is beyond our present scope, the
behavior of the nonlinear model for steeper slopes also changes. The value
of slope for which the next bifurcation occurs decreases dramatically as the

value of p is increased from zero (see Botsford (1991) for details). Thus, the addition of age structure is stabilizing in that (1) instability requires a steeper slope of the stock-recruitment function, and (2) other exotic nonlinear behavior such as period doubling and chaos also require steeper slopes. The additional realism attained by the explicit addition of age structure shows that instability and chaotic behavior are not as likely in iteroparous populations.

An even more realistic approach is to allow the survival to have any reasonable age-dependent form, include both sexes, and allow the population to be either closed,

$$R_t = B_t f[C_t] \tag{4a}$$

or open

$$R_t = B_e f[C_t] \tag{4b}$$

where B_t is total annual reproduction which is a weighted sum over age structure (i.e., $\Sigma b_a n_{a,t}$), C_t is a similar weighted sum reflecting the effect of older individuals on density-dependent recruitment (i.e., $\Sigma c_a n_{a,t}$), and $f[.]$ is recruitment survival, a nonlinear function decreasing from a value of 1.0 to 0.0 (Botsford and Wickham 1978, Botsford 1986a, Roughgarden, et al. 1985). For the closed population, the weighting function b_a is fecundity at age, while for the open population B_e is a constant. In both cases, the weighting function c_a is the relative effect of an individual at each age on recruitment (e.g., cannibalism rate at age). The latter may be based on physiological considerations or behavioral observations.

Most analyses of this or similar models have used linearization about equilibrium to show that populations can be locally unstable with nonlinear cyclic behavior (Allen and Basasibwaki 1974, Botsford and Wickham 1978, Frauenthal 1975, Rorres 1976, Cushing 1980, Botsford 1982, McKelvey et al. 1980, Levin and Goodyear 1980, Roughgarden et al. 1985, Diekmann et al. 1986, Hastings 1987, Bence and Nisbet 1989). Most of the results can be stated in terms of three characteristics of expected behavior: (1) a larger decline in recruitment with an increase in density (i.e., more negative normalized slope of recruitment survival function) is destabilizing, hence conducive to cycles; (2) cycles are more likely to occur when the influence of adult individuals on recruitment is more narrowly concentrated at a few older age classes (i.e., as the population becomes more semelparous), and (3) if locally unstable, the period of cycles is approximately twice the mean age of density-dependent influence of older individuals.

While this behavior near the bifurcation at which the linearized model becomes locally unstable is fairly well understood, the behavior of the nonlinear model beyond this point has been studied much less thoroughly. Guckenheimer et al. (1977) and Levin (1981) have explored the different kinds of nonlinear behavior that can occur in these age-structured models. Tuljapurkar (1987) has recently described periodic solutions to a similar nonlinear model in continuous time. Also, for a similar model of Tribolium, Hastings (1987) has recently shown that as the duration of the recruiting life stage becomes longer, cycles are also more likely.

The realism of these models has allowed them to be directly compared to life history characteristics of individuals in research programs designed to determine causes of certain behaviors of populations. The above characteristics of expected behavior of these models have been used to study behavior of an intertidal invertebrate, the barnacle *Balanus glandula* (Roughgarden et al. 1985, Bence and Nisbet 1989), as well as to specifically test mechanisms responsible for cycles in populations of the Dungeness crab, *Cancer magister*. To accomplish the latter, researchers incorporated each proposed mechanism into an age or size structured model, then tested to see whether it met conditions necessary to cause cycles (i.e., conditions 1 and 2 above; Hobbs and Botsford 1989) and whether the period of the cycles produced by the model matched the observed period (i.e., condition 3 above; McKelvey, et al. 1980, Botsford 1986a, Botsford and Hobbs 1991). Mechanisms that could not produce the observed cycles were rejected (Botsford 1981a, McKelvey and Hankin 1981). Clearly, these descriptions of the explicit dependence of population behavior on life history would not have been possible without explicit consideration of population structure, and it would not have been possible to test the various mechanisms proposed to be responsible for specific population behavior without using realistic models.

This type of analysis does not just yield information about a single special kind of population behavior, such as cycles, but also about issues such as the influence of various life history characteristics on general levels of variability to be expected from random environmental influences on recruitment. For example, Horwood and Shepherd (1979) and Reed (1983) showed that age structured models yield different behavior in response to environmental variability than the previously used logistic model (May et al., 1978). Gurney and Nisbet (1980) also described the response of age structured populations to random variability. Botsford (1986b), using age, size, and sex structured models, showed how changes in life history patterns, single-sex harvesting, and slope of the density-dependent recruitment relationship changed the frequency content and variance of random variability in recruitment and catch. For example, a population with a male-only fishery was more sensitive to a random environment (affecting recruitment) than one with a two-sex fishery, because the latter reduced the slope of the recruitment survival function at equilibrium, while the former did not.

The next step in this effort to understand recruitment in marine popula-

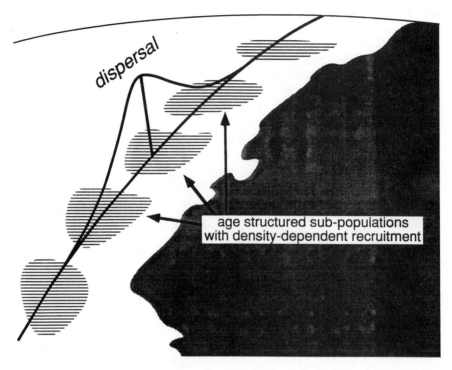

Figure 10.2. A schematic view of a meroplanktonic metapopulation distributed along a coast. Age structured subpopulations with density-dependent recruitment are inter-connected by dispersal of meroplanktonic larvae among them.

tions is to explore the effects of larval transport among spatially distributed populations on population stability and persistence (Fig. 10.2). Roughgarden and Iwasa (1986) have analyzed the consequences of several age structured models sharing a common larval pool (cf. Roughgarden et al. 1988, Possingham and Roughgarden 1990). Hobbs and Botsford (in prep.) modified the above models (equation (4)) for a single location by adding dispersal of the larval phase along a single spatial dimension (i.e., a coastline) and allowing density-dependence in recruitment both before and after dispersal:

$$R_t(x) = f_{\text{post}} [C_t(x);x] \int_{x_1}^{x_u} p(x, y) f_{\text{pre}}[W_t(y);y] B_t(y) \, dy \qquad (5)$$

where R_t, C_t, and B_t are now functions of spatial position x, x_1 and x_u are the lower and upper possible spatial positions $p(x, y)$ is the portion of larvae hatched in the interval $(y, y + dx)$ that settle in the interval $(x, x + dx)$, and $f_{post}[.]$ and $f_{pre}[.]$ are post-dispersal density-dependence and pre-dispersal density-dependence, respectively (Hobbs and Botsford, in prep.). Note that this type of model obviates the need for having both open and closed models (i.e., as in equations (4a) and (4b)), since whether a population is treated as open or closed is simply a matter of spatial scale.

Analysis of this model has extended the above interpretation of the behavior of populations with age structure and density-dependent recruitment by showing how symmetrical dispersal over varying distances, and varying slopes of pre-dispersal and post-dispersal density-dependence (i.e., rate of decline in density-dependent recruitment with increasing density) affect stability and spatial coherence of a population distributed along a coast (Hobbs and Botsford in prep.). An unexpected result is that increasing dispersal among populations distributed along a coast can either have a stabilizing, synchronizing effect or a destabilizing, desynchronizing effect depending on parameter values (Hobbs and Botsford, in prep.).

Analysis of the dynamic behavior of models such as this, and the effects of the spatio-temporally varying environment on them, will provide useful practical results and increased understanding not possible without the spatial dimension(s). Constant, symmetrical dispersal is of, course, not very realistic and is just the first step. More realistic models will explicitly include the effects of real physical forcing based on physical oceanographic considerations. The impact of variability in oceanographic conditions on larval transport and survival will be incorporated in a time-varying dispersal kernel $p(x, y, t)$ in Eq. (5). An example is the effect of variable spring winds in transporting Dungeness crab larvae to areas favorable for successful settlement (Hobbs et al. 1991). These models will then provide the potential for more responsive fisheries management, a more complete understanding of recruitment, and prediction of the impacts of global climate change on our coastal oceans.

Levels of Integration

Another aspect of individual state structured models which makes them a powerful tool in research programs designed to determine the cause(s) of behavior of specific populations is the fact that they connect two levels of integration, the individual level and the population level (e.g., Huston et al. 1988, Nisbet et al. 1989). The advantages of this characteristic are typically put forth in terms of a general statement noting that each level offers unique problems and insights, and each level finds its explanations of mechanisms in the levels below, and its significance in the levels above. However, its value can be more convincingly illustrated by describing its use in a specific problem such as understanding population variability in spatially distributed marine populations.

A substantial impediment to the understanding of recruitment in marine populations of meroplanktonic species (and many other problems in population biology) is that important events occur over a wide range of temporal and spatial scales (cf., Haury et al. 1978, Smith 1978, Walsh 1978, Levin 1987, Judge et al. 1988). For example, Dungeness crab populations cycle synchronously with a period of 10 years, over a range of several thousand kilometers of coastline, and larvae extend offshore a couple of hundred kilometers (Fig. 10.3). Mechanisms responsible for these cycles probably occur relatively rapidly over small areas (e.g., intraspecific interactions and individual feeding over days and meters, mesoscale oceanographic forcing over hundreds of meters and weeks). Thus, a straightforward, complete study of this system would require sampling on a daily basis and a one meter grid for a period of several decades over half a million square kilometers, obviously an impossible amount of sampling.

The use of individual state structured models may aid in reducing the amount of sampling required to draw conclusions regarding the mechanisms controlling population variability. Biological oceanographers (and other ecologists) have realized that events which occur at *different trophic levels* occur over different temporal and spatial scales (Steele 1978, Haury et al. 1978). An extension of this idea, which provides useful interpretation for meroplanktonic marine organisms at least, is the idea that events which occur at *different levels of integration* occur on different temporal and spatial scales (e.g., Botsford, et al. 1989). Populations both vary autonomously and are sensitive to environmental forcing on time scales on the order of several years (i.e., one or two times the generation time, the inverse of instantaneous fishing mortality rate; Horwood and Shepherd 1979, Reed 1983, Botsford 1986b). On the other hand, interactions between individuals, such as cannibalism or juvenile competition, occur on time scales of days. This general notion, to the extent that it holds true, allows realistic population models containing individual structure to be used to substantially reduce the amount of sampling required to understand population behavior.

When population models are formulated as the collective behavior of individuals, the conditions under which certain population behaviors of interest (e.g., irruptions, cycles, local extinction, multiple equilibria) arise can often be stated in terms of conditions on the behavior of individuals (e.g., predatory responses). Because processes involving individuals typically occur on shorter time and space scales, they can be sampled with an achievable level of effort. For example, in the work on the Dungeness crab mentioned above, conditions under which, say, cannibalism can cause population cycles (and the period of the cycles) depend on the behavior of individual crabs.

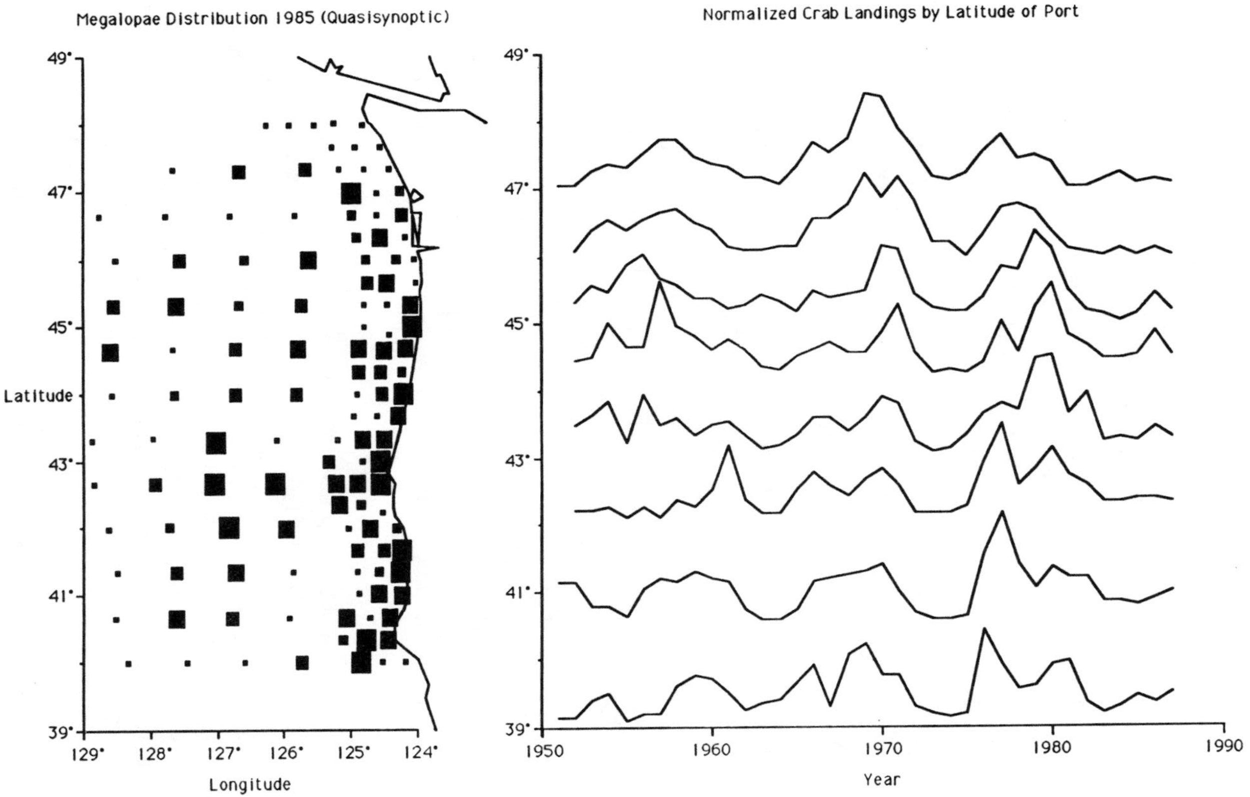

Figure 10.3. A composite view of the spatial distribution of meroplankton of Dungeness crab larvae in a typical year. On the left, abundance is proportional to square size. The temporal variability of abundance on inter-annual time scales, as reflected in catch, is shown on the right.

The relative voraciousness of cannibals at each age determines the weighting function used to compute effective population size C_t from age structure. This can be determined from a study of gut content over a short time as a test of the hypothesis that cannibalism is the cause of the cycles (e.g., Botsford and Hobbs 1991). In this way, populations that span ranges of spatial and temporal scales that are impossible to sample completely can still be studied. One simply uses models with individual state structure to connect conditions at the individual level, which involve spatial and temporal scales that can be sampled, to the behaviors of interest at the population level, which involve variability on prohibitively large scales. Note that this approach does not require that all population-level phenomena can be explained on the basis of individual-level phenomena. We are only testing whether specific explanations that involve a possible influence of the individual level on the population level can cause observed behavior.

Problems with Descriptions of Distribution

Any research program that attempts to use realistic, individual state structured models to better understand or manage populations eventually meets a problem of such complexity that the approach of describing populations as distributions over state variables is either very difficult, inefficient, or impossible. As the mathematical descriptions (i.e., partial differential equations, delay differential equations, or matrix equations) become more complex, meaningful analytical solutions are extremely limited, analysis of qualitative behavior is more difficult, and even numerical solutions are sometimes not possible. Various simplifications have been used, such as lumping a range of size, age, or spatial classes together. Also, special numerical techniques have been developed, such as keeping track of size distributions of cohorts rather than the complete distribution over size and age (e.g., Botsford 1984), and the escalator boxcar train (de Roos 1988, de Roos, et al. 1991). However, even these may not be completely satisfactory.

For reasons of efficiency, simplicity or actual necessity, several researchers have recently turned to the approach of simulating a large number of individuals, a development that was the central focus of this meeting (Huston et al. 1988). For example, I have made use of simulations of individuals in situations where interactions between small numbers of individuals was important (e.g., in a single cohort size simulator applicable to growth of plant monocultures or cohorts of fish). The chief advantage of this approach is that it is conceptually easier. It requires much less mathematical background and mathematical work, though it may require considerable programming expertise. The chief disadvantages are that (1) one will not know, with as much confidence, whether the population level results are a consequence of collective individual behavior, or a subtle quirk of the way in which the simulation was programmed, and (2) one can quickly outstrip the capability of even the fastest computer. The former results from essentially discarding the foundation of a substantial body of mathematical knowledge regarding solutions to differential or difference equations. The latter may not be immediately obvious when thinking about single runs of a model, but is inescapable when the number of runs necessary to establish sensitivity to various management tactics, biological uncertainty, or initial conditions, is considered.

The choice of whether to take the individual simulation approach can be made by weighing the advantages and disadvantages in cases for which either is appropriate; but in others, one must use the individual simulation approach because the distribution-based approach will not work. Exactly when this occurs is not completely clear, but it is more likely to occur as characteristics such as local smallness, the presence of interactions among individuals, and the lack of mixing increase, as indicated above in the discussion of adequate descriptions of individual state (cf. Metz and Diekmann 1986).

Discussion

Over the past couple of decades there has been an increasing awareness of the importance of individual heterogeneity in life history process to population dynamics. Models that explicitly reflect individual differences in age, space, size, and stage have led to many advances in our understanding of population behavior. These models have primarily been distribution-based models, but they may in the future involve the approach of simulating large numbers of individuals.

Because it appears likely that the individual simulation approach will be used more frequently in the future, the consequences of that shift should be carefully weighed (hence this conference). To avoid finding ourselves in a situation ten years from now, in which we have a large number of specific simulation results in which we may not have complete confidence, but few advances in general understanding, we must proceed cautiously.

If we are to continue our analyses of the population consequences of individual heterogeneity, while at the same time continuing to build a useful general understanding of population dynamics (i.e., a theory), several guidelines appear to be necessary. The first is that we continue to make use of qualitative analysis, as well as analytical and numerical solutions to equations describing distributions over individual state variables. This will allow us to continue to take advantage of a large, extant body of existing knowledge. Although the individual simulation approach has been developed primarily by groups with considerable experience with distribution-based models (e.g., Huston et al. 1988), it is accessible and appealing to other types of quantitatively oriented biologists with less mathematical interest and training. Hence, this guideline may require that a biologist seek out a mathematical collaborator. The second is that when we adopt an individual simulation approach, we move away from results established using analytically tractable models as slowly as possible. Our simulation models should initially overlap the mathematically analyzable models, so that we can verify results. They should then differ in small steps, and we should make sure we completely understand differences in behavior before proceeding with further steps (i.e., assure ourselves that they do not arise from artifacts of the simulation). The third is that we continue to seek a general understanding of population dynamics that is consistent across a number of specific simulation results. Mathematical analyses of distribution-based models have allowed us to see how specific characteristics of population behavior depend on characteristics of individuals (e.g., which characteristics of mortality, growth, and reproductive rates and which individual interactions lead to population behavior such as cycles, synchrony over space, heterogeneity in spatial distribution, etc.). Because we will not be able to inspect analytical solutions with a simulation approach, these kinds of generalities will be less obvious, but they are necessary if we are to avoid re-inventing the wheel for each specific case.

An important area of practical application to which the individual simulation approach seems particularly suited is the dynamics of small, endangered populations. The characteristics of local smallness, individual interactions and low mixing describe the problems in this area well. The randomness associated with a low number of discrete events (e.g., demographic stochasticity) are omnipresent in this area, and even if the population is large enough that they are not present, the question of whether to subdivide the population, which would then lead to local smallness, is one which is often asked (i.e., the SLOSS (single large or several small) question; see, for example, Quinn and Hastings 1987, Soulè 1987). The fact that meaningful results regarding low probability events (extinction) are desired implies that a large number of simulation runs will be required in most cases, and underscores the problem of computing limitations (e.g., Harris et al. 1987). There are some analytical results, obtained by treating the elements of a Leslie matrix as probabilities for binomial trials (e.g., Sykes 1969, Tuljapurkar 1989). This impediment has led researchers away from realistic, age-structured models to more simplified versions which do not explicitly reflect individual processes (e.g., Goodman 1987).

A shift to the individual simulation approach will require additional future research. One of the areas which I see as requiring further translation of abstract considerations into practical guidelines is the question of when one must turn to an individual simulation approach. A second area would be the development of more efficient numerical techniques for the individual simulation approach. In our experience in my laboratory, populations are often locally small only part of the time and locally large the rest of the time. As a consequence, we need not mimic mortality as binomial trials all of the time, but rather can use a hybrid model which shifts to the computation intensive individual simulation approach only when needed.

In addition to these areas of research on the individual simulation approach, there is still much to do on individual state structure in general. For example, most of the focus thus far has been on the forward problem: how we should describe populations on the basis of specific individual characteristics. However, there are many situations in which we have aggregate distribution data and wish to know individual characteristics. An example is the determination of growth and mortality rates from sequential samples of size distribution data of larval fish (Hackney and Webb 1978, Banks et al. 1988, 1991). A second area is determination of the effect of various techniques for simplifying individual state-structured models. It would be useful to know more about the influence on accuracy and qualitative behavior of practices such as lumping over individual state variables.

The increasing trend toward the development of models and theories that depend on individual-ecological information, as derived from behavioral ecology or physiological energetics, provides considerable optimism for future progress in the field of ecology. Schoener (1986) argues that a mechanistic approach based on hierarchical study of the individual, population, and community levels will enhance theoretical understanding of biological communities. Models with individual state structure will probably provide a means for formulating and evaluating the appropriate linkages between the different levels of integration. This will require further development of individual-level constructs such as physiological energetics or animal behavior in providing information on interactions between individuals (cf., Nisbet et al. 1989, Hallam et al. 1990). Greater empirical support from studies of individual characteristics, and careful development of new approaches to individual-based models, may substantially facilitate the continuing construction of a general, useful view of population and community dynamics.

Acknowledgements

This research was supported in part by a grant from the A. P. Sloan Foundation. I am also very grateful for comments provided by J. F. Quinn and R. E. Hobbs.

Literature Cited

Allen, R. L. and P. Basasibwaki. 1974. Properties of age structure models for fish populations. *J. Fish. Res. Board Can.* **31**:1119–1125.

Bagenal, T. B. 1978. Aspects of fish fecundity. In *Ecology of Freshwater Fish Production.* S. B. Gerking (ed.), pp. 75–101. Wiley, New York.

Banks, H. T., L. W. Botsford, F. Kappel, and C. Wang. 1988. Modeling and estimation in size structured population models. In *Mathematical Ecology, Proceedings of the Autumn Course Research Seminars,* T. G. Hallam, L. J. Gross, and S. A. Levin (eds.), pp. 521–541. World Scientific, Singapore.

Banks, H. T., L. W. Botsford, F. Kappel, and C. Wang. 1991. Estimation of growth and survival in size-structured cohort data: an application to larval striped bass (*Morone saxatilis*). *J. Math. Biol.,* in press.

Bence, J. R. and R. M. Nisbet. 1989. Space-limited recruitment in open systems: the importance of time delays. *Ecology* **70**:1434–1441.

Botsford, L. W. 1981a. Comment on cycles in the northern California Dungeness crab population. *Can. J. Fish. Aquat. Sci.* **38**:1295–1296.

Botsford, L. W. 1981b. More realistic fishery models: cycles, collapse, and optimal policy. In *Renewable Resource Management,* T. L. Vincent and J. Skowronski (eds.), pp. 6–20. Springer-Verlag, New York.

Botsford, L. W. 1982. Age- and size-specific models in the Dungeness crab fishery. In *Proceedings of the International Conference on Population Biology,* H. I. Freeman and C. Strobeck (eds.), pp. 394–400, Lecture Notes in Biomathematics 52. Springer-Verlag, New York.

Botsford, L. W. 1984. Effect of individual growth rates on expected behavior of the northern California Dungeness crab (*Cancer magister*) fishery. *Can. J. Fish. Aquat. Sci.* **41**:99–107.

Botsford, L. W. 1986a. Population dynamics of the Dungeness crab (*Cancer magister*). In *North Pacific Workshop on Stock Assessment and Management of Invertebrates,* G. S. Jamieson and N. Bourne (eds.), pp. 140–153. Can. Spec. Publ. Fish. Aquat. Sci. 92.

Botsford, L. W. 1986b. Effects of environmental forcing on age-structured populations: northern California Dungeness crab (*Cancer magister*) as an example. *Can. J. Fish. Aquat. Sci.* **43**:2345–2352.

Botsford, L. W. 1987. Participant's Comment. In *Modeling and Management of Resources under Uncertainty,* T. L. Vincent, Y. Cohen, W. J. Grantham, G. P. Kirkwood, and J. M. Skowronski (eds.), pp. 137–138, Proceedings of the Second U.S.-Australia Workshop on Renewable Resource Management. Springer-Verlag, New York.

Botsford, L. W. 1991. Further analysis of Clark's delayed recruitment model. Bulletin of Mathematical Biology **54**:275–293.

Botsford, L. W., Armstrong, D. A., and Shenker, J. M. 1989. Oceanographic Influences on the Dynamics of Commercially Fished Populations. In *Coastal Oceanography of Oregon and Washington,* M. R. Landry and B. M. Hickey (eds.), pp. 511–565. Elsevier, The Netherlands.

Botsford, L. W. and R. C. Hobbs. 1991. Population dynamics of the Dungeness crab (*Cancer magister*) II. Proceedings, International Council for the Exploration of the Sea (submitted).

Botsford, L. W. and D. E. Wickham. 1978. Behavior of age-specific, density-dependent models and the northern California Dungeness crab (*Cancer magister*) fishery. *J. Fish. Res. Board Can.* **35**:833–943.

Caswell, H. 1989. *Matrix Population Models.* Sinauer Associates, Sunderland, MA.

Caswell, H., H. E. Koenig, J. A. Resh, and Q. E. Ross. 1972. An introduction to systems science for ecologists. In *Systems Analysis and Simulation in Ecology,* B. C. Patten (ed.), pp. 3–78. Academic Press, New York.

Clark, C. W. 1976. A delayed-recruitment model of population dynamics, with an application to baleen whale populations. *J. Math. Biol.* **3**:381–391.

Cushing, J. M. 1980. Model stability and instability in age structured populations. *J. Theor. Biol.* **86**:709–730.

deRoos, A. M. 1988. Numerical methods for structured population models: the escalator boxcar train. *Num. Meth. for Part. Diff. Equations.* **4**:173–195.

deRoos, A. M., O. Diekmann, and J. A. J. Metz. 1992. Studying the dynamics of structured population models: a versatile technique and its application to Daphnia. *Am. Nat.,* in press.

Diekmann, O., R. M. Nisbet, W. S. C. Gurney, and F. van den Bosch. 1986. Simple mathematical models for cannibalism: a critique and a new approach. *Math. Biosci.* **78**:21–46.

Euler, L. 1760 (1970). A general investigation into the mortality and multiplication of the human species. Translated by N. B. Keyfitz. *Theor. Popul. Biol.* **1**:307–314.

Frauenthal, J. C. 1975. A dynamic model for human population growth. *Theor. Popul. Biol.* **8**:64–73.

Goodman, D. 1987. Consideration of stochastic demography in the design and management of biological reserves. Natural Resource Modeling **1**:205–234.

Guckenheimer, J., G. Oster, and A. Ipaktchi. 1977. The dynamics of density-dependent population models. J. Math. Biol. **4**:101–147.

Gurney, W. S. C. and R. M. Nisbet. 1980. Age dependent population dynamics in static and variable environments. *Theor. Popul. Biol.* **17**:321–344.

Hackney, P. A. and J. C. Webb. 1978. A method for determining growth and mortality rates of ichthyoplankton. In *Fourth National Workshop on Entrainment and Impingement.* L. D. Jensen (ed.), pp. 115–124. Ecological Analysts, Melville, N.Y.

Hallam, T. G., R. R. Lassiter, J. Li, and L. A. Suarez. 1990. Modelling individuals employing an integrated energy response: Application to Daphnia. *Ecology* **71**:938–954.

Harris, R. B., L. A. Maguire, and M. L. Shaffer. 1987. Sample sizes for minimum viable population estimation. *Conserv. Biol.* **1**:72–76.

Hastings, A. 1987. Cycles in cannibalistic egg-larval interactions. *J. Math. Biol.* **24**:651–666.

Hastings, A. 1988. When should you include age structure? In *Community Ecology,* A. Hastings (ed.), pp. 25–34. Springer-Verlag, New York.

Haury, L. R., J. A. McGowan, and P. H. Wiebe. 1978. Patterns and processes in the time-space scales of plankton distribution. In *Spatial Pattern in Plankton Communities*, J. H. Steele (ed.), pp. 277–328. Plenum Press, New York.

Hobbs, R. C. and L. W. Botsford. 1989. Dynamics of an age-structured prey with density- and predation-dependent recruitment: the Dungeness crab and a nemertean egg predator worm. *Theor. Popul. Biol.* **36**:1–22.

Hobbs, R. C., L. W. Botsford, and A. Thomas. 1992. Influence of hydrographic conditions and wind forcing on the distribution and abundance of Dungeness crab, *Cancer magister* larvae. Can. J. Fish. and Aquat. Sci., (in press).

Holling, C. S. 1964. The analysis of complex population processes. *Can. Entomol.* **96**:335–347.

Horwood, J. W. and J. A. Shepherd. 1979. The sensitivity of age-structured populations to environmental variability. *Math. Biosci.* **57**:59–82.

Huston, M., D. DeAngelis, and W. Post. 1988. New computer models unify ecological theory. *Bioscience* **38**:682–691.

Judge, M. L., J. F. Quinn, and C. Wolin. 1988. Variability in recruitment of *Balanus glandula* along the central California coast. *J. Exp. Mar. Biol. Ecol.* **119**:235–251.

Leslie, P. H. 1945. On the use of matrices in certain population mathematics. *Biometrika* **33**:183–212.

Levin, S. 1981. Age structure and stability in multiple-age spawning populations. In *Renewable Resource Management*, T. L. Vincent and J. Skowronski (eds.), pp. 21–45, Springer-Verlag, New York.

Levin, S. 1987. Scale and predictability in ecological modeling. In *Modeling and Management of Resources under Uncertainty*, T. L. Vincent, Y. Cohen, W. J. Grantham, G. P. Kirkwood, and J. M. Skowronski (eds.), pp. 2–10, Proceedings of the Second U.S.-Australia Workshop on Renewable Resource Management. Springer-Verlag, New York.

Levin, S. and C. P. Goodyear. 1980. Analysis of an age-structured fishery model. *J. Math. Biol.* **9**:245–274.

Levins, R. 1966. The strategy of model building in population biology. *Am. Sci.* **54**:421–431.

Lotka, A. J. 1907. Relation between birth rates and death rates. *Science* **26**:21.

Ludwig, D. 1987. Computer-intensive method for fisheries stock assessment. In *Modeling and Management of Resources under Uncertainty*, T. L. Vincent, Y. Cohen, W. J. Grantham, G. P. Kirkwood, and J. M. Skowronski (eds.), p. 114–124, Proceedings of the Second U.S.-Australia Workshop on Renewable Resource Management. Springer-Verlag, New York,

Ludwig, D. and C. J. Walters. 1985. Are age-structured models appropriate for catch-effort data? *Can. J. Fish. Aquat. Sci.* **42**:1066–1072.

May, R. M. 1981. The role of theory in ecology. *Am. Zool.* **21**:903–910.

May, R. M., J. R. Beddington, J. W. Horwood, and J. G. Shepherd. 1978. Exploiting natural populations in an uncertain world. *Math. Biosci.* **42**:219–252.

May, R. M. and G. F. Oster. 1976. Bifurcations and dynamic complexity in simple ecological models. *Am. Nat.* **110**:573–599.

McKelvey, R. and D. Hankin. 1981. Reply to comment on cycles in the northern California Dungeness crab population. *Can. J. Fish. Aquat. Sci.* **38**:1296–1297.

McKelvey, R., D. Hankin, K. Yanosko, and C. Snygg. 1980. Stable cycles in multistage recruitment models: an application to the northern California Dungeness crab (*Cancer magister*) fishery. *Can. J. Aquat. Sci.* **37**:2323–2345.

Metz, J. A. J. and O. Diekmann. 1986. *Dynamics of Physiologically Structured Populations*. Lecture Notes in Biomathematics 68. Springer-Verlag, Berlin.

M'Kendrick, A. G. 1926. Application of mathematics to medical problems. *Proc. Edinb. Math. Soc.* **44**:98–130.

Nisbet, R. M. and W. S. C. Gurney. 1982. *Modelling Fluctuating Populations*. Wiley, New York.

Nisbet, R. M., W. S. C. Gurney, W. W. Murdoch, and E. McCauley. 1989. Structured population models: a tool for linking effects at individual and population level. *Biol. J. Linn. Soc.* **37**:79–99.

Popper, K. 1935. *The Logic of Scientific Discovery*. [A translation of *Logik der Forschung*, 1935] Harper and Row 1959, New York.

Possingham, H. P. and J. Roughgarden. 1990. Spatial population dynamics of a marine organism with a complex life cycle. *Ecology* **71**:973–985.

Quinn, J. F. and A. E. Dunham. 1984. On hypothesis testing in ecology and evolution. In *Ecology and evolutionary biology: a round table on research*. G. W. Salt (ed.), pp. 22–37. University of Chicago Press, Chicago.

Quinn, J. F. and A. Hastings. 1987. Extinction in subdivided habitats. *Conserv. Biol.* **1**:198–208.

Reed, W. J. 1983. Recruitment variability and age structure in harvested animal populations. *Math. Biosci.* **65**:239–268.

Ricker, W. E. 1954. Stock and Recruitment. *J. Fish. Res. Board Can.* **11**:559–623.

Rorres, C. 1976. Stability of an age-specific population with density-dependent fertility. *Theor. Popul. Biol.* **10**:26–46.

Roughgarden, J., S. Gaines, and H. Possingham. 1988. Recruitment dynamics in complex life cycles. *Science* **241**:1460–1466.

Roughgarden, J. and Y. Iwasa. 1986. Dynamics of a metapopulation with space-limited subpopulations. *Theor. Popul. Biol.* **29**:235–261.

Roughgarden, J., Y. Iwasa, and C. Baxter. 1985. Demographic theory for an open marine population with space-limited recruitment. *Ecology* **66**:54–67.

Salmon, W. C. 1973. *Logic*. Prentice-Hall, Englewood Cliffs, New Jersey.

Schoener, T. W. 1986. Mechanistic approaches to community ecology: a new reductionism? *Am. Zool.* **26**:81–106.

Smith, P. E. 1978. Biological effects of ocean variability: time and space scales of biological response. *Rapp. P. V. Reun. Cons. Int. Explor. Mer.* **173**:117–127.

Soulè, M. E. (ed.). 1987. *Viable Populations for Conservation*. Cambridge University Press, Cambridge, England.

Steele, J. H. (ed.) 1978. *Spatial Pattern in Plankton Communities*. Plenum Press, New York.

Sykes, Z. M. 1969. Some stochastic versions of the matrix model for population dynamics. J. Amer. Stat. Assoc. **64**:111–130.

Tuljapurkar, S. 1987. Cycles in nonlinear age-structured models I. Renewal Equations. *Theor. Popul. Biol.* **32**:26–41.

Tuljapurkar, S. 1989. An uncertain life: demography in random environments. *Theor. Popul. Biol.* **35**:173–184.

Walsh, J. J. 1978. The biological consequences of interaction of the climatic, El Nino, and event scales of variability in the eastern tropical Pacific. *Rapp. P. V. Reun. Cons. Int. Explor. Mer.* **173**:182–192.

Werner, E. E. and J. F. Gilliam. 1984. The ontogenic niche and species interactions in size-structured populations. *Ann. Rev. Ecol. Syst.* **15**:393–425.

Wooton, R. J. 1979. Energy costs of egg production and environmental determinants of fecundity in teleost fishes. In *Fish Phenology: Anabolic Adaptiveness in Teleosts*. P. J. Miller (ed.), pp. 133–160. Academic Press, London.

11

Empirical and Theoretical Approaches to Size-Based Interactions and Recruitment Variability in Fishes

Larry B. Crowder, James A. Rice, Thomas J. Miller, and Elizabeth A. Marschall

ABSTRACT. Traditional age-based population models have been widely used both in ecology and fisheries management. But age-based models have been difficult to apply to fishes and other organisms where survivorship and fecundity depend more strongly on size than age. In fishes, recruitment variation is often high and has important population implications. The mechanisms controlling survival and recruitment of fishes appear to operate at the level of the individual, so modeling this process may best be done using individual-based models. Individual-based models have been widely used in fish behavioral and physiological ecology, but the population implications of variation among individuals have not been carefully examined.

Our studies of recruitment mechanisms of Lake Michigan bloater (*Coregonus hoyi*) have shown the importance of body size and growth rate variation to recruitment. Based on a literature review, body size and growth variation in larval fishes seems to be of widespread importance. An individual-based model for predation on larval bloaters suggests that variation among individuals in growth rate can influence cohort survival and size structure. Size-dependent mechanisms controlling recruitment in fishes may be successfully integrated using individual-based models.

Introduction

Population Modeling

Population modeling in ecology has traditionally based classes or catagories on age. This probably derives from the dependence of early population modelers on human demographic models (Hutchinson 1978). But humans are not necessarily representative of other animal populations; they are long lived, show little variation in offspring number and are extremely age-specific in rates of fecundity and survival. Most animals, including many

237

vertebrates, are relatively short-lived, show high variation in the number of offspring produced, and vital rates may be more directly linked to body size than age *per se* (Kirkpatrick 1984, Sauer and Slade 1987, Sebens 1987).

Frequently, these models combine many individuals into the same group that is modeled with a single state variable, such as population size or the number of individuals in age class three. In some cases this approach may be appropriate, but in many cases, the biology dictates a lower level of aggregation in the model. For example, differences among individuals within a class due to either genetics or environment may force the consideration of finer classes and ultimately an individual-based approach (Huston et al. 1988). In addition, most models exclude localized effects and assume all interactions among individuals are homogeneously distributed. For sessile organisms, in particular, this may be an inappropriate assumption. This also leads to a finer scale of resolution in the model (Huston et al. 1988). In the final analysis, the appropriate scale of aggregation in a model of population dynamics must depend upon the questions being asked and the systems under examination.

Modeling fish population dynamics is a practical enterprise that derives from an interest in managing commercial stocks of fishes. In particular, one seeks to model fish populations to understand how to enhance the yield of the fishery while sustaining the population's capacity to produce that yield. Models of fish population dynamics also enhance our ecological insights regarding how these populations function. Because fishes have highly plastic growth rates, traditional age-based models might have seemed inappropriate, *a priori*, but their application has a long and venerable history in fisheries (Gulland 1977).

The link between fisheries and population modeling in basic ecology is an old one, but ecological and fisheries theories have tended to develop separately (Kerr 1982, Werner 1982). Our goal here is to summarize the major approaches taken in modeling fish populations, and in this context to review our work over the last 10 years.

Demographic techniques initially used in fisheries management as well as for other organisms were "borrowed" from human demography (Hutchinson 1978). Early population models in general ecology treated all individuals as identical (as in the classic exponential and Pearl-Verhulst logistic models of population growth). Early fishery yield modeling did the same (as in the surplus-yield model, Schaefer 1954). Later models from both general ecology and fisheries considered age classes explicitly (in fisheries, the dynamic pool model, Beverton and Holt 1957). The data necessary for such models could be organized in a life table that outlined the age-specific survivorship and age-specific fecundity typical of those populations. Age could be formally accomodated using the population projection matrix approach (Leslie 1945).

While this approach often proved acceptable for mammals and birds, where key parameters of fecundity and survivorship are fairly predictable (but see Sauer and Slade 1987), age-based models have been difficult to apply to fishes, insects, other invertebrates, and many plants where survivorship and fecundity at age are much less predictable (Sebens 1987). These organisms generally produce large numbers of offspring which experience high mortality during their early life history, and highly variable survival.

To resolve this problem, the population projection approach can be extended to include stages rather than ages as the unit of aggregation in the models (for a review see Caswell 1989). These models are particularly appropriate when different ontogenetic stages of organisms with complex life histories have substantially different population parameters, or spend variable periods of time within a stage (Hartshorn 1975, Werner and Caswell 1977, Caswell 1986, Crouse et al. 1987, Werner 1988). The mathematics and analyses of matrix population models are now well developed.

Fishes also have highly plastic growth and may span a wide range of sizes at a single age. In fishes, age is relatively easy to estimate although this can be difficult in the tropics (Pauly 1980). However, survivorship and fecundity depend more directly on size than on age *per se* due to variability in growth rates (Beyer 1989). Because a wide variety of processes can affect growth rates and size structures of fish populations (e.g., food, temperature, population density, predators, exploitation), age specific survival and fecundity can change dramatically from one period to the next. This requires re-parameterizing the age-based model for the "new" conditions. An alternative, only recently being explored, is to develop parallel size-based population models (e.g., VanSickle 1977, DeAngelis et al. 1979, 1984, Kirkpatrick 1984, Ebenman and Persson 1988, Beyer 1989).

Recruitment Variation

One of the key characteristics of fish populations, which challenges our ability to understand or manage populations effectively, is variation in survival and recruitment. Most fisheries biologists consider that this variation occurs due to processes in the early life history (e.g., egg-larva-juvenile stages). By recruitment, fisheries biologists refer to the addition of fish to the harvestable population. One may also think of recruitment to the population (e.g., year class strength is determined during the first year of life).

Variable recruitment is a big problem from two perspectives. From an ecological view, variable recruitment is important because the population dynamics and community structure we see in adults are often determined by events very early in the life history. Populations with variable survival in their early life history are common, and their variable recruitment has been studied from ecological and life history perspectives (Strong 1984, Sale 1990). In fisheries, understanding the mechanisms underlying variable recruitment is one of the major research problems (Steele et al. 1980, Rothschild and Rooth 1982, Fritz et al. 1990). From a management perspective, variable recruitment is a serious issue, because uncertainty in the amount of fish available for harvest makes management uncertain, and errors can have significant political and economic implications (Beddington et al. 1984, Steele 1984, Walters 1984). For this reason, many fisheries managers would prefer to believe in classic fishery models where simple deterministic functions relate adult stock size to number of recruits. Put quite simply, this approach has failed. It is possible that the traditional approaches to recruitment variability and fish population modeling will not lead us to either greater understanding of the recruitment process or to improved ability to predict recruitment (Larkin 1977, Beyer 1989).

In the last 10–15 years, the emphasis in recruitment studies has shifted from predicting recruitment based on regression or stock/recruit relationships toward understanding the mechanisms underlying recruitment variability. Nevertheless, most recruitment studies still have two problems. First, many researchers still focus on estimating mortality in the early life history stages. However, estimating mortality with the precision required is extremely difficult (Smith 1981). Second, most programs still look at recruitment as a population or cohort-level phenomenon. The mechanisms governing survival and recruitment (and ultimately evolution) operate at the level of the individual. When individuals differ substantially, the results we see at the population level may derive from a small minority of atypical individuals. For example, survivors (recruits) from a cohort of spawned larvae are probably not average fish (Sharp 1987). When this is the case, interpretations based on modeling the average individual are likely to be misleading.

Individual-Based Models in Fish Ecology

Ecologists have long been interested in modeling the behavior and physiology of animals. These frequently are individual-based models rather than population-based models, in the sense that what is being modeled is individual behavior or physiology rather than population level effects of these behaviors. Models of foraging behavior of individuals have been widely studied for 25 years (c.f. Pyke 1984, Real and Caraco 1986, Stephens and Krebs 1986). Foraging in fish has been extensively modeled (Werner and Hall 1974, Mittelbach 1981, Werner, Mittelbach, et al. 1983, Crowder 1985), although all of these attempts have dealt with how an *average* individual may be expected to choose prey (or patches of food) given a particular encounter rate with prey of differing utility. In general, these models can successfully predict food size choice and habitat switching based on the relative value of prey resources (Mittelbach 1981, Werner, Mittelbach, et al. 1983).

Habitat choice can also be modified by predation risk (Werner, Gilliam, et al. 1983, Gilliam and Fraser 1987, Abrahams and Dill 1989). More elaborate models of habitat choice involve tradeoffs of predation risk and growth in various habitats (Gilliam 1982, Werner and Gilliam 1984, Gilliam and Fraser 1987). These models are also based on an "average" individual. Some data are beginning to suggest, however, that individuals may differ substantially in their foraging behaviors, and thus not all individuals conform to the theoretical predictions (Marschall et al. 1989).

Another well-regarded set of individual-based models of fishes are the bioenergetic-growth models of Kitchell and colleagues (Kitchell et al. 1977, Stewart et al. 1981, Rice and Cochran 1984, Kitchell 1983, Stewart and Binkowski 1986, Hewett and Johnson 1987). In these models, the energy budget of an *average* individual is simulated. By knowing something about the physiology of fishes as a function of their body size and temperature, one can estimate either cumulative consumption from growth *or* growth from consumption. These models have been thoroughly tested and validated (Rice and Cochran 1984), but they are based on an average individual. In order to extrapolate to the population level, one must multiply the consumption by individuals in each size (age) class by the number of individuals in that class. This approach has been successfully used to assess the effects of free ranging predators on their prey resources in Lake Michigan (Stewart et al. 1981, Kitchell and Crowder 1986). While this approach adequately predicts cohort level consumption or average growth of individuals in a cohort, it does not necessarily predict growth or consumption of any specific individual. Further, it does not deal with causes or consequences of variation in growth rate among individuals.

Individual-based models in fish ecology have been pursued when the process of interest occurs on an individual level or when aggregation at a higher level (age/size groups) was not considered representative of the process. Foraging models clearly fall into the former category, whereas recruitment models probably fall into the latter category. In recruitment to a fish population, small differences among individuals may have big effects on their probability of being represented among the survivors of a cohort. Modeling recruitment by modeling the average fish is problematical because the average fish dies in less than a week (Sharp 1987).

Recruitment Mechanisms in Bloater (*Coregonus Hoyi*)

For the past decade, we have been studying recruitment mechanisms of fishes. Through work with the bloater, our thinking has focused on the importance of body size, size-dependent interactions and the characteristics of individuals in determining survival.

Bloaters are Coregonine fishes, one of a suite of seven ciscoes that co-existed in Lake Michigan until serious abiotic and biotic modifications of that system from the 1900s to 1960s led to local extinctions of all species other than bloater (Smith 1970, Wells and McLain 1973). Bloaters spawn primarily during January through March, depositing their eggs on the bottom at 70–100 m depth. Their eggs are relatively large, about 2 mm in diameter; larvae hatch in late spring or early summer at about 9.5–10 mm body length (Wells 1966). After hatching, the larve spend 5–10 days in the hypolimnion and then migrate to the surface where they feed on zooplankton during their first summer (Rice, Crowder, and Holey 1987). Late in the summer, the juveniles migrate to the hypolimnion and feed on benthic prey for the remainder of their life (Crowder and Crawford 1984).

Bloaters experienced dramatic reductions in recruitment success beginning in the mid 1960s (Brown 1970), and were placed on the threatened species list in Michigan by the mid 1970s. This decline seems to have been due to both direct and indirect effects of alewife, (*Alosa pseudoharengus*), an exotic fish which invaded Lake Michigan, establishing large populations by the mid 1960s (Crowder 1980, Crowder et al. 1987, Luecke et al. 1990, Miller et al. 1990). In studying bloater recruitment, we wanted to consider the range of possible mechanisms and use an approach which would allow us to reduce the list from all possible recruitment mechanisms to those which are most likely. From a thorough review of the literature on marine fish larvae, we outlined the commonly hypothesized mechanisms thought to control recruitment. Starvation in some critical life stage (e.g., first feeding) was popular (Hjort 1914, Lasker 1975, 1978), as was physical environmental variation (e.g., failed transport or retention in an appropriate habitat, Smith 1981, Sinclair 1989). Predation was also popular, particularly as an alternative to starvation (Crowder 1980, Hunter 1981).

We adopted a "strong inference" approach to screen among the alternative hypotheses and eliminate one or more (Platt 1964). We sought to turn the problem around—instead of studying sources of mortality to explain why 99+% of the fish die, we wanted to test the null hypothesis that survivors (i.e., recruits) are drawn at random from the cohort of spawned offspring. If recruits are just average fish, there is no point to look for particular explanatory hypotheses based on individual differences. Rather than looking at the relative success of cohorts across years (as in stock/recruit relations), we focused on which individuals within a cohort survived. We also let their unique characteristics tell us what was important, thus narrowing the scope of hypotheses we needed to explore.

Fortunately, larval fishes carry a detailed record of their birthdate and growth rates in their otoliths. From daily rings in these small inner ear bones, we could estimate birthdate, growth rates and often detect periods of low growth or other "stress" (Rice et al. 1985). By examining the otoliths of larvae caught at successively later points in their early life history, distri-

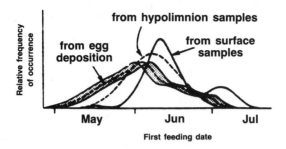

Figure 11.1 Comparison of combined distributions of first-feeding dates observed for larval bloaters in 1983 surface and hypolimnion samples with expected distributions of first-feeding dates from egg deposition and incubation temperatures. Because absolute abundances at each stage are unknown, all curves are scaled to the same area. Mismatches between curves identify periods of differential relative mortality within the season. From Rice, Crowder, and Holey 1987.

butions increasingly differed (Fig. 11.1); we found that survivors to one month of age were drawn disproportionately from those which were spawned and hatched later (Rice, Crowder, and Holey 1987). Furthermore, daily growth rates of larvae caught in the field and of larvae from laboratory studies (Rice, Crowder, and Binkowski 1987) confirmed that starvation was unimportant as a direct source of mortality for bloater. We found that growth rate *was* important—early larvae grew half as fast as later larvae and were underrepresented among the survivors. Stress marks also correlated with those fish that failed to survive and did not correlate with specific dates or environmental events.

These results suggested that recruitment success was related to size or growth dependent mortality. Subsequent experiments in the laboratory identified alewife as one of the most important predators and showed that their predation on bloaters was strongly size-dependent (Luecke et al. 1990). In other experiments we showed that reductions in zooplankton size due to intense planktivory (e.g., by alewife) can indirectly reduce bloater growth rates, particularly in the early juvenile stage, potentially reducing recruitment (Crowder et al. 1987, Miller et al. 1990). Thus, bloater recruitment success seems to depend on the size and growth dynamics of larvae and juveniles. Direct predation is strongly size-dependent, and reduced food or temperatures can prolong exposure to predators, leading to reduced survival. Clearly, the physical milieu and the biotic mechanisms *interact* to influence ultimate recruitment. Furthermore, it is particular individuals that are successful, and their condition (size, growth rate, etc.) is decidedly not average.

Our approach allowed us to narrow the range of possibilities and focus our efforts on the key hypotheses. It also confirmed our notion that the survivors tend to be atypical rather than average individuals. This implies that our understanding of recruitment might be enhanced most rapidly by focusing on the unique characteristics of individual survivors rather than on estimating mortality at the population level (Rice, Crowder, and Holey 1987). The approach of comparing the characteristics of individual survivors to those of earlier samples has been applied successfully to other fishes (Methot 1983, Crecco and Savoy 1985), but will be difficult to apply to organisms other than fishes (e.g. marine invertebrates) unless one can identify structures that contain a record of characteristics important to an individual's recruitment.

Larval Size and Recruitment Success

Bloaters did not fit some of our expectations based on the marine fish larvae literature. They were robust and highly resistant to starvation relative to typical marine fish larvae (time to 50% mortality for starved larvae was 25 days vs. less that 7 for typical marine larvae!). Body size scaling is well known from the ecological and physiological literature (Thompson 1917, Haldane 1927, McMahon and Bonner 1983, Peters 1983, Calder 1984), but had never been thoroughly examined for larval fish. Further, body size and growth dynamics in the presence of size-selective predators can have complex population level implications (Ebenman and Persson 1988).

Fishes grow through 4–5 orders of magnitude of length in their life history, and often 2–3 orders of magnitude of this growth occurs in the first year of life (Werner and Gilliam 1984). Furthermore, sizes of larval fishes at hatching span at least one order of magnitude; even within a species, size at hatching may vary by more than 50% (Blaxter and Hempel 1963). Variation in body size among species may account for some of the substantial ecological differences observed across larval fishes (Miller et al. 1988).

In our recent paper (Miller et al. 1988), we reviewed various aspects of the larval ecology of 72 species of marine, freshwater and anadromous fishes with respect to hatching size and larval growth. The review included fish which hatch between 1.6 mm and 17 mm in length. Physiological response times were standardized to 15 C using a Q_{10} of 2.3 (Checkley 1984). Times to yolk absorption, 50% mortality for starving larvae and point of no return (irreversible starvation) all increased linearly with hatching size. The "window of opportunity" to feed (time to point of no return minus time to first possible feeding) also increased with larval size (Fig. 11.2). Even small increases in body size at hatching confer large benefits in terms of flexibility at first feeding. For every 0.1 mm increase in length at hatching, larvae gain about 6 hours in which to find food; each 1 mm increases the "window of opportunity" 2.5 days (Fig. 11.2).

In addition, vulnerability to predators was found to be size-specific. Probability of capture per encounter declines with increasing larval prey size,

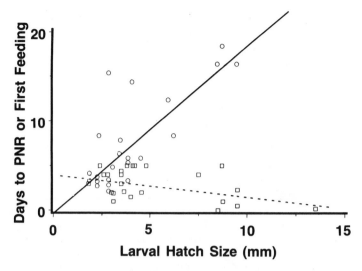

Figure 11.2   Time to point of no return (solid line) or to first possible feeding (dotted line) as a function of total larval length at hatching. The regression equation for the difference between these lines for the six paired data points on the same species is $y = 2.46$ (length) $- 4.81$, $r^2 = 0.89$. From Miller et al. 1988.

and the slopes of these functions are steepest for small predators and shallowest for large predators. In fact, it was possible to fit an equation to predict capture success per encounter as a function of predator-prey size ratio (Fig. 11.3). This equation seemed to describe probability of capture reasonably well whether the predators were fishes, jellies or euphausids! Probability of capture increases from zero at a predator-prey size ratio of about 2.5, to 1.0 at a predator-prey ratio of 15. A consideration of body size does much to explain differences among species in ecological performance.

Individual-Based Modeling of Larval Fish Recruitment

We came by our interest in individual-based modeling from an appreciation of the biology of the recruitment process in fishes. It is *individuals* that survive to recruit; the unique characteristics of individuals, and not population averages, determine which individuals survive. Individual-based models are not only interesting, but are perhaps the only logical way to model these processes.

Not only are many of the individual processes that determine recruitment size dependent, but they also interact. Slow growth leads to prolonged exposure to a size-dependent predator. To examine the population level im-

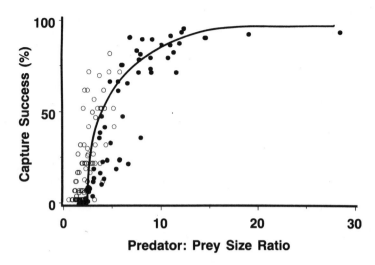

Figure 11.3 Capture success (CS) as a function of predator-prey length ratio. Points are original data from various invertebrate (open circles) and fish (closed circles) predators. The equation for the fitted line is CS = 100 − ((ration + 3.37)/44.76)$^{-2.28}$. From Miller et al. 1988.

plications of these processes, we formulated a conceptual model summarizing these interactions (Fig. 11.4). All steps in the model are size-dependent, interact, and *occur at the level of the individual*, but the results of these interacting factors are not always intuitively obvious. We are collecting data both on Great Lakes fishes and on fishes from southeastern US estuaries to enhance this model, but the basic form of the model has already been developed independently by Don DeAngelis, Kenny Rose and others at Oak Ridge National Laboratories. We have been working closely with them to test, enhance and exercise it.

This individual-based model structure provides an ideal framework for evaluating the recruitment implications of individual variability and size-dependent interactions. For example, the strong size-dependence of predation on larval and juvenile fishes (Miller et al. 1988) suggests that variation in growth rate among individuals, which is often observed, could have a substantial impact on the number and characteristics of fish surviving exposure to predation. We have explored this hypothesis using the predation component of the model, parameterized for size-dependent predation by yearling alewife on bloater larvae and juveniles (Luecke et al. 1990). Simulations tracked the growth and survival of each individual in a cohort of bloater larvae as they were exposed to alewife predation over the first 60 d of life.

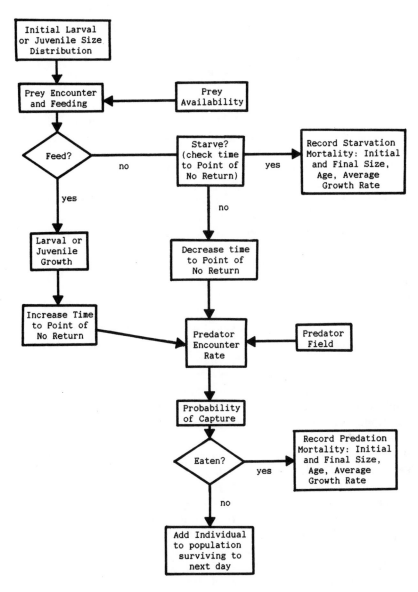

Figúre 11.4 Structure of the conceptual model for larval and juvenile fish recruitment. Each step has a size-dependent component.

When cohorts of larvae with the same initial mean growth rate, but different levels of variance in growth rate among individuals, were exposed to size-dependent predation, there were substantial effects on the number and characteristics of survivors. Our results suggest that survival could be 3–4× higher for cohorts with high variance in growth rate among individuals than for low variance cohorts, with all other conditions held constant. Also, selection for fast growing (and therefore larger) individuals increased with increasing variance in growth rate; at high variance almost all survivors were drawn from the upper 25% of the initial growth rate distribution. (Rice et al. ms.) Clearly, survivors are not average individuals!

In nature, prey are often exposed to growing predators or to different sized predators as they grow (Wilbur 1988), small changes in hatching size, growth rate, or predator size can all have major effects on survival. As a result, the recruitment process can by highly dynamic, and the results may not be intuitively obvious. We see individual-based models as a powerful and flexible approach for evaluating the outcome of these interactions.

Discussion and Conclusions

Size-dependent interactions provide a unifying framework for understanding mechanisms governing survival and recruitment in fishes. We suspect that this may also be true for the many other organisms with high fecundity and variable survival (Strong 1984). Our understanding of the interacting size-dependent factors controlling recruitment has already been enhanced through individual-based models. Unfortunately, one can apply individual-based models only to populations from which individual-based data can be obtained.

In recruitment, key processes occur at the level of the individual. While we have stressed body size as an individual characteristic, we acknowledge that a variety of other factors can also contribute to differential survival. Maternal contributions of yolk, lipids, proteins and growth hormones, may influence individual embyro quality and subsequent larval growth (and survival) without changing egg size *per se*. Furthermore, these higher quality or larger eggs may derive from females which are themselves decidedly non-average; for example, larger and older females may contribute eggs which have a differential chance for survival (Bagenal 1973, Moodie et al. 1989, Zastrow et al. 1989).

Not all individual-based models examine the consequences of variability among individuals. For example, most individual-based models in behavioral ecology treat an average individual and do not address directly the population level implications of such behaviors (Łomnicki 1988). Optimal foraging models deal with individual behaviors and choices, but they do so for an average individual. The same is true of individual-based bioenergetics models. Neither model examines the consequences of a population of foragers, for which each has a different encounter rate with prey, hunger levels, or experience, for example. Some data are beginning to suggest that variation among individuals may be important for these processes (Marschall et al. 1989), but it has not been explicitly modeled. Whether foraging at the population level can be represented by the sum of "average" individuals depends on the variance among individuals. Some very interesting work has been done both empirically and via simulation to examine the effects of variance in size on cannibalism within a cohort (DeAngelis et al. 1979) and in fish predator-prey interactions (Adams and DeAngelis 1987).

One advantage of considering the individual as the unit for modeling is that natural selection also acts on individuals. If we are considering the evolution of body size, hatching size, and potential growth trajectories (Kirkpatrick 1984, 1988, Werner 1988), it may be advantageous to examine the amount of variance in these parameters. Theoretical work is required on how body size affects life history strategies, particularly with reference to variation in body size within and among cohorts of the same species, or from the same mother under different environmental or physiological conditions.

Individual-based modeling has already proven a useful technique in ecology. These models offer a heuristic advantage, allowing biologists to understand the processes being modeled, and seem best applied when the biological situation dictates their use. This depends, however, on the representativeness of the average individual for the process to be modeled. In a behavioral ecology model of foraging, the average individual may be an appropriate unit to model. In fish recruitment, as in plant competition (Huston et al. 1988), the contrary is true, and some individuals are more equal than others in terms of initial body size, growth rate or spatial location and so must be modeled with an individual-based approach, taking into account variation among individuals.

More aggregated models are better known in terms of their dynamical behavior and should be used in preference to individual-based models when possible. Caswell distinguished two types of individual-based models: the *i-state configuration* models where each individual is modeled explicitly (as in our larval survival model), and the *i-state distribution* models where individuals are represented by distributions of some characteristic (e.g., size or stage). The mathematics of "*i*-state distribution" models usually include partial differential equations (e.g., the Von Forester equation) or population projection matrices. While these approaches might initially seem more difficult to biologists than the "*i*-state configuration" approach, they are much simpler and their dynamics better understood mathematically.

What, then, is the appropriate level of aggregation in a model? Why or when do we need to include individual state ("i-state") structure in a model (Botsford, Caswell, Metz, this volume)? To answer this question, one must then ask, at a given scale of aggregation, is the average individual representative? If behavior, physiology or other vital rate functions vary importantly among individuals, then the model would benefit from an individual-based approach. When individuals can be aggregated into larger catagories, the modeling is simplified and can employ better-known techniques.

Individual-based models are attractive tools for ecologists, but like all tools they must be handled carefully to produce the desired results. We suggest that apprentice modelers (i.e., most biologists) find an expert under which to learn the trade. Biologists also have much to bring to this process, particularly in identifying the key questions and observations, and in establishing the appropriate level of aggregation for modeling the biological process of interest. For some modelers, "the real world is but a special case," but it is this world biologists hope to understand and manage.

Acknowledgements

We thank the organizers of this workshop for their kind invitation. Sea Grant has supported our recruitment related research for nearly a decade (University of Wisconsin Sea Grant Projects R/LR-22, R/LR-33, R/LR-37; University of North Carolina Sea Grant Projects R/MER-12, R/MG8903, R/MER-18). Recently, the Electric Power Research Institute has provided support for graduate fellowships for TJM and EAM. We also thank the UNC Marine Science Council, the U.S. Forest Service and the Agricultural Research Service for supporting our research. Numerous colleagues contributed to the work outlined here; Fred Binkowski, Don DeAngelis, Chris Luecke, Kenny Rose, Annet Trebitz and Steve Yeo deserve our special thanks.

Literature Cited

Abrahams, M. V. and L. M. Dill. 1989. A determination of the energetic equivalence of the risk of predation. *Ecology* **70**:999–1007.

Adams, S. M. and D. L. DeAngelis. 1987. Indirect effects of early bass-shad interactions on predator population structure and food web dynamics. In *Predation: Direct and Indirect Impacts on Aquatic Communities*, W. C. Kerfoot and A. Sih (eds.), pp. 103–117. University Press of New England, Hanover, NH.

Bagenal, T. B. 1973. Fish fecundity and its relations with stock and recruitment. Rapp P-V *Reun. Cons. Int. Explor. Mer* **164**:186–198.

Beddington, J. R. et al., 1984. Management under uncertainty. In *Exploitation of Marine Communities*. R. M. May (ed.), pp. 227–244. Springer-Verlag, Berlin.

Beverton, R. J. H. and S. J. Holt. 1957. *On the dynamics of exploited fish populations*. Fish Investigations Series II:No. 19. H. M. Stationery Office, London.

Beyer, J. E. 1989. Recruitment stability and survival—simple size-specific theory with examples from the early life dynamics of marine fish. *Dana* **7**:45–147

Blaxter, J. H. S. and G. Hempel. 1963. The influence of egg size on herring larvae (*Clupea harengus* L.). *Rapp P. V. Reun. Cons. Int. Explor. Mer* **28**:211–240.

Brown, E. H., Jr. 1970. Extreme female predominance in the bloater (*Coregonus hoyi*) of Lake Michigan in the 1960s. In *Biology of the Coregonid Fishes*, C. C. Lindsay and C. S. Woods (eds.), pp. 501–514. University of Manitoba Press, Winnepeg.

Calder, W. A. III. 1984. *Size, Function and Life History.* Harvard University Press, Cambridge, MA.

Caswell, H. 1986. Life cycle methods for plants. *Lect. Math. Life Sci.* **18**:171–233.

Caswell, H. 1989. *Matrix Population Models.* Sinauer Associates, Sunderland, MA.

Checkley, D. M. Jr. 1984. Relation of growth to ingestion for larvae of Atlantic herring *Clupea harengus* and other fish. *Mar. Ecol. Prog. Ser.* **18**:215–224.

Crecco, V. A. and T. F. Savoy. 1985. Effects of biotic and abiotic factors on growth and relative survival of young American shad, *Alosa sappidissima*, in the Connecticut River. *Can. J. Fish. Aquat. Sci.* **42**:1640–1648.

Crouse, D. T., L. B. Crowder, and H. Caswell. 1987. A stage-based model for loggerhead sea turtles and implications for conservation. *Ecology* **68**:1412–1423.

Crowder, L. B. 1980. Alewife, rainbow smelt and native fishes in Lake Michigan: Competition or predation? *Environ. Biol. Fishes* **5**:225–233.

Crowder, L. B. 1985. Optimal foraging and feeding mode shifts in fishes. *Environ. Biol. Fishes* **12**:57–62.

Crowder, L. B. and H. L. Crawford. 1984. Ecological shifts in resource use by bloaters in Lake Michigan. *Trans. Am. Fish. Soc.* **113**:694–700.

Crowder, L. B., M. E. McDonald, and J. A. Rice. 1987. Understanding recruitment of Lake Michigan fishes: The importance of size-based interactions between fish and zooplankton. *Can. J. Fish. Aquat. Sci.* **44(II)**:141–147.

DeAngelis, D. L., S. M. Adams, J. E. Breck, and L. J. Gross. 1984. A stochastic predation model: Application to largemough bass observations. *Ecol. Modell.* **24**:21–41.

DeAngelis, D. L., D. C. Cox, and C. C. Coutant. 1979. Cannibalism and size dispersal in young-of-the-year largemouth bass: experiments and model. *Ecol. Modell.* **8**:133–148.

Ebenman, B. and L. Persson (eds.). 1988. *Size-structured Populations: Ecology and Evolution.* Springer-Verlag, Berlin.

Fritz, E. S., L. B. Crowder and R. C. Francis. 1990. The National Oceanic and Atmospheric Administration plan for recruitment fisheries oceanography research. *Fisheries* **15**:25–31.

Gilliam, J. F. 1982. Habitat use and competitive bottlenecks in size-structured fish populations. Ph.D. Dissertation, Michigan State University.

Gilliam, J. F. and D. F. Fraser. 1987. Habitat selection under predation hazard: test of a model with foraging minnows. *Ecology* **68**:1856–62.

Gulland, J. A. 1977. *Fish Population Dynamics.* Wiley, London.

Haldane, J. B. S. 1927. On being the right size. In *Possible Worlds and Other Essays*, Chatto and Windrus (eds.), pp. 1–8. Harper, London.

Hartshorn, G. S. 1975. A matrix model of tree population dynamics. In *Tropical Ecological Systems*, F. B. Golley and E. Medina (eds.), pp. 41–51. Springer-Verlag, Berlin.

Hewett, S. W. and B. L. Johnson. 1987. *A generalized bioenergetics model of fish growth for microcomputers.* University of Wisconsin Sea Grant Institute, Madison, WI.

Hjort, J. 1914. Fluctuations in the great fisheries of northern Europe viewed in the light of biological research. *Rapp. P. V. Reun. Cons. Int. Explor. Mer* **2**:1–228.

Hunter, J. R. 1981. Feeding ecology and predation of marine fishes larvae. In *Marine Fish Larvae: Morphology, Ecology and Relation to Fisheries*, R. Lasker (ed.), pp. 33–79. University of Washington Press, Seattle.

Huston, M., D. DeAngelis, and W. Post. 1988. New computer models unify ecological theory. *BioScience* **38**:682–691.

Hutchinson, G. E. 1978. An Introduction to Population Ecology. Yale University Press, New Haven, CT.

Kerr, S. R. 1982. Niche theory and fisheries ecology. *Trans. Am. Fish. Soc.* **109**:254–257.

Kirkpatrick, M. 1984. Demographic models based on size, not age, for organisms with indeterminate growth. *Ecology* **65**:1874–1884.

Kirkpatrick, M. 1988. The evolution of size in size-structured populations. In *Size-structured Populations: Ecology and Evolution*, B. Ebenman and L. Persson (eds.), pp. 13–28. Springer-Verlag, Berlin.

Kitchell, J. F. 1983. Energetics. In *Fish Biomechanics*, P. W. Webb and D. Weihs (eds.), pp. 312–338. Praeger, New York.

Kitchell, J. F. and L. B. Crowder. 1986. Predator-prey interactions in Lake Michigan: Model predictions and recent dynamics. *Environ. Biol. Fishes* **16**:205–211.

Kitchell, J. F., D. J. Stewart, and D. Weininger. 1977. Applications of a bioenergetics model to yellow perch (*Perca flavescens*) and walleye (*Stizostedion vitreum vitreum*). J. Fish. Res. Board Can. **34**:1922–1935.

Larkin, P. A. 1977. An epitaph for the concept of MSY. *Trans. Amer. Fish. Soc.* **107**:1–11.

Lasker, R. 1975. Field criteria for survival of anchovy larvae: the relation between inshore chlorophyll maximum layers and successful first feeding. *U.S. Nat. Mar. Fish. Serv. Fish. Bull.* **73**:453–462.

Lasker, R. 1978. The relation between oceanographic conditions and larval anchovy food in the California current: identification of factors contributing to recruitment failure. *Rapp. P. V. Reun. Cons. Int. Explor. Mer* **173**:212–230.

Leslie, P. H. 1945. On the use of matrices in certain population mathematics. *Biometrika* **33**:183–212.

Łomnicki, A. 1988. *Population ecology of individuals.* Princeton University Press, Princeton, NJ.

Luecke, C., J. A. Rice, L. B. Crowder, S. F. Yeo, and F. P. Binkowski. 1990. Recruitment mechanisms of bloater in Lake Michigan: An analysis of the predatory gauntlet. *Can. J. Fish. Aquat. Sci.* **47**:524–532.

Marschall, E. A., P. L. Chesson, and R. A. Stein. 1989. Foraging in a patchy environment: prey encounter rate and residence time distributions. *Anim. Behav.* **37**:444–454.

McMahon, T. A. and J. T. Bonner. 1983. *On Size and Life.* Scientific American Inc., New York.

Methot, R. D. Jr. 1983. Seasonal variation in survival of larval *Engraulis mordax* estimated from the age distribution of juveniles. *U.S. Nat. Mar. Fish. Serv. Fish. Bull.* **81**:741–750.

Miller, T. J., L. B. Crowder, J. A. Rice, and E. A. Marschall. 1988. Larval size and recruitment mechanisms in fishes: Toward a conceptual framework. *Can. J. Fish. Aquat. Sci.* **45**:1657–1670.

Miller, T. J., L. B. Crowder and F. P. Binkowski. 1990. The effect of changes in the zooplankton assemblage on growth and recruitment success of bloater. *Trans. Am. Fish. Soc.* **119**:483–491.

Mittelbach, G. G. 1981. Foraging efficiency and body size: a study of optimal diet and habitat use by bluegills. *Ecology* **62**:1370–1386.

Moodie, G. E. E., N. L. Loadman, M. D. Wiegand, and J. A. Malhias. 1989. Influence of egg characteristics on survival, growth and feeding in larval walleye (*Stizostedion vitreum*). *Can. J. Fish. Aquat. Sci.* **46**:516–521.

Pauly, D. 1980. On the interrelationships between natural mortality, growth parameters, and mean environmental temperature in 175 fish stocks. *Rapp. P. V. Reun. Cons. Int. Explor. Mer* **39**:175–192.

Peters, R. H. 1983. *The ecological implications of body size.* Cambridge University Press, Cambridge, England.

Platt, J. R. 1964. Strong inference. *Science* **146**:347–353.

Pyke, G. H. 1984. Optimal foraging theory: a critical review. *Ann. Rev. Ecol. Syst.* **15**:523–575.

Real, L. and T. Caraco. 1986. Risk and foraging in stochastic environments. *Ann. Rev. Ecol. Syst.* **17**:371–390.

Rice, J. A. and P. A. Cochran. 1984. Independent evaluation of a bioenergetics model for largemouth bass. *Ecology* **65**:732–739.

Rice, J. A., L. B. Crowder and F. P. Binkowski. 1985. Evaluating otolith analysis for bloater *Coregonus hoyi*: Do otoliths ring true? *Trans. Am. Fish. Soc.* **114**:532–539.

Rice, J. A., L. B. Crowder, and F. P. Binkowski. 1987. Evaluating potential sources of mortality for larval bloater (*Coregonus hoyi*): Starvation and vulnerability to predation. *Can. J. Fish. Aquat. Sci.* **44**:467–472.

Rice, J. A., L. B. Crowder, and M. E. Holey. 1987. Exporation of mechanisms regulating larval survival in Lake Michigan bloater: A recruitment analysis based on characteristics of individual larvae. *Trans. Am. Fish. Soc.* **116**:703–718.

Rice, J. A., T. J. Miller, K. A. Rose, L. B. Crowder, D. E. DeAngelis, E. A. Marschall, and A. S. Trebitz. ms. Growth rate variation and larval survival: Implications of an individual-based size-dependent model. *Can. J. Fish. Aquat. Sci.*, submitted.

Rothschild, B. J. and C. G. H. Rooth. 1982. *Fish Ecology III*. University of Miami Technical Report No. 820028. Miami, FL.

Sale, P. F. 1990. Recruitment of marine species: Is the bandwagon rolling in the right direction? *Trends Ecol. Evol.* **5**:25–27.

Sauer, J. R. and N. A. Slade. 1987. Size-based demography of vertebrates. *Ann. Rev. Ecol. Syst.* **18**:71–90.

Schaefer, M. B. 1954. Some aspects of the dynamics of populations important to the management of commercial marine fishes. *Bull. Inter-Am. Trop. Tuna Comm.* **1**:27–56.

Sebens, K. P. 1987. The ecology of indeterminate growth in animals. *Ann Rev. Ecol. Syst.* **18**:371–408.

Sharp, G. D. 1987. Averaging the way to inadequate information in a varying world. *Am. Inst. Fish. Res. Biol. Briefs* **16**:3–4.

Sinclair, M. 1989. *Marine Populations: An Essay on Population Regulation and Speciation*. University of Washington Press, Seattle.

Smith, P. E. 1981. Fisheries on coastal pelagic schooling fish. In *Marine Fish Larvae: Morphology, Ecology, and Relation to Fisheries*, R. Lasker (ed.), pp. 1–32. University of Washington Press, Seattle.

Smith, S. H. 1970. Species interactions of the alewife in the Great Lakes. *Trans. Am. Fish. Soc.* **99**:754–765.

Steele, J. H. 1984. Kinds of variability and uncertainty affecting fisheries. In *Exploitation of Marine Communities*, R. M. May (ed.), pp. 245–262. Springer-Verlag, Berlin.

Steele, J., C. Clark, P. Larkin, R. Lasker, R. May, B. Rothschild, E. Ursin, J. Walsh, and W. Wooster. 1980. *Fisheries Ecology: Some Constraints that Impede our Understanding*. Ocean Science Board, National Academy of Sciences, Washington, D.C.

Stephens, D. W. and J. R. Krebs. 1986. *Foraging Theory*. Princeton University Press, Princeton, N.J.

Stewart, D. J. and F. P. Binkowski. 1986. Dynamics of consumption and food conversion by Lake Michigan alewives: An energetics-modeling synthesis. *Trans. Am. Fish. Soc.* **115**:643–661.

Stewart, D. J., J. F. Kitchell, and L. B. Crowder. 1981. Forage fishes and their salmonid predators in Lake Michigan. *Trans. Am. Fish. Soc.* **110**:751–763.

Strong, D. R. 1984. Density-vague ecology and liberal population regulation in insects. In *A New Ecology: Novel Approaches to Interactive Systems*, P. W. Price, G. N. Slobodchikoff and W. S. Gaud (eds.), pp. 313–329. Wiley, New York.

Thompson, D'A. W. 1917. *On Growth and Form*. Cambridge University Press, Cambridge, England.

VanSickle, J. 1977. Analysis of a distributed-parameter population model based on physiological age. *J. Theor. Biol.* **64**:571–586.

Walters, J. 1984. Managing fisheries under biological uncertainty. In *Explotation of Marine Communities*, R. M. May (ed.), pp. 263–274. Springer-Verlag, Berlin.

Wells, L. 1966. Seasonal depth distribution of larval bloaters (*Coregonus hoyi*) in southeastern Lake Michigan. *Trans. Am. Fish. Soc.* **95**:388–396.

Wells, L. and A. L. McLain. 1973. Lake Michigan: man's effects on native fish stocks and other biota. *Great Lakes Fish. Comm. Tech. Rep.* 20.

Werner, E. E. 1982. Niche theory in fisheries ecology. *Trans. Am. Fish. Soc.* **109**:257–260.

Werner, E. E. 1988. Size, scaling, and the evolution of complex life cycles. In *Size-Structured Populations: Ecology and Evolution*, B. Ebenman and L. Persson (eds.), pp. 60–84. Springer-Verlag, Berlin.

Werner, P. A. and H. Caswell. 1977. Population growth rates and age versus stage-distribution models for teasel (*Dipsacus sylvestris* Huds.) *Ecology* **58**:1103–1111.

Werner, E. E. and J. F. Gilliam. 1984. The ontogenetic niche and species interactions in size-structured populations. *Ann. Rev. Ecol. Syst.* **15**:393–426.

Werner, E. E., J. F. Gilliam, D. J. Hall, and G. G. Mittelbach. 1983. An experimental test of the effects of predation risk on habitat use in fish. *Ecology* **64**:1540–48.

Werner, E. E. and D. J. Hall. 1974. Optimal foraging and size selection of prey by bluegill sunfish (*Lepomis macrochirus*). *Ecology* **55**:1042–1052.

Werner, E. E., G. G. Mittelbach, D. J. Hall, and J. F. Gilliam. 1983. Experimental tests of optimal habitat use in fish: the role of relative habitat profitability. *Ecology* **64**:1525–39.

Wilbur, H. M. 1988. Interactions between growing predators and growing prey. In *Size-structured Populations: Ecology and Evolution*. B. Ebenman and L. Persson (eds.), pp. 157–172. Springer-Verlag, Berlin.

Zastrow, C. E., E. D. Houde, and E. H. Saunders. 1989. Quality of striped bass (*Morone saxatilis*) eggs in relation to river source and female weight. *Rapp. P. V. Reun. Cons. Int. Explor. Mer* **191**:34–42.

12

A Vitality-Based Stochastic Model for Organism Survival

James Jay Anderson

ABSTRACT. Survivorship curves, describing the fraction of a cohort alive as a function of age, have three typical forms; linear, convex, and concave depending on whether mortality is constant over all life or is greatest in early or late life. A successful survivorship model must be able to describe these forms. In addition, model parameters should be biologically meaningful in a self-consistent and rigorous framework. Few models, if any, have achieved these goals. In this paper I present a novel individual-based survivorship model that fits virtually all survivorship data tested and has parameters with clear intuitive meanings. Survivorship curves of organisms are described in terms of an individual's vitality and accidental mortality rate. Mortality is thus partitioned into two parts: one dependent on an individual's history and another independent of it. The balance of processes is characterized by a ratio of population half-life from accident only and vitality-related death only. Vitality dynamics are described by a random walk through two parameters: initial vitality and vitality rate of change. Accidental mortality is described by a Poisson process characterized by average time between accidental deaths. Parameters are estimated by fitting the model to survivorship data using a Marquardt nonlinear least squares technique. The model fits survivorship curves well, including examples of plants, insects, molluscs, fish, birds, mammals, and humans. Initial vitality varied by two orders of magnitude and increased linearly with adult weight for homeotherms. The rate of vitality change and accidental mortality rate varied by one and three orders of magnitude respectively. For different races of an insect, parameters varied within a factor of four.

Introduction

The pursuit of a mathematical understanding of mortality began in the seventeenth century with the work of John Graunt (1662), who published the first demographic study for the City of London. Now, as then, a practical need exists for a quantitative understanding of mortality in populations if we are to sustain our world and its ecological balance. In pursuit of laws of mortality a concept has emerged that survival is structured by a force of mortality describing a cohort decline in terms of the fraction of the cohort alive. The idea originated with Gompertz (1825), who assumed the force of mortality exponentially increases with age. The idea of a driving force has been so appealing that it has taken on a stature nearly equal to the forces of physics. Statistical and mechanistic models of mortality invariably begin with the force and attempt to explain and model survivorship curves in terms of changes in the force of mortality with organism age. Statistical models generally do not establish a biological basis for the force of mortality. Their utility is in how well they fit survivorship curves. Reviews of statistical models by Manton and Stallard (1984), Keyfitz (1982), and Pollard (1973), for example, are abundant with algebraic equations describing the force of mortality. Mechanistic models, in comparison, look for deeper meanings to this force, and for four decades applied mathematicians have searched for an underlying law of mortality. Economos (1982), in a review of the history of this search, noted that a seminal paper by Simms (1942) set the foundation for a series of models in which the force of mortality was described with stochastic models of an organism's physiological ability to maintain its homeostatic equilibrium during aging and the influence of exogenous processes (Sacher and Trucco 1962; Strehler 1960; Strehler and Mildvan 1960, Sacher 1966, Woodbury and Manton 1977, Yashin et al. 1985, Yashin et al. 1989). Another approach, taken by Economos (1982) and extended by Piantanelli (1986), considers the change of an organism's vitality with age in relation to its physiology. Mortality occurs when vitality drops below a critical threshold.

From these studies, general laws of mortality have not emerged, although there is a commonality of approach in stochastically relating mortality to organism physiology. Although stochastic models provide good fits to survivorship curves, they are mathematically complex and their biological foundations have been questioned (Economos 1982). Vitality based models along the lines of Economos also fit observations, but their treatment of vitality dynamics is piecemeal in that the rate of change of vitality with age and its statistical distributions do not evolve from a single underlying dynamic equation.

In this paper I apply the concept of vitality in a new and more integrated manner through a stochastic equation. It differs from earlier models in that it does not use the force of mortality. Instead it is based on an abstract concept of organism vitality that has no exact connection with physiology or the concept of homeostasis. This model gives a simple partition of mortality resulting from vitality related causes and accidental causes in terms of three parameters.

The Concept

I begin by categorizing mortality into two classes: that dependent on the past history of the organism and that independent of the past history. The first type is designated *vitality-related mortality*. Vitality is an abstract holistic property that can be accumulated and lost. It changes with moment-to-moment experiences of the organism and is composed of a combination of factors. An organism's physiological resistance to disease, level of stress, behavior, success and failure in feeding, predator attacks, mating, parental care, and habitat choice all provide incremental changes in vitality. In this model, vitality is stochastic and mortality occurs if vitality reaches zero. Mortality can also occur independent of an organism's vitality. This will be referred to as *accidental mortality*. Examples include harvest of a population, and catastrophic events. The important feature is that accidental morality rate is independent of past or present organism condition. Predation where predator success depends in part on prey body size, would be categorized as vitality-related mortality. Indiscriminate predation on a group of prey would be categorized as accidental mortality.

The model can be envisioned as a modified form of the gamblers ruin process (Karlin and Taylor 1975). Assume a gambler plays roulette at a casino. Beginning with an initial sum of money his wealth incrementally increases and decreases as he wins and looses with each bet. The gambler is compulsive and will not stop on his own accord. But since the odds favor the casino he will eventually stop when he loses all his money. He may also stop if his friend finds him. We can determine the probability of when the gambler stops as a function of probabilities of losing his money and his friend finding him. The probability of his bankruptcy can be described by a random walk that depends on his initial bankroll, the value of his bets, and the betting odds. The probability of his friend finding him is independent of these things.

The model for mortality follows the same processes. An individual plays the game of survival against nature. It begins with an initial vitality and in each instant of time its actions either increase or decrease its vitality. Death ends the game, and this happens either when the individual's vitality reaches zero or it dies from accidental causes. The probability of survival, P, can be expressed in terms of the product of the probability of not dying from vitality related causes, P_v, and the probability of not dying from accidental causes, P_a, and is expressed:

$$P(t) = P_v(t)P_a(t). \tag{1}$$

We now develop these two probabilities separately.

Vitality-Related Mortality

For vitality-related survival assume vitality changes from a combination of individual processes according to the equation:

$$\frac{dv}{dt} = \Sigma r_i, \tag{2}$$

where v is organism vitality, t is its age, and r_i is vitality rate of change due to the ith process. Vitality is a dimensionless quantity and r has a dimension of t^{-1}. The combination of processes affecting vitality is assumed to be additive. For example, natural aging is assumed to have a contribution to vitality rate of change that adds to the contribution of disease and stress. The assumption of rate additivity is analogous to additivity of physical forces, and its justification, as with the physical counterpart, is based on how well the model describes the data.

Processes that change vitality must be variable in time. Disease occurs randomly, feeding activity and predation occur mostly with a diurnal periodicity. Thus, we can assume the combined effect of all processes may be represented by an average rate and a fluctuating rate characterizing moment-to-moment variations about the average. The combined vitality rates are expressed in terms of deterministic and stochastic rates by the equation:

$$\Sigma r_i = r + W(t), \tag{3}$$

where r is the sum of the deterministic rates, and $W(t)$ is a white noise process describing rapid fluctuations of the rate. White noise has a zero mean value and is uncorrelated so:

$$E[W(t)] = 0, \tag{4}$$
$$E[W(t)W(t')] = \sigma^2\delta(t - t'), \tag{5}$$

where $\delta(t - t')$ is a Dirac delta function and σ^2 is the incremental variance or intensity of the white noise process and has a dimension of t^{-1} (for reference see: Karlin and Taylor 1981, Goel and Richter-Dyn 1974, Gardiner 1985).

The stochastic differential equation for vitality rate of change becomes:

$$\frac{dv}{dt} = r + W(t), \tag{6}$$

where organism vitality as a function of time is a random variable with a continuous sample path. With this model, vitality is a Brownian process in which individual paths of v over time are continuous but random. The equation for the vitality rate of change relates past events with the present environment. Since vitality is a random variable, each path is different, so in a common environment each member of a cohort will have a different vitality. The statistical properties of an individuals's vitality is thus considered through the probability distribution of vitality with age. The probability is conditional, such that the probability density of v at time t is dependent on the vitality, v_0, at time t_0. This is denoted:

$$p = p(v, t|v_0, t_0). \tag{7}$$

The conditional probability of a random variable with dynamics defined by a stochastic differential equation can be expressed by a Fokker-Planck equation, which describes change in probability of measure v according to the rate of growth of its mean and variance. Details on Fokker-Planck equations can be obtained from a number of sources including: Karlin and Taylor 1981, Goel and Richter-Dyn 1974, Gardiner 1985, Risken 1984, Wax 1954. Assuming that r and σ are constant over the interval 0 to t the Fokker-Planck equation for the rate of change of the conditional probability density of a process with dynamics described by Eq. (6) is:

$$\frac{\partial p}{\partial t} = -r \frac{\partial p}{\partial v} + \frac{\sigma^2}{2} \frac{\partial^2 p}{\partial v^2}. \tag{8}$$

The equation describes the change in probability density of v in terms of the rate of change of the mean of v due to the deterministic rate and change in the variance of v due to the stochastic rate. To solve Eq. (8) we require boundary conditions. Mortality is incorporated into the model by expressing the lower boundary condition as an absorbing boundary so that the process stops when it reaches the boundary. Since vitality is an abstract and relative variable we can define the absorbing boundary condition at $v = 0$. The absorbing boundary is then specified:

$$p(v, t|0, t_0) = 0. \tag{9}$$

This merely states that the probability of reentering the region, $v > 0$, from the boundary is zero. In our terms, once dead forever dead. Assuming no maximum value to vitality the upper boundary condition is set at infinity so:

$$p(\infty, t|v_0, t_0) = 0. \tag{10}$$

With these boundary conditions and the assumption that r and σ are constant in time the solution of the above Fokker-Planck equation is (Cox and Miller 1965):

$$p(v, t|v_0, t_0) = \frac{1}{\sqrt{2\pi t\sigma^2}} \left(exp \frac{-(v - v_0 - rt)^2}{2t\sigma^2} \right.$$
$$\left. - exp\left\{ \frac{-(v + v_0 - rt)^2}{2t\sigma^2} - \frac{2rv_0}{\sigma^2} \right\} \right), \tag{11}$$

where v_0 is the initial vitality at $t_0 = 0$, r is the deterministic rate of change of vitality, and σ^2 is the intensity of the stochastic rate of change of vitality. In this model Eq. (11) describes the probability density of vitality over an interval of age in which the vitality rate is constant in a stochastic sense. Thus, by definition, a life state is an interval in which processes determining mortality are constant. In many situations the stage can be taken as the organism's life span. To describe survivorship curves in greater detail several stages may be required. For example, an organism with distinct larval and adult stages might require different parameters for each stage. In addition, parameters might also change due to environmental influences independent of an organism's life stage. In this situation stages would need to be defined in terms of both time and age. The present analysis assumes a single life stage model.

The probability that an organism is alive at a given time is expressed by the cumulative probability of vitality. This is obtained by integration of Eq. (11) over possible values of vitality so:

$$P_v(t) = \int_0^\infty p(v, t|v_0, 0)dv, \tag{12}$$

where for convenience we let $t_0 = 0$ be the initial time for the life stage. If we let:

$$\xi_1 = \frac{v - v_0 - rt}{\sqrt{2t\sigma^2}} \quad \xi_2 = \frac{v + v_0 - rt}{\sqrt{2t\sigma^2}}, \tag{13}$$

then the integration of Eq. (11) is represented as:

$$P_v(t) = \frac{1}{\sqrt{\pi}} \int_{\xi_1(0)}^\infty e^{-\xi_1}d\xi_1 - \frac{1}{\sqrt{\pi}} exp\left(-\frac{2rv_0}{\sigma^2} \right) \int_{\xi_2(0)}^\infty e^{-\xi_2}d\xi_2. \tag{14}$$

Noting the complementary error function is defined:

$$erfc(z) = \frac{2}{\sqrt{\pi}} \int_z^\infty e^{-\xi^2} d\xi, \tag{15}$$

then Eq. (14), expressing the probability of survival related to vitality is:

$$P_v(t) = \frac{1}{2} (erfc(u_1) - erfc(u_2)e^{-u_3}), \tag{16}$$

where the coefficients are defined:

$$u_1 = -\frac{v_0 + rt}{\sqrt{2t\sigma^2}} \quad u_2 = \frac{v_0 - rt}{\sqrt{2t\sigma^2}} \quad u_3 = \frac{2rv_0}{\sigma^2}. \tag{17}$$

Accidental Mortality

The effect on survival of mortality that is accidental and independent of an organism's vitality, is described by a Poisson process. This assumes that the probability of accidental mortality is small and events are randomly distributed. The probability that an organism has not experienced accidental mortality is the probability of observing zero events in time t and can be expressed:

$$P_a(t) = e^{-kt}, \tag{18}$$

where k is the rate coefficient for accidental mortality and has a dimension of t^{-1}. The average time to accidental mortality is k^{-1}.

Survival Probability

The survival probability to age t is defined from Eq. (1), Eq. (16), and Eq. (18). Since the parameter σ always appears in conjunction with v_0 or r, the survival equation can be normalized by σ^2 to give:

$$P(t) = \frac{1}{2} \left[erfc \frac{-V_0 - Rt}{\sqrt{2t}} - erfc \frac{V_0 - Rt}{\sqrt{2t}} e^{-2V_0R} \right] e^{-kt}, \tag{19}$$

where the *normalized initial vitality* and the *normalized vitality rate* of change are defined:

$$V_0 = \frac{v_0}{\sigma} \quad R = \frac{r}{\sigma}, \tag{20}$$

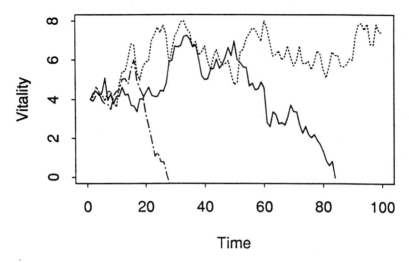

Figure 12.1. Three examples of random trajectories of vitality over time.

and k is the rate of accidental mortality. The parameter V_0 has a dimension of $t^{1/2}$ and R has a dimension of $t^{-1/2}$.

Cohort Survival

For cohort survival consider that each organism has a different initial vitality. Describing the initial distribution of normalized vitality by the function $I(V_0)$, cohort survivorship is expressed by the integral:

$$S(t) = \int_0^\infty I(V_0)P(t)dV_0. \tag{21}$$

In simple situations initial vitality is constant in all organisms, so $I(V_0') = \delta(V_0 - V_0')$ is a Dirac delta function and Eq. (21) reduces to Eq. (19). In this paper Eq. (19) represents cohort survival.

Model Characteristics

The model describes survival and mortality in terms of vitality loss and accidental events. Vitality evolves over time according to Eq. (6). Since this is a stochastic equation, each sample path is random and is described by a Wiener process, which is continuous but very kinky and is nowhere differentiable (Fig. 12.1). Paths start at V_o and stop when $V = 0$, representing mortality. Accidental mortality stops a path before it reaches zero. Path characteristics are set by V_o and R.

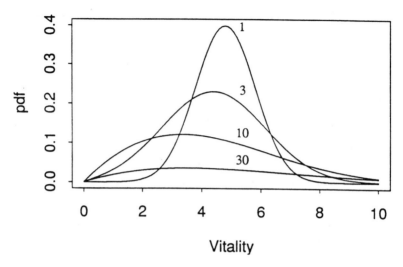

Figure 12.2. Probability density function of vitality at four times. Initial vitality 5.

The vitality probability density function changes with time according to Eq. (11). Over time the function spreads, the mode, and area under the curve decrease (Fig. 12.2) as organisms gain or lose vitality. Area is proportional to number of organisms surviving without accidental mortality.

The probability of survival as described by Eq. (19) is dependent on three parameters: V_0, R, and k (initial vitality, rate of change of vitality, and the coefficient of accidental mortality). For selected values of the parameters (Table 12.1) concave, linear, and convex survival curves can be generated (Fig. 12.3). Strongly concave shapes (curve A), are generated when R/V_0 ~ -1. These curves have a short initial stage with high survival followed by an exponential-like decrease in survival resulting from a loss of vitality. Highly convex shapes (curve D) are produced when initial vitality is large and vitality related mortality occurs later in life. Curves B and C represent situations where both accidental- and vitality-related mortalities occur. Curve

Table 12.1. Model parameters for curves illustrated in Fig. 12.3

Curve	V_0	R	k
A	2	−2	0
B	5	−2	0.2
C	15	−2	0.2
D	15	−2	0
E	1	0.5	0

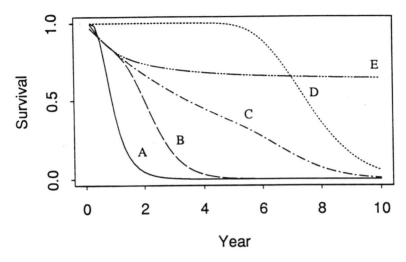

Figure 12.3. Survivorship curve for different model parameters.

B has a clear inflection point at about 2 years where vitality-related mortality dominates over accidental mortality. In curve C accidental mortality dominates over the majority of the organism life. Curve E represents a situation where accidental mortality is zero and the rate of change of vitality is positive. In this situation there is an initial loss as organisms with high vitality loss rates die off. Remaining organisms suffer little mortality. This situation is observed in dose-response experiments where a toxicant kills only weak organisms while strong ones survive.

If accidental mortality is eliminated, so $k = 0$, several characteristics of the survivorship curve can be approximated by simple empirical relationships dependent on V_0 and R. For the domain $R < 0$, the population half-life, that is the time, T_{50v}, for survival to drop to 50% from the effect of vitality-related mortality only, can be approximated by the equation:

$$T_{50v} \approx \frac{V_0^2}{0.5 - V_0 R}. \tag{22}$$

From an analysis of residuals the estimate of T_{50v} in Eq. (22) was determined to be within 8% of the correct value for $R < 0$ and within 1% for $R < -1$. The slope of the survivorship curve at T_{50v} for $R < 0$ can be approximated by the equation:

$$\left. \frac{dP}{dt} \right|_{P=0.5} = -0.4 V_0. \tag{23}$$

The asymptotic level of survival when $R > 0$ and $k = 0$ can be developed from. Eq. (19) by setting $t = \infty$ and noting that $erfc(-\infty) = 2$ and $erfc(\infty) = 0$. The asymptotic survival is:

$$P(\infty) = 1 - e^{-2V_0 R}. \tag{24}$$

The above relationships are useful for developing initial estimates of model parameters for fitting functions when accidental mortality is zero.

The population half-life from accidental causes only is:

$$T_{50a} = -\frac{\log 0.5}{k} \tag{25}$$

The ratio of the population half-life from accidental mortality to the population half-life from vitality-related mortality provides a measure of the importance of the two processes.

Fitting The Model to Data

The model defined by Eq. (19) can be fit to survivorship curves using a standard nonlinear least squares fitting routine, the Marquardt Method (Press et al. 1988). The model to be fitted is defined by Eq. (19):

$$P = P(t, \varphi), \tag{26}$$

where $\varphi = \{V_0, R, k\}$.

The method defines a chi-square merit function and finds model parameters that minimize the function. The merit function is defined:

$$\chi^2(\varphi) = \sum_{i=1}^{N} \left[\frac{P_i - P(t_i, \varphi)}{\sigma_i} \right]^2, \tag{27}$$

where N is the number of data points for our case and P is the probability of survival defined by Eq. (19), P_i is an observed survival data point, and σ_i is the standard deviation about the point. The method searches the parameter space to find the set of parameters φ that minimize χ^2.

To search the parameter space the method requires the Hessian, which for Eq. (19) is defined by the following equations:

$$\frac{\partial P}{\partial V} = \left[\frac{e^{-U_1^2}}{\sqrt{2\pi t}} + e^{-2V_0 R} \cdot \left(\frac{e^{-U_2^2}}{\sqrt{2\pi t}} + R \cdot erfc(U_2) \right) \right] e^{-kt}, \qquad (28)$$

$$\frac{\partial P}{\partial R} = \left[\frac{te^{-U_1^2}}{\sqrt{2\pi t}} - e^{-2V_0 R} \cdot \left(\frac{te^{-U_2^2}}{\sqrt{2\pi t}} - V_0 \cdot erfc(U_2) \right) \right] e^{-kt}, \qquad (29)$$

and

$$\frac{\partial P}{\partial k} = -ke^{-kt}P, \qquad (30)$$

where:

$$U_1 = -\frac{V_0 + Rt}{\sqrt{2t}} \quad U_2 = \frac{V_0 - Rt}{\sqrt{2t}}. \qquad (31)$$

Uncertainties in model parameters are given in the fitting method and are a function of σ_i. In general, the standard deviation about the data is not exactly known for survivorship data used in the examples in this paper. In this case we can obtain relative error estimates by assuming all measurement errors have the same standard deviation and that the model describes the survival process. Then we can assign an arbitrary standard deviation to the data, $\sigma_i = 1$, and fit the model parameters by minimizing χ^2. Next we estimate the actual standard deviation of model fit to data, the residuals of the fit, with the formula:

$$s^2 = \frac{1}{N} \sum_{i=1}^{N} [P_i - P(t_i, \varphi)]^2. \qquad (32)$$

Uncertainties of model parameters, u_φ, are derived from the square roots of the diagonal elements of the covariance matrix. These uncertainties are computed in the fitting process. Using the model fit residuals given by Eq. (32) an estimate of parameter standard deviation is obtained with:

$$\sigma_\varphi = u_\varphi \cdot s. \qquad (33)$$

Although this approach prohibits an independent assessment of the goodness-of-fit, it does allow some kind of error to be assigned to the points (Press et al. 1988).

The fitting procedure requires initial model parameter estimates. One approach to obtain initial estimates is to assume $k = 0$ and use Eq. (22) and Eq. (23) to estimate the parameters as:

$$V_0 = 2.5P_t - 0.8 \quad R = \frac{1}{2} - \frac{V_0}{T_{50v}}, \tag{34}$$

where P_t is the slope of P with respect to t at $P = 0.5$. A second approach is to use initial parameters selected from species with similar life strategies (See Table 12.2 for examples). The selection of initial parameters also can be guided with model characteristics discussed above.

Examples

The model gives a good fit to survivorship curves of a variety of organisms including plants, insects, mollusks, fish, birds, mammals, and humans (Fig. 12.4–12.8 and Table 12.2). Tables 12.2 and 12.3 give parameter estimates for a variety of survival curves using the Marquardt method. Model fit residuals are small and have normal distributions. Parameter uncertainties generally are less than 10%.

The model fits a large variety of plant types typified by nearly linear survivorship curves. To achieve this shape both vitality-related and accidental mortality must occur with equal rates. For the survivorship curve of the grass illustrated in Fig. 4 the half-life ratio of accidental to vitality-related mortalities is near one (Table 12.2). For buttercups vitality-related mortality is dominant as reflected by the half-life ratio of 5 (Table 12.2).

Survivorship curves for insects often have an initial linear decrease in survival, followed by a rapid mid-life decline and a leveling off at older ages (Fig. 12.5). The model fits this data with high vitality loss and accidental mortality rates. The half-life ratio of the adult mosquito *Aedes aegypti* (L.) was 4.9 (Table 12.2). This suggests vitality-related mortality dominated. A high accidental mortality rate may not be expected in this data since the survivorship curve was determined for caged insects.

Survivorship curves for birds generally decrease in an exponential-like manner (Fig. 12.6). The model fits this feature with a low initial vitality and vitality loss rate. Accidental mortality is also relatively important. A half-life ratio of 3 suggests vitality-related mortality dominates (Table 12.2).

Fish also exhibit an exponential-like decline in survival (Fig. 12.7). The model reproduces this pattern with a low initial vitality. The half-life ratio is 280 suggesting that, in fish, vitality-related mortality is much larger than accidental mortality (Table 12.2).

In mammals survivorship curves often decrease rapidly in old age. This is illustrated for a 1969 census of Australian men (Fig. 12.8). The initial

Table 12.2. Fit of Eq. (19) to survivorship data using the Marquardt method. The residual or standard deviation of the fit, s, is defined according to Eq. (32). Uncertainties, in parentheses ( ), are defined from Eq. (33), and N is the number of data points. Units for V_0, R, and k, are $yr.^{1/2}$, $yr.^{-1/2}$, and $yr.^{-1}$, respectively. T_{50v} and T_{50a} are population half-lives in years according to Eq. (22) and Eq. (25).

Name	V_0	R	k	N	s	T_{50v}	T_{50a}	T_{50a}/T_{50v}
plant	5.50	−0.76	0.137	8	0.025	6.5	5.1	0.8
Trichachne[1]	(0.98)	(0.15)	(0.01)					
plant	1.54	−0.45	0.066	9	0.014	2.0	10.5	5.3
Ranunuculus[2]	(0.08)	(0.14)	(0.11)					
rotifer	0.49	−23.58	2.258	12	0.033	0.02	0.3	15.4
Foloscularia[3]	(0.04)	(1.9)	(2.5)					
insect	1.21	−7.26	0.90	11	0.017	0.16	0.8	4.9
mosquito[4]	(0.07)	(0.36)	(0.20)					
mollusk	12.94	−1.03	0.14	8	0.016	12.1	4.9	0.4
Patella[5]	(3.04)	(0.27)	(0.01)					
fish	0.20	−0.33	0.035	20	0.005	0.07	19.8	280.2
Oncorhynchus[6]	(0.01)	(0.11)	(0.14)					
bird	1.36	−0.42	0.131	11	0.016	1.7	5.3	3.1
Vanellus[7]	(0.14)	(0.46)	(0.27)					
rodent	2.21	−2.84	0.341	13	0.019	0.7	2.0	2.8
Microtus[8]	(0.15)	(0.17)	(0.07)					
sheep	38.67	−3.52	0.26	8	0.028	10.9	2.7	0.2
Ovis[9]	(4.65)	(0.39)	(0.01)					
human	51.9	−0.75	0.002	20	0.019	68.3	346.6	5.1
Homo sapien[10]	(2.34)	(0.03)	(0.0002)					

[1]a grass in southern Arizona (see Fig. 12.4). (Sarukhan and Harper 1973).
[2]European buttercup *R. acis*. (Sarukhan and Harper 1973).
[3]Sessile rotifer *F. conifera* near New Haven Conn. (Edmondson 1945)
[4]female (see Fig. 12.5). (Crovello and Hacker 1972).
[5]South African limpet *P. cochlear*. (Branch 1975)
[6]juvenile sockeye salmon *Oncorhynchus nerka* (see Fig. 12.7) (Foerster 1938).
[7]lapwing greater than 6 months old, *V. vanellus* (see Fig. 12.6) (Kraak et al. 1940).
[8]European field vole, *M. agrestis* (Leslie and Ransom 1940).
[9]Dall sheep *O. dalli dalli*, greater than 1 year old (Caughley 1966).
[10]Australian males, census in 1961 (see Fig. 12.8) (Pollard 1973).

linear decline is mainly from accidental death. The model achieves these properties with a high initial vitality and a small accidental death rate. The accidental mortality rate from the model is $0.002 \ yr^{-1}$. The accidental death rate of young men in the 1961 census was $0.0017 \ yr^{-1}$ (Pollard 1973). Thus, model and observed accidental death rates are in close agreement. The half-life ratio of 5 (Table 12.2) suggests vitality-related mortality dominates. In contrast, in one year plus Alaskan sheep (Table 12.2) accidental mortality

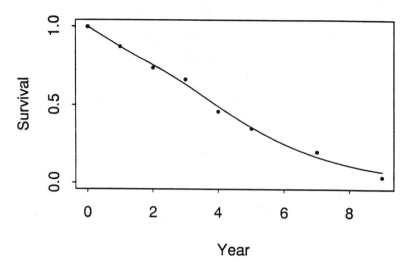

Figure 12.4. Survivorship curve for grass *Trichachne* (Saruknan and Harper 1973).

dominates (half-life ratio 0.2). The high accident rate in this species may result from hunting or natural accidents.

Table 12.3 illustrates how model parameters may vary for different strains of *Drosophila*. Initial vitality and vitality rate of change are essentially identical in three strains while accidental mortality rates are different. In two other strains initial vitality and vitality rate are lower, but the accidental rates

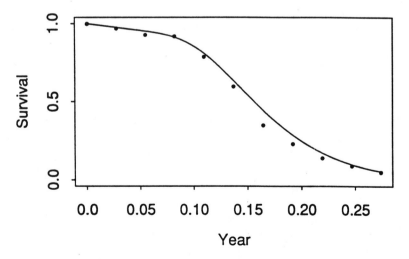

Figure 12.5. Survivorship curve for mosquito (Crovello and Hacker 1972).

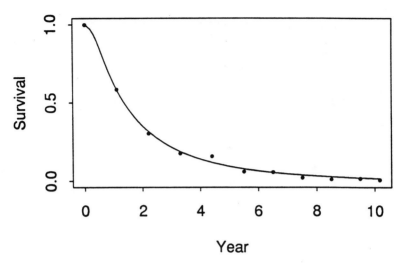

Figure 12.6. Survivorship curve for lapwing (Kraak et al. 1940)

are higher. In all strains accidental and vitality-related mortalities are of about equal significance.

Discussion

The separation of vitality-related and accidental mortality gives several

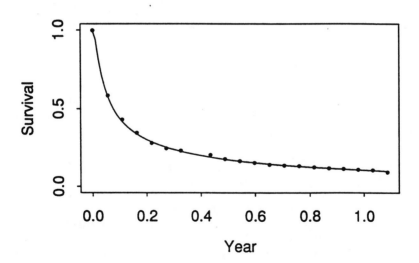

Figure 12.7. Survivorship curve for juvenile sockeye salmon (Foerster 1938).

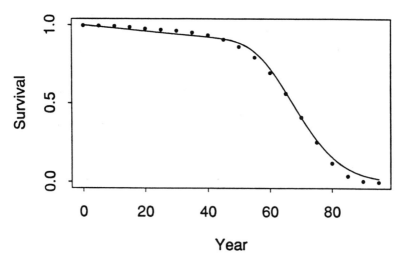

Figure 12.8 Survivorship curve for Australian men (Pollard 1973).

interesting results. With a human population the observed rate of accidental death and that predicted by the model were in agreement. This is evidence that, at least for this population, the model separates the two forms of mortality. In many species vitality-related mortality is dominant. For fish it appears to be very dominant. At first glance these results seems unrealistic since predation can be a significant cause of juvenile fish mortality. This result is supported by studies indicating survival is correlated with early life body size in fish (Miller et al. 1988, Werner and Gilliam 1984). The salient feature is that survival related to any past or current characteristic of the

Table 12.3. Model parameters for different geographical races of *Drosophila serrata* and *D. birchi*. See text and Table 12.1 for explanation of symbols. (From Birch et al 1963).

Race	V_0	R	k	s	T_{50v}	T_{50a}	T_{50a}/T_{50v}
D. serrata	1.92	−7.24	1.16	0.027	0.26	0.60	2.3
N. of Sydney	(0.17)	(0.62)	(0.21)				
D. serrata	1.96	−7.69	1.97	0.012	0.25	0.35	1.4
Brisbane	(0.09)	(0.36)	(0.10)				
D. birchi	1.93	−8.36	2.65	0.015	0.22	0.26	1.2
Cairns	(0.14)	(0.60)	(0.15)				
D. birchi	0.70	−3.00	3.66	0.016	0.19	0.19	1.0
Rabaul	(0.06)	(0.29)	(0.38)				
D. birchi	0.40	−2.26	4.08	0.033	0.11	0.17	1.5
Morsby	(0.08)	(0.65)	(2.67)				

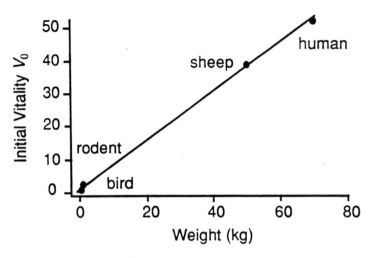

Figure 12.9 Initial vitality vs. adult weight in homeotherms. Vitality data from Table 12.2.

organism, such as growth or body size, is vitality-related by definition. In this context is it not surprising that fish have a high rate of vitality-related mortality. Evidence on growth derived from otolith rings of juvenile fish (Crowder et al., this volume) suggests that survivors grew more rapidly in early life than did the population as a whole. If the specific growth rate is proportional to vitality the relationship between growth and survival observed by Crowder et al. is what would be predicted by the vitality model.[1] Capture of fish in nets is an exception, though, since nets indiscriminately collect both weak and strong fish. In contrast to fish, the model indicated limpet survivorship curves were dominated by accidental mortality. This would result if predators attacked prey equally with high and low vitality. Thus it seems reasonable that a partition between the two mortality forms in part depends on the nature of predator-prey interactions. In situations where outcome of an interaction depends on prey escape response, vitality should affect the mortality rate. In situations where predation is random and the prey has no defense against the predator, mortality would be categorized as accidental mortality.

An additional result is suggested for homeotherms. It appears that initial vitality increases with adult weight in a linear manner (Fig. 12.9). It is well known that many physiological and ecological rates also vary with body size (Calder 1984). Some of these well established empirical relationships may provide further evaluation of this model.

---

[1]A manuscript is in preparation using the vitality model to couple growth and survival.

Conclusions

Survivorship curves, describing the fraction of a cohort alive as a function of age, have three typical forms: linear, convex, and concave. With linear curves the rate of mortality is constant for all ages. With concave curves mortality is greatest in early life stages and is approximately proportional to the number of organisms surviving, and with convex curves mortality is low in early life stages and increases with age. Many survivorship curves exhibit a mixture of these patterns. For example, in many species early life mortality rates are low, increase in mid-life and decrease again in late life to produce an inverse-sigmoid form. A successful survivorship model must be able to describe these diverse patterns. In addition, a successful model should provide meaningful biological interpretations to its parameters through a self-consistent and rigorous framework. Few models, if any, have achieved these goals. In this paper I present a new survivorship model that fits virtually all survivorship data tested and is derived from a dynamic model that provides clear and intuitive meanings to the model parameters.

The model treats two types of mortality: one depends on organism vitality and the other is the result of accidents. An essential feature of vitality is that it changes over time according to an individual's experience and, as such, it reflects the dependence of the past on an individual's ability to survive. Change in vitality is treated as a random walk process, the Wiener process, and depends on two parameters: an initial vitality, and the rate of vitality change. Accidental mortality, by definition, is random and independent of past events. It is described by a Poisson process and is characterized by the average time between accidents. The model thus describes survivorship curves in terms of two processes using three parameters. It seems evident that every successful survivorship curve must in some way encompass these two processes. It also follows that a survivorship curve must have at least three parameters. Accidental mortality can be described by one parameter, the average time between accidents. But the process dependent on past events requires a minimum of two parameters; an initial value and a rate of change over time. Two parameter Gompertz-type models are insufficient to describe the two mortality processes. Many higher parameter models often do not provide significantly better fits to data, and their biological relevance is often unclear.

This model is unique from other vitality based models in that it does not apply a homeostasis, in which vitality is interpreted as the ability to return to homeostasis. The model is also unique in not explicitly incorporating energetics, a foundation of ecological theory. Individuals with more vitality are better able to survive than individuals with less vitality. Vitality is gained and lost, like money, thus the focus is in describing its dynamics. Although this approach is novel the vitality model is the most reduced form of an individual based model in that it combines into a single measure all elements that distinguish the individual. If an ecological theory can be developed in terms of this single variable the task of relating growth, mortality, fecundity, and other basic ecological processes might be immeasurably simplified. It remains to be seen if this level of simplification is possible. It is clear the model fits data well, and the next step is to relate the parameters describing the dynamics of vitality to measurable properties of individuals such as growth, fecundity, toxic exposure, population density, and predator number. The limited analysis thus far indicates that for mammals, initial vitality increases with body weight. Also, a relationship between vitality rate and toxicant levels can be demonstrated (manuscript in preparation). These initial studies suggest at least partial success, and have not ruled out the possibility of developing a simplified ecological theory based on the dynamics of vitality.

Acknowledgments

This work was supported by the US Environmental Protection Agency under Grant R811348, the University of Washington College of Ocean and Fishery Sciences, and the Bonneville Power Administration.

Literature Cited

Birch, L. C., T. Dobzhansky, P. O. Elliott, and R. C. Lewontin. 1963. Relative fitness of geographical races of *Drosophila serrata*. *Evolution* **17**:72–83.

Branch, G. M. 1975. Interspecific competition in *Petella cochlear* Born. *J. Anim. Ecol.* **44**:263–281.

Calder, W. A. 1984. Size, Function and life history. Harvard Univ. Press, Cambridge p. 396.

Caughley, G. 1966. Mortality patterns in mammals. *Ecology* **47**:906–918.

Cox, D. R. and H. D. Miller. 1965. *The Theory of Stochastic Processes*. Wiley, New York.

Crovello, T. J. and C. S. Hacker. 1972. Evolutionary strategies in life table characteristics among fernal and urban strains of *Aedes aegypti* (L.). *Evolution* **26**:185–196.

Economos, A. C. 1982. Rate of aging, rate of dying, and the mechanism of mortality. *Arch. Gerontol. Geriatr.* **1**:3–27.

Edmondson, W. T. 1945. Ecological studies of sessile *Rotatoria:* II. Dynamics of populations and social structures. *Ecol. Monogr.* **15**:141–172.

Foerster, R. E. 1938. Mortality trend among young sockeye salmon (*Oncorhynchus nerka*) during various stages of lake residence. *J. Fish. Res. Board Can.* **4(3)**:184–191.

Gardiner, C. W. 1985. *Handbook of Stochastic Methods for Physics, Chemistry and the Natural Sciences.* Springer-Verlag, Berlin.

Goel, N. S. and N. Richter-Dyn. 1974. *Stochastic Models in Biology.* Academic Press, New York.

Gompertz, B. 1825. On the nature of the function expressive of the law of human mortality. *Philos. Trans. of the Royal Society* **115**:513–585.

Graunt, J. 1662. *Natural and Political Observations Mentioned in a Following Index and Made upon the Bills of Mortality, by John Graunt Citizen of London . . . London.* Printed by Tho: Roycroft, for John Martin, James Allestry, and Tho: Dicas at the sign of the bell in St. Paul's churchyard. MDCLXII.

Karlin, S. and H. M. Taylor. 1975. *A First Course in Stochastic Processes.* Academic Press, New York.

Karlin, S. and H. M. Taylor. 1981. *A Second Course in Stochastic Processes.* Academic Press, New York.

Keyfitz, N. 1982, Choice of function for mortality analysis: Effective forecasting depends on a minimum parameter representation. *Theor. Popul. Biol.* **21**:329–352.

Kraak, W. K., G. L. Rinkel, and J. Hoogenheide. 1940. Oecologische bewerking van de Europese ringgegevens van der Kievit *Vanellus vanellus* (L.). *Ardea* **29**:151–157.

Leslie, P. H. and R. M. Ransom. 1940. The mortality, fertility, and ratio of natural increase in the vole (*Microtus agrestis*) as observed in the laboratory. *J. Anim. Ecol.* **9**:27–52.

Manton, K. G. and E. Stallard. 1984. *Recent Trends in Mortality Analysis.* Academic Press, Orlando, FL.

Miller, T. J., L. B. Crowder, J. A. Rice, and E. A. Marshall. 1988. Larval size and recruitment mechanisms in fishes: Towards a conceptual framework. *Can. J. Fish. Aquat. Sci.* **45**:1657–1670.

Piantanelli, L. 1986. A mathematical model of survival kinetics. I. Theoretical basis. *Arch. Gerontol. Geriatr.* **5**:107–118.

Pollard, J. H. 1973. *Mathematical Models for the Growth of Human Populations.* Cambridge University Press, Cambridge, England.

Press, W. H., B. P. Flannery, S. A. Teukolsky, and W. T. Vetterling. 1988. *Numerical Recipes in C: The Art of Scientific Computing.* Cambridge University Press, Cambridge, England.

Risken, H. 1984. *The Fokker-Planck Equation.* Springer, Berlin.

Sacher, G. A. 1966. The Gompertz transformation in the study of the injury-mortality relationship: application to late radiation effects and aging. In *Radiation and Aging*, P. L. Lindop and G. A. Sacher, (eds.), pp. 411–445. Taylor and Francis Limited, London.

Sacher, G. A. and E. Trucco. 1962. The stochastic theory of mortality. *Ann. New York Acad. Sci.* **96**:985–1007.

Sarukhan, J. and J. L. Harper. 1973. Studies on plant demography: *Ranunculus repens* L. and R. *acris* L. I. Population flux and survivorship. *J. Ecol.*, **61**:675–716.

Simms, H. S. 1942. The use of a measurable cause of death (hemorrhage) for the evaluation of aging. *J. Gen. Physiol.* **26**:169–178.

Strehler, B. L. 1960. Fluctuating energy demands as determinants of the death process. (A parsimonious theory of the Gompertz function). In *The Biology of Aging*, B. L. Strehler, (ed.), pp. 309–314. American Institute of Biological Sciences, Washington D.C., USA.

Strehler, B. L. and A. S. Mildvan. 1960. General theory of mortality and aging. *Science* **132**:14–21.

Wax, N. 1954, *Selected Papers on Noise and Stochastic Processes*. Dover, New York.

Werner, E. E. and J. F. Gilliam 1984. The ontogenetic niche and species interactions in size-structured populations. *Ann. Rev. Ecol. Syst.* **15**:393–425.

Woodbury, M. A. and K. G. Manton. 1977. A random walk model of human mortality and aging. *Theor. Popul. Biol.* **11**:37–48.

Yashin, A. I., K. G. Manton, and E. Stallad. 1989. The propagation of uncertainty in human mortality processes operating in stochastic environments. *Theor. Popul. Biol.* **35**:119–141.

Yashin, A. I., K. G. Manton, and J. W. Vaupel. 1985. Mortality and aging in a heterogeneous population: A stochastic process model with observed and unobserved variables. *Theor. Popul. Biol.* **27**:154–175.

[1]A manuscript is in preparation using the vitality model to couple growth and survival.

13

Evaluating the Size Effects of Lampreys and Their Hosts: Application of an Individual-Based Model

Neil A. MacKay

ABSTRACT. An individual-state configuration model was constructed that considers a size distribution of lampreys instead of an average size. Model simulations vary only in lamprey host size and bracket the observed field data. The model predicts that lamprey-induced mortality is seasonal and strongly pulsed, consistent with independent field studies and previous modeling work. Lamprey growth increases with host size. The model suggests that increases in lamprey size lead to nonlinear increases in lamprey-induced mortality, while increases in lamprey host size lead to nonlinear decreases in host mortality. The doubling in lamprey size that has occurred in the last twenty-five years could have resulted in 3 to 15 times greater lamprey-induced mortality. By allowing variance in lamprey size using the individual-based approach, predictions of lamprey-induced mortality increased by a factor of 2 or more over predictions derived from a lamprey of average size. Model modifications are suggested that would allow the model to estimate the functional response of lampreys in Lake Michigan.

Introduction

The sea lamprey (*Petromyzon marinus*) has been credited with reducing its prey populations, particularly salmonids, since it invaded the upper Great Lakes in the 1930's (Smith 1971, Smith et al. 1974, Smith and Tibbles 1980, Wells 1980). Prior to the lamprey's arrival, Lake Michigan supported large sport and commercial fisheries of lake trout (*Salvelinus namaycush*). By the mid 1950's a combination of exploitation pressure and lamprey predation had extirpated the lake trout from Lake Michigan. The discovery in 1958 of the lampricide TFM (3-trifluoromethyl-4-nitrophenol) allowed control of the sedentary, stream-dwelling larval lampreys (called ammocoetes), and hence controlled the adult population at 5–10% of its peak abundance (Smith and Tibbles 1980, Torblaa and Westman 1980, Hanson and Swink

1989). Since that time, little progress has been made on alternative control methods. Lamprey control still relies heavily upon the chemical treatment of streams (Lamsa et al. 1980, Sawyer 1980, Smith and Tibbles 1980). Meanwhile, the Lake Michigan sport fishery recovered and is now flourishing (Smith and Tibbles 1980, Talhelm 1988). Lake trout have been reintroduced and extensive stocking of Pacific salmonids that began in the 1960's has continued to the present (Kitchell 1990).

Though the fishery is economically strong (Talhelm and Bishop 1980, Talhelm 1988), many fish are still vulnerable to lamprey attack (Walters, Steer, and Spangler 1980). Forecasting lamprey-induced mortality can assist in determining salmonid stocking policy and fishery regulations. Normally, estimates of future mortality are predicted by linear extrapolation from present estimates. Modeling work, however, has shown that increases in lamprey size lead to nonlinear increases in lamprey-induced mortality (Kitchell 1990). Since the full implementation of the TFM control program in Lake Michigan in 1965 (Smith and Tibbles 1980), lamprey sizes have doubled (Heinrich et al. 1980), perhaps resulting in four or five fold increases in lamprey-induced mortality (Kitchell 1990). Small increases in average size and/or the relative abundance and size of "above average" individuals can have very large effects on host mortality rates. Decreases in the relative density of hosts can also lead to nonlinear increases in mortality. In their lamprey-lake trout simulation model, Lett et al. (1975) found that lamprey-induced mortality increased geometrically as lamprey population size increased linearly.

Prediction of host mortality is also thwarted by a threshold dilemma. As lamprey numbers increase, they may reach a point where they lead to nonlinear increases in host mortality, due primarily to the depletion of the larger, hardier hosts (Christie 1972, Farmer and Beamish 1973, Lett et al. 1975). As size-selective parasites, lampreys choose the largest hosts (Hall and Elliott 1954, Berst and Waino 1967, Davis 1967, Budd et al. 1969, Farmer and Beamish 1973, Cochran 1985, Henderson 1986), upon which they usually act as a parasite, not a predator (Christie 1974). In the absence of large hosts, a lamprey will attack small hosts, and will therefore act as a predator. This theory has been used to explain the collapse of the Lake Ontario lake trout fishery (Christie 1972). Walters, Steer, and Spangler (1980) suggested that depensatory lamprey-induced mortality may lead to catastrophic declines in host stocks. Nonlinearities such as these reinforce the dangers of linear extrapolation.

Another assumption of extrapolations is that there is a good estimate upon which to base predictions. This assumption may not be met in the case of lamprey-induced mortality, due to the difficulty in assessing their host's mortality (Walters, Spangler et al. 1980). Host mortality has been difficult to assess because: (1) while lamprey-induced mortality is most certainly size-related, field estimates of lamprey size are usually limited to times when the lampreys enter the lake as parasitic juveniles and when they leave the lake to ascend streams and spawn (12–18 months later), (2) dead hosts sink to the bottom of the lake where they are difficult to enumerate, and (3) mortality predictions based on average lamprey size would be inaccurate since lamprey-induced mortality is nonlinearly related to lamprey size.

The first problem can be addressed with bioenergetics modeling. The size of the lamprey throughout its parasitic phase can be estimated by knowing the initial and final mass of a lamprey, its diet, and its thermal regime (Hewett and Johnson 1987).

The second problem was addressed by Bergstedt and Schneider (1988) in their four year trawling survey for lake trout carcasses in Lake Ontario. This study is the only direct measurement of lamprey-induced mortality documented in the literature. The costs of such intense sampling preclude its adoption into the current Great Lakes monitoring protocol.

The third problem (effects of lamprey size on mortality rates) can be solved by considering variance in the size distribution of lampreys instead of just an average lamprey size.

A size distribution of lampreys can be modeled using an i-state configuration model (sensu Caswell, this volume). These models have proven useful in simulating size dependent interactions such as predator-prey processes in fish (DeAngelis et al. 1984, Adams and DeAngelis 1987, Huston et al. 1988, Trebitz 1989, Madenjian and Carpenter 1991, Madenjian et al. 1991). While lampreys are not gape-limited predators like the species previously modeled, their interactions with their hosts are size-based and therefore amenable to individual-based modeling.

I-state configuration models are useful in situations where data are lacking but mechanisms are better understood, because these models rely on fewer parameters. The predictions they provide are in easily measurable (and hence readily available) variables, such as size distributions. Their flexibility and the ease with which behavioral rules can be added allow the realism of stochasticity. Most importantly, they allow for variability at the level of the individual in key parameters such as size, which has an enormous impact on predator-prey interactions.

Methods

The objective of this study was to develop an individual-state configuration model that could be used to evaluate the effects of lamprey and host size on host mortality. Lamprey growth was represented with the bioenergetics equations of Kitchell and Breck (1980) set within the construct of a random encounter model. Simplicity was stressed throughout the design of the model to aid the interpretability of the results. The stochasticity of the simulation model created lamprey size distributions which could then be compared to field observations.

The bioenergetics of the sea lamprey are reasonably well known (Beamish 1973, Farmer et al. 1975, Farmer et al. 1977, Farmer 1980) and the bioenergetics parameters used here have been used successfully to predict the seasonal pattern of lamprey-induced mortality (Kitchell and Breck 1980) and the seasonal vulnerability of different sizes of lamprey hosts (Kitchell 1990). For a listing of the bioenergetics equations and parameters used in the model, see Kitchell and Breck (1980) or Hewett and Johnson (1987).

As lamprey growth is very sensitive to water temperature (Farmer et al. 1977; Kitchell and Breck 1980), seasonal temperature data are used for the simulation. Lamprey growth is the integrator of consumption in the bioenergetics model. A change in growth rate translates into a change in host blood consumed (Hewett and Johnson 1987).

The simulation model runs from January 1, which is the approximate time the lamprey enters the lake and begins feeding, until December 31, when feeding stops and the transformation to sexual maturity occurs (Applegate 1950; Hardisty and Potter 1971; Johnson and Anderson 1980). A cohort of 2000 lampreys begins the simulation at 10 grams, which is the average size at which transformation to parasitic feeding occurs. For purposes of simplicity, host size is uniform within a simulation, but varied between runs of the model. As a result, some of the nonlinear dynamics that occur in natural systems cannot be represented. The smallest host size is 650 grams, corresponding to the smallest hosts that lampreys select in the field (Bergstedt and Schneider 1988). This size corresponds to an age 3 lake trout in Lake Michigan (Stewart et al. 1983). The other host sizes modeled were 1000 g, 2000 g, and 3000 g, which span the majority of lake trout sizes attacked by the sea lamprey in Lake Michigan.

The Flowchart

The simulation model runs on a daily timestep, as illustrated in Figure 13.1. Each day the model selects each lamprey, one at a time. If the lamprey was not attached to a host at the end of the previous day, the individual searches for a host. If the lamprey doesn't find a host, then it loses mass according to the bioenergetic rules. If the lamprey finds a host, it attaches, rasps a hole in the host, and commences feeding. At the end of the day, the model calculates whether the rate of feeding and the recent history of blood removal are enough to kill the host. If so, the lamprey releases from the host and must search for a new host on the following day. If the host does not die, then the lamprey 'decides' (see below) whether or not it will feed on the host the next day or drop off immediately. If the lamprey drops off, then it must search for a new host the following day. If it remains on the host, then it begins the following day feeding on the host.

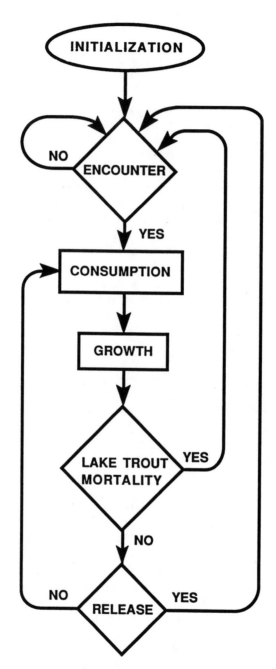

Figure 13.1. The simulation model flowchart. An arrow that points upward signals the start of the next day.

Stochastic Aspects of Model

Two processes in the model are stochastic: the search for a host, and the decision to drop off a host before the host dies. Every non-feeding lamprey has a 10% chance each day of finding a host. This value was assigned arbitrarily (for simplicity) but in fact should depend on densities of lampreys and their hosts in the field. Every attached lamprey has a 10% chance each day of deciding to detach from the host upon which it is feeding. Barring the premature death of the host, this rule results in a mean attachment time of ten days. In their simulation model, Lett et al. (1975) allowed the lampreys to find a new host (as long as one was available) every two days and set the feeding interval function as a normal distribution with a mean attachment time of ten days (again barring host death). While most hosts make no attempt to avoid approaching lampreys (small hosts sometimes do) (Farmer and Beamish 1973), hosts have often been reported to attempt to remove lampreys by brushing against the substrate (Farmer and Beamish 1973; Cochran 1986). This model does not allow hosts to force the separation, however.

Host Mortality

Host mortality was calculated using the regression of Farmer et al. (1975):

$$y = 2046 \cdot x^{-1.533},$$

where x is the percent of host blood removed each day by the lamprey, and y is the days until death. The host blood was estimated as 3.5% of the wet weight of the fish (Conte et al. 1963, Smith 1966). The blood removed daily is a function of lamprey size and the water temperature (Farmer et al. 1977, Kitchell and Breck 1980). Host death is defined as the time when the energy content of the host blood is only 16% of the original energy content, given that host blood volume remains constant due to osmotic influx of water (Farmer 1980). The model predicts host mortality only when the lamprey has been feeding on the host longer than the regression prediction of days until death.

Host Size and Availability

Because this version of the model is designed to evaluate the effects of lamprey size, the model contains a single host size per run. Host size effects are assessed by a series of independent simulations. Also, there is no limit on host availability and therefore no feedback of lamprey predation on host availability. These factors constrain the model from representing all of the nonlinearities and feedbacks that are present in this system, but aid the interpretability of the results.

Model Output

The model produces the size-frequencies of lampreys given a host size, and reports lamprey-induced host mortality at daily intervals. The model also records the number of hosts each lamprey kills, which can be used to determine the effect of lamprey size on host mortality, and conversely, to determine the effect of host size on lamprey growth.

Comparisons with Field Data

The size distribution of spawning run lampreys collected from several different tributaries of Lake Michigan in 1989 by the U.S. Fish and Wildlife Service was compared with the model-generated size distributions calculated with different host sizes.

Results and Discussion

The size distributions predicted by the model appear similar to the size distribution of spawning run lampreys collected from Lake Michigan in spring 1989 (Figs. 13.2, 13.3). Characterizations of the distribution confirm that model results resemble field data (Table 13.1), although no objective criteria were used. Simulations with larger hosts resulted in more dispersed distributions (larger standard deviation) than those with smaller hosts (Fig. 13.2, Table 13.1).

Lampreys exhibited better growth as the size of the host increased (Fig. 13.2). This is not unexpected since larger hosts can provide a food source for a longer duration than can smaller hosts that perish more quickly. Once the hosts dies, the lamprey must begin searching for a new host.

The model predicted that lamprey-induced mortality should be seasonally pulsed, being lowest in spring and early summer and maximal in late summer/early fall (Fig. 13.4). This result is consistent with field studies (Christie and Kolenosky 1980, Spangler et al. 1980, Bergstedt and Schneider 1988), laboratory studies (Farmer 1980, Hanson and Swink 1989), and other modeling studies (Lett et al. 1975, Kitchell and Breck 1980, Kitchell 1990). Because lamprey consumption is highly temperature and size dependent (Farmer et al. 1977), the late summer early fall period that combines large lamprey size with optimal feeding temperatures results in maximal consumption rates and therefore the highest levels of host mortality (Kitchell and Breck 1980).

Increases in lamprey size led to non-linear increases in lamprey-induced mortality at the larger host sizes (Fig. 13.5). As lampreys increased in size, they became more like predators than parasites. Their large size enabled them to endanger even the largest hosts. Smaller lampreys could not kill the largest hosts alone, and this model did not allow multiple attacks on a single host. These results are consistent with the predictions of Kitchell (1990), who estimated the period of host vulnerability (termed the scope for mortality) of different sized hosts along a gradient of increasing lamprey size.

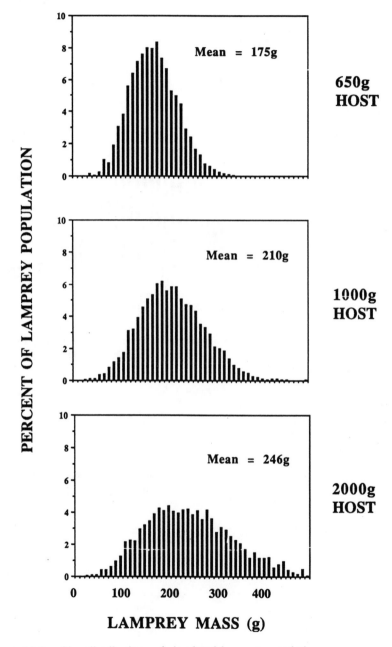

Figure 13.2. Size distributions of simulated lamprey populations.

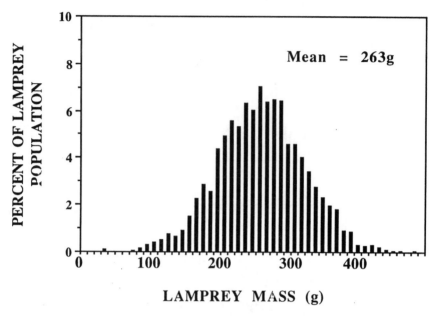

LAMPREY SIZE FREQUENCY DISTRIBUTIONS
LAKE MICHIGAN, 1989 FIELD DATA

Figure 13.3. The size distribution of spawning run lampreys collected from various tributaries of Lake Michigan in spring 1989 (J. Heinrich, U.S. Fish and Wildlife Service, Marquette, MI).

Table 13.1. Descriptive statistics of simulated and observed lamprey populations. Field data were for Lake Michigan lampreys collected in the spring spawning run of 1989 (J. Heinrich, U.S. Fish and Wildlife Service, Marquette, MI).

Host weight (g)	Standard Deviation	Mean lamprey weight (g)	Coefficient of Variation
650	49	175	0.28
1000	66	210	0.32
2000	90	246	0.36
3000	99	249	0.40
Field	61	263	0.23

LAMPREY-INDUCED HOST MORTALITY
THROUGH THE YEAR

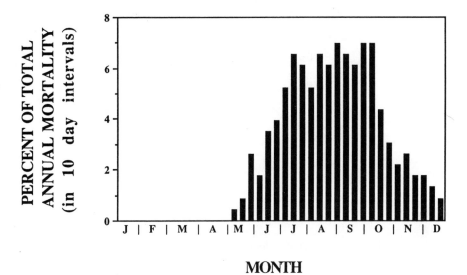

Figure 13.4. Predicted percentage of total annual mortality over the calendar year. Predictions based on individual-based model simulation. Bars represent mortality occurring in 10 day intervals.

As lamprey sizes have doubled over the last 25 years (Heinrich et al. 1980), this increase in size may have resulted in 3–15 times greater lamprey-induced host mortality. As presented in Table 13.2, mortality is dependent on the size of the host. This model considers only direct mortality, so large hosts are nearly impervious to attacks by small lampreys (i.e., cannot be killed). As the average host size increases, the relationship between lamprey-induced mortality and lamprey size becomes increasingly nonlinear (Fig. 13.5).

Comparing the model predictions of lamprey-induced mortality attributed to a lamprey of mean size to the average mortality caused by all lampreys in the size distribution reveals the benefit of the individual-based approach. If the relationship between lamprey-induced mortality and lamprey size is linear (as it is for small host sizes), then the individual-based approach would not provide any extra information because the variance in lamprey weight would not be significant. Because this relationship is nonlinear for large host sizes, considering a size distribution of lampreys results in twice the mortality predicted by modeling an average individual (Table 13.3).

The model showed that increases in host size led to nonlinear decreases

LAMPREY INDUCED MORTALITY VS. LAMPREY WEIGHT OVER DIFFERENT HOST SIZES

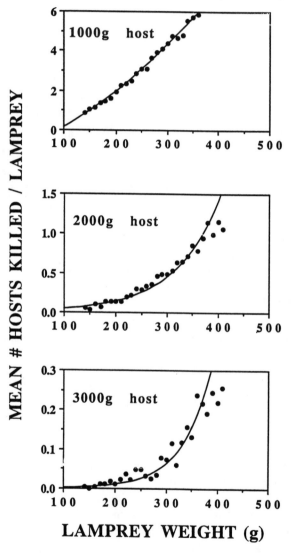

Figure 13.5. Predicted average number of host fishes killed per lamprey for hosts of different sizes. Predictions based on individual-based simulation model. A second order polynomial model is fit to the data in the 1000 g host panel. The curves for the 2000 g and 3000 g hosts are generated using exponential models.

Table 13.2. Comparison of the number of hosts killed annually predicted by the model over different host sizes as a result of the doubling in mean size of lamprey over the last 25 years. The factor column is the ratio of the predicted mortality of a 140 g lamprey to the predicted mortality of a 280 g lamprey.

| Host weight (g) | Hosts killed/lamprey Lamprey weight | | Factor |
	140 g	280 g	
650	3.08	9.62	3.1
1000	0.799	3.78	4.7
2000	0.045	0.425	9.4
3000	0.004	0.0625	15.6

in lamprey-induced mortality. The larger the host/lamprey size ratio, the less vulnerable the host, due to its larger blood volume and greater regenerative capabilities. If the lamprey is removing less than ten percent of the host's blood per day, then the host could survive for more than 60 days (Farmer et al. 1977). This knowledge led researchers to consider that lampreys may not be dangerous to fish populations if there is a sufficient number of large fishes to serve as a buffer for the smaller fish (Christie 1972, Lett et al. 1975).

This model considered direct mortality—that is, removing blood faster than the host can replenish it for a duration sufficient to kill the fish (Farmer

Table 13.3. Comparison of the predatory impact predicted by considering the whole size distribution of lampreys to the impact predicted by extrapolating mortality from the mean lamprey weight. The mean host mortality is the average number of hosts killed per lamprey for all of the lampreys. The host mortality of the mean lamprey is the number of hosts killed per average-sized lamprey. The factor column is the ratio of the two values. A ratio near 1 suggests that mortality is linear for that simulation. This factor is the underestimate of total mortality that would occur by modeling an average individual instead of considering a size distribution.

Host weight (g)	Mean host mortality (# killed/lamprey)	Host mortality of mean lamprey (# killed/lamprey)	Factor
650	4.40	4.44	0.99
1000	2.21	2.17	1.02
2000	0.365	0.292	1.25
3000	0.090	0.040	2.25

et al. 1977)—as equaling total lamprey-induced mortality, when in fact there is a substantial indirect component to lamprey-induced mortality. Laboratory studies have found that indirect mortality, here defined as host mortality following the release of the lamprey, could be as high 50% of the total mortality (Swink and Hanson 1986, 1989). This indirect mortality was often caused by secondary fungal infection and had been noted previously in the literature (Parker and Lennon 1956). As the model predicts that the number of attacks per lamprey increases as a function of lamprey size, indirect mortality could also be non-linearly related to lamprey size. If this model considered indirect mortality, even large hosts could have been vulnerable.

Future Considerations

Now that the basic relationships between lampreys and their hosts are better understood, future work could address more specific concerns. By limiting the number of hosts in the system, using an array of host sizes instead of a single host size per run, and creating a mechanistic encounter function, this model could predict lamprey functional response surfaces, a need specified by the Sea Lamprey International Symposium (Walters, Spangler et al. 1980) and the Workshop to Evaluate Sea Lamprey Populations (Heinrich et al. 1987). Based on individual-based model simulations presented here, the predation rate would depend not only on lake trout density but on lamprey size. Hence, a functional response surface would be appropriate.

The release function of the model also could be modified, taking into account optimal foraging arguments suggested by several papers on lampreys (Kitchell and Breck 1980, Lamsa et al. 1980, Cochran and Kitchell 1986, 1989). The model could then be calibrated with population estimates of lampreys and their prey in Lake Michigan. The percentage of hosts marked (i.e., bearing lamprey scars), a data set available for many years for Lake Michigan, would provide an independent measure of the accuracy of the model predictions.

If validated (Caswell 1976), this model could then be used to predict the consequences of various lamprey control strategies and their effects on host populations. The value of this tool might be increased if the model predictions were tested in conjunction with proposed lamprey control programs such as that currently planned for Lake Champlain.

Acknowledgments

I would like to thank Drs. Jim Kitchell and Chuck Madenjian, who advised me in model development and reviewed this manuscript. John Heinrich kindly provided the field data. I would also like to thank Brett Johnson and two anonymous reviewers whose comments strengthened this paper. Chris Carr provided programming assistance. This research was funded by the National Oceanic and Atmospheric Administration, Department of Commerce, Office of Sea Grant, through an institutional grant to the University of Wisconsin.

Literature Cited

Adams, S. M. and D. L. DeAngelis. 1987. Indirect effects of early bass-shad interactions on predator population structure and food web dynamics. In *Predation: Direct and Indirect Impacts on Aquatic Communities,* W. C. Kerfoot and A. Sih (ed.), pp. 103–117. University Press of New England, Hanover, N.H.

Applegate, V. C. 1950. Natural history of the sea lamprey, *Petromyzon marinus,* in Michigan. Special Sicence Report. *U.S. Fish Wildl. Serv. No.* **55**:1–237.

Beamish, F. W. H. 1973. Oxygen consumption of adult *Petromyzon marinus* in relation to body weight and temperature. *J. Fish. Res. Board Can.* **30**:1367–1370.

Bergstedt, R. A. and C. P. Schneider. 1988. Assessment of sea lamprey (*Petromyzon marinus*) predation by recovering of dead lake trout (*Salvelinus namaycush*) from Lake Ontario, 1982–85. *Can. J. Fish. Aquat. Sci.* **45**:1406–1410.

Berst, A. H. and A. A. Waino. 1967. Lamprey parasitism of rainbow trout in southern Georgian Bay. *J. Fish. Res. Board Can.* **24**:2539–2548.

Budd, J. C., F. E. J. Fry, and P. S. M. Pearlstone. 1969. Final observations of the survival of planted lake trout in South Bay, Lake Huron. *J. Fish. Res. Board Can.* **26**:2413–2424.

Caswell, H. 1976. The validation problem. In *Systems Analysis and Simulation in Ecology*. Vol. IV, B. C. Patten, (ed.), pp. 313–325. Academic Press, New York.

Christie, W. J. 1972. Lake Ontario: effects of exploitation, introductions, and eutrophication on the salmonid community. *J. Fish. Res. Board Can.* **29**:913–929.

Christie, W. J. 1974. Changes in the fish species composition of the Great Lakes. *J. Fish. Res. Board Can.* **31**:827–854.

Christie, W. J. and D. P. Kolenosky. 1980. Parasitic phase of the sea lamprey (*Petromyzon marinus*) in Lake Ontario. *Can. J. Fish. Aquat. Sci.* **37**:2021–2038.

Cochran, P. A. 1985. Size-selective attack by parasitic sea lampreys: consideration of alternate null hypotheses. *Oecologia* **67**:137–141.

Cochran, P. A. 1986. Attachment sites of parasitic lampreys: comparisons among species. *Environ. Biol. Fish.* **17**:71–79.

Cochran, P. A. and J. F. Kitchell. 1986. Use of modeling to investigate potential feeding strategies of parasitic lampreys. *Environ. Biol. Fish.* **16**:219–223.

Cochran, P. A. and J. F. Kitchell. 1989. A model of feeding by parasitic lampreys. *Can. J. Fish. Aquat. Sci.* **46**:1845–1852.

Conte, F. P., H. H. Wagner, and T. O. Harris. 1963. Measurement of blood volume of the fish, *Salmo gairdneri. Am. J. Physiol.* **205**:533–540.

Davis, R. M. 1967. Parasitism by newly transformed anadromous sea lampreys on landlocked salmon and other fishes in a coastal Maine lake. *Trans. Am. Fish. Soc.* **96**:11–16.

DeAngelis, D. L., S. M. Adams, J. E. Breck, and L. J. Gross. 1984. A stochastic predation model: Application to largemouth bass observations. *Ecol. Model.* **24**:25–41.

Farmer, G. J. 1980. Biology and physiology of feeding in adult lampreys. *Can. J. Fish. Aquat. Sci.* **37**:1751–1761.

Farmer, G. J. and F. W. H. Beamish. 1973. Sea lamprey (*Petromyzon marinus*) predation on freshwater teleosts. *J. Fish. Res. Board Can.* **30**:601–605.

Farmer, G. J., F. W. H. Beamish, and P. F. Lett. 1977. Influence of water temperature on the growth rate of the landlocked sea lamprey (*Petromyzon marinus*) and the associated rate of host mortality. *Can. J. Fish. Aquat. Sci.* **34**:1373–1378.

Farmer, G. J., F. W. H. Beamish, and G. A. Robinson. 1975. Food consumption of the adult landlocked sea lamprey, *Petromyzon marinus* L. *Comp. Biochem. Physiol.* **50A**:753–757.

Hall, A. E. and O. R. Elliott. 1954. Relationship of length of fish to incidence of sea lamprey scars on white suckers, *Catostomus commersoni*, in Lake Huron. *Copeia* **1**:73–74.

Hanson, L. H. and W. D. Swink. 1989. Downstream migration of recently metamorphosed sea lampreys in the Ocqueoc River, Michigan, before and after treatment with lampricides. *North Am. J. Fish. Manag.* **9**:327–331.

Hardisty, M. W. and I. C. Potter. 1971. The general biology of adult lampreys. In *The Biology of Lampreys. Vol. I*, M. W. Hardisty and I. C. Potter, (ed.), pp. 127–206. Academic Press, London.

Heinrich, J. W., J. G. Seelye, and B. G. H. Johnson. 1987. Proceedings of the workshop to evaluate sea lamprey populations (WESLP) in the Great Lakes, August 1985. In *Evaluation of Sea Lamprey Populations in the Great Lakes: Background Papers and Proceedings of the August 1985 Workshop*, B. G. H. Johnson, (ed.), pp. 1–31. Great Lakes Fishery Commission Special Publication 87-2. Great Lakes Fishery Commission, Ann Arbor, MI.

Heinrich, J. W., J. G. Weise, and B. R. Smith. 1980. Changes in biological characteristics of the sea lamprey (*Petromyzon marinus*) as related to lamprey abundance, prey abundance, and sea lamprey control. *Can. J. Fish. Aquat. Sci.* **37**:1861–1871.

Henderson, B. A. 1986. Effect of sea lamprey (*Petromyzon marinus*) parasitism on the abundance of white suckers (*Catostomus commersoni*) in South Bay, Lake Huron. *J. Appl. Ecol.* **23**:381–389.

Hewett, S. H. and B. L. Johnson. 1987. A generalized bioenergetics model of fish growth for microcomputers. *Univ. Wisc. Sea Grant Inst. Tech. Rep.* WIS-SG-87-245.

Huston, M., D. DeAngelis, and W. Post. 1988. New computer models unify ecological theory. *Bioscience* **38**:682–691.

Johnson, B. G. H. and W. C. Anderson. 1980. Predatory-phase sea lamprey (*Petromyzon marinus*) in the Great Lakes. *Can. J. Fish. Aquat. Sci.* **37**:2007–2020.

Kitchell, J. F. 1990. The scope for mortality caused by sea lamprey. *Trans. Am. Fish. Soc.* **119**:642–648.

Kitchell, J. F. and J. E. Breck. 1980. Bioenergetics model and foraging hypothesis for sea lamprey (*Petromyzon marinus*). *Can. J. Fish. Aquat. Sci.* **37**:2159–2168.

Lamsa, A. K., C. M. Rovainen, D. P. Kolenosky, and L. H. Hanson. 1980. Sea lamprey (*Petromyzon marinus*) control—Where to from here? Report of the SLIS control theory task force. *Can. J. Fish. Aquat. Sci.* **37**:2175–2192.

Lett, P. F., F. W. H. Beamish, and G. J. Farmer. 1975. System simulation of predatory activities of sea lampreys (*Petromyzon marinus*) on lake trout (*Salvelinus namaycush*). *Can. J. Fish. Aquat. Sci.* **32**:623–631.

Madenjian, C. P. and S. R. Carpenter. 1991. Individual-based model for growth of young-of-the year walleye: a piece of the recruitment puzzle *Ecological Applications* 1, 268–279.

Madenjian, C. P., B. M. Johnson, and S. R. Carpenter. 1991. Stocking strategies for fingerling walleyes: an individual-based approach *Ecological Applications* 1, 280–288.

Parker, P. S. and R. E. Lennon. 1956. Biology of the sea lamprey in its parasitic phase. *U.S. Dept. Int. Res. Rep.* 44.

Sawyer, A. J. 1980. Prospects for integrated pest management of the sea lamprey (*Petromyzon marinus*). *Can. J. Fish. Aquat. Sci.* **37**:2081–2092.

Smith, B. R. 1971. Sea lampreys in the Great Lakes of North America. In *The Biology of Lampreys. Vol. I.*, M. W. Hardisty and I. C. Potter, (ed.), pp. 207–247. Academic Press, London.

Smith, B. R. and J. J. Tibbles. 1980. Sea lamprey (*Petromyzon marinus*) in Lakes Huron, Michigan, and Superior: history of invasion and control, 1936–78. *Can. J. Fish. Aquat. Sci.* **37**:1780–1801.

Smith, B. R., J. J. Tibbles, and B. G. H. Johnson. 1974. Control of the sea lamprey (*Petromyzon marinus*) in Lake Superior, 1953–70. *Great Lakes Fish. Comm. Spec. Rep.* 28.

Smith, L. S. 1966. Blood volumes of three salmonids. *J. Fish. Res. Board Can.* **23**:1439–1446.

Spangler, G. R., D. A. Robson, and H. A. Regier. 1980. Estimates of lamprey-induced mortality in whitefish, (*Coregonus clupeaformis*). *Can. J. Fish. Aquat. Sci.* **37**:2146–2150.

Stewart, D. J., D. Weininger, D. V. Rottiers, and T. A. Edsall. 1983. An energetics model for lake trout, *Salvelinus namaycush:* application to the Lake Michigan population. *Can. J. Fish. Aquat. Sci.* **40**:681–698.

Swink, W. D. and L. H. Hanson. 1986. Survival from sea lamprey (*Petromyzon marinus*) predation by two strains of lake trout (*Salvelinus namaycush*). *Can. J. Fish. Aquat. Sci.* **43**:2528–2531.

Swink, W. D. and L. H. Hanson. 1989. Survival of rainbow trout and lake trout after sea lamprey attack. *North Am. J. Fish. Manag.* **9**:35–40.

Talhelm, D. R. 1988. *The International Great Lakes Sport Fishery of 1980*. Great Lakes Fishery Commission Special Publication 88-4. Great Lakes Fishery Commission, Ann Arbor, MI.

Talhelm, D. R. and R. C. Bishop. 1980. Benefits and costs of sea lamprey (*Petromyzon marinus*) control in the Great Lakes: some preliminary results. *Can. J. Fish. Aquat. Sci.* **37**:2169–2174.

Torblaa, R. L. and R. W. Westman. 1980. Ecological impacts of lampricide treatments on sea lamprey (*Petromyzon marinus*) ammocoetes and metamorphosed individuals. *Can. J. Fish. Aquat. Sci.* **37**:1835–1850.

Trebitz, A. S. 1989. Development and application of an individual-based population model for first-year largemouth bass. M.S. thesis, University of Tennessee.

Walters, C. J., G. Spangler, W. J. Christie, P. J. Manion, and J. F. Kitchell. 1980. A synthesis of knowns, unknowns, and policy recommendations from the Sea Lamprey International Symposium. *Can. J. Fish. Aquat. Sci.* **37**:2202–2208.

Walters, C. J., G. Steer, and G. Spangler. 1980. Responses of lake trout (*Salvelinus namaycush*) to harvesting, stocking, and lamprey reduction. *Can. J. Fish. Aquat. Sci.* **37**:2133–2145.

Wells, L. 1980. Lake trout (*Salvelinus namaycush*) and sea lamprey (*Petromyzon marinus*) populations in Lake Michigan, 1971–78. *Can. J. Fish. Aquat. Sci.* **37**:2047–2051.

14

Simulating Populations Obeying Taylor's Power Law

R. A. J. Taylor

ABSTRACT. **The distribution of animal population density varies with the mean density in a systematic fashion, known as Taylor's power law. It has been proposed that the specificity and repeatability of this relationship is a consequence of the specific behavior of the members of the population. Assuming that this is so, investigations of the population dynamic consequences of behavioral interactions between individuals, or small groups, should have simulated organisms placed in relation to each other in realistic juxtapositions. This requires an individually-based simulation in which each individual's Cartesian coordinates may be generated according to a prescribed power law exponent. Generating realistic spatial distributions which cover the full range of densities encountered in Nature is not easy, even for such a "simple" spatial distribution as the Poisson. An algorithm is presented which places organisms on a plane in realistic spatial distributions according to any defined power law exponent within the observed range $(.5 < \beta < 3.5)$. Difficulty in generating extremely aggregated distributions suggests that distributions with extremely high spatial variances at high density may be the result of an "amplification" process (West and Shlesinger 1990).**

Introduction

In what is arguably the most influential publication on population dynamics theory, May (1974) defined analytical or "strategic" models at one end and pragmatic or "tactical" models at the other end of a continuum. Tactical models describe systems in detail, and are intended to answer specific, usually applied, questions, while strategic models sacrifice detail in favor of generalizations to provide a conceptual framework for investigating general principles. He wrote of the strategic models: "There is no explicit evolution in our ecosystems . . . we consider isolated communities which are uniformly unvarying in space; time is the only independent variable" (May 1974). These two basic premises, no evolution and no movement, still hold in many analytical population models. In essence the theory deals with whole species populations in which all movement is internal and its dynamical consequences are relevant only insofar as they influence birth and death. With the constraints imposed by the paradigm of population change in time alone, population dynamics literally lacks a fundamental dimension-space. The main reason for this deficiency is that space is notoriously difficult to incorporate into analytical models, requiring, as it does, the cataloguing of the positions of organisms relative to each other in that space. One way round this problem is to split the space up into a number of cells with defined relative positions, but within which the positions of the modeled organisms are not recorded (e.g. Hassell and May 1973, 1974). This approach represents an intermediate between group-based models and individual-based models: the degree of individuality (spatial resolution) can be adjusted as needs dictate.

Behavioral interactions are generally fairly short-ranged and occur between a limited number of individuals, so that lumping together even small groups into cells loses an important component of the behavioral detail. If the fine detail of behavior has little influence on the overall behavior of populations, the loss of detail within cells would be irrelevant. But it is the fine detail of behavior which governs host-finding ability, foraging success, the outcomes of aggressive interactions, and reproductive success; in short, survival success on the individual, ecological and evolutionary time scales. Ultimately, therefore, only models in which each individual's behavior is simulated can the population consequences of animal behavior be investigated, for it is the variability between individuals which determines their success or failure. The consequences of success or failure at higher levels of organization must also depend on the distribution of variability within the population. Making assumptions of uniformity, such as are made in connection with the gas laws, removes the essential attribute of behavioral variability which distinguish animals from gas molecules.

The analytical approach employed in investigations of the gas laws has met with some success, however. Tractable differential equations have been developed by Turchin (1989) to investigate the population consequences of aggregation behavior using models incorporating attraction and repulsion of conspecifics. His analytical models are a more rigorous treatment of density-dependent spatial behavior than Taylor's (1981a, 1981b) simulation experiments, but lack the spatial frame that Taylor and Taylor (1977) advocated. Turchin's models are constructed in one spatial dimension, but are readily expanded into two dimensions. His objective was to develop an analytical framework in which behavioral parameters could be identified and measured experimentally. His approach, as is usual with differential equation models, emphasized equilibrium conditions and examined strategic questions. Tactical questions, such as how much variation in abundance about an equilibrium, either spatial or temporal, could be expected while maintaining long-term stability of the population were not addressed.

To investigate this sort of question, Perry (1988) introduced an explicit spatially coordinated arena with his stepping-stone model to investigate spatial variability. He concluded that moderately aggregated populations could be generated by demographic rate processes, but failed to account for extremely high levels of aggregation, even in the stepping-stone model. This study of behaviorally-induced migration built on his earlier investigation of the negative binomial as a generator of behavioral variation (Perry and Taylor 1986). In the earlier work it was found that the conclusions arrived at by May (1978) on the stability criteria for his phenomenological model (see below) were not met in real interacting populations. Perry and Taylor (1986) suggested an alternative way of building heterogeneity into the negative binomial model by allowing parasites and hosts to respond to each other by movement. They showed how host aggregation may be generated by avoidance behavior. Chesson and Murdoch (1986) also examined the phenomenological model as a special case of a generalized host-parasite model involving aggregation. They concluded that the phenomenological model is too restrictive because it does not capture the parasitoids' behavioral response to host density realistically.

A common thread running through population dynamics is the concept of density-dependence: it is the raw material of the feedback processes which must be at the heart of the long term persistence most populations exhibit. Many of the questions population models are constructed to answer have as a common denominator the question, "how does density-dependence operate?" This old question has been receiving renewed attention recently. Experimental manipulation of predation has revealed spatial density-dependence where it had hitherto been obscured in temporal data (Gould et al. 1990), but it is not even certain that we can detect and measure density-dependence efficiently (see Solow and Steele 1990 and references therein).

The problem is that density is fundamentally a spatial concept, consequently density-dependence is also essentially a spatial phenomenon, but it takes time to operate and become apparent. Thus investigations of density-dependence must take proper notice of both the spatial and temporal dimensions. In their investigation of the temporal consequences of spatial density-dependence, Stewart-Oaten and Murdoch (1990) found that, contrary to conventional wisdom, spatial heterogeneity is more likely to be destabilizing than stabilizing: an increase in clumping beyond that seen with the Poisson is more likely to lead to instability. Yet populations are clumped over a wide range of scales, and stable over differing spatial and temporal periods. Even when a population is globally stable for long periods of time, clumps within the population range may be highly unstable (Taylor and Taylor 1979). Because intrinsic behavioral and extrinsic environmental factors may act site-specifically to influence distribution and abundance, population dynamic models investigating density-dependence may require a spatial frame.

In this essay I argue that the behavioral responses of animals, many of which are density-dependent, are an intimate part of the dynamics of natural populations, that individual behavior is expressed at the population level in the distribution of individuals in space relative to each other as well as to the obvious heterogeneities in their environment. To model populations, therefore, whether for strategic or tactical purposes, the short range of behavior mandates an individual approach. To obviate the need to incorporate each and every behavioral interaction which may contribute to the population structure, I present an algorithm for generating realistic spatial distributions at any predetermined level of spatial resolution. This algorithm can then be applied to generate populations at a resolution appropriate for the investigation at hand.

Argument

Density and Distribution

Population density is a measure of how close together individuals are packed, an attribute we generally determine by sampling. We can imagine the sampling process as a count of individuals portrayed in a set of photographs of the population taken simultaneously at different places. This instantaneous record of the population provides information on the number of animals per photograph, or relative density, and their relative positions. Knowing the area of each photograph we can also determine the absolute density (Taylor et al. 1991). The number of organisms per photo, X_i, forms a frequency distribution characterized by its moments. The moments most commonly computed are the mean, $M[X]$, and variance, $V[X]$. But even within a single photograph, the density experienced by each individual will be slightly different. Using their Cartesian coordinates to divide the picture up into a Voronoi tessellation we can estimate the density experienced by each individual (Fig. 14.1).

Defining $Y_i = 1/X_i$, where y_i is the proportion of the picture occupied by the ith individual and using the standard approximations for obtaining the mean and variance of a function of a random variable (Kendall et al. 1983) we obtain:

$$E[Y] = 1/E[X]$$

$$V[Y] = V[X]\left\{\frac{dY}{dX}\right\}^2 = \frac{V[X]}{M[X]^4}. \tag{1}$$

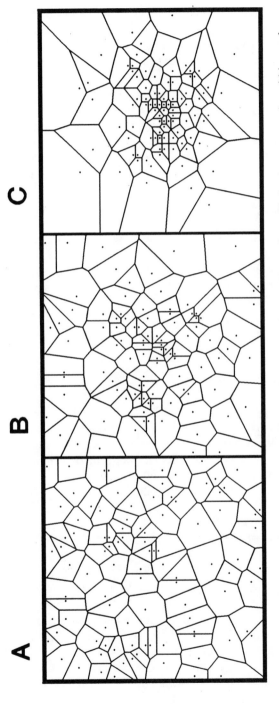

Figure 14.1 The area occupied by a population of individuals divided into a Voronoi tessellation has every point within a polygon closer to its corresponding individual than to any other. The tesselations illustrate how population intensity varies within and between photographs of populations experiencing different aggregations: A, random (Poisson); B, aggregated (Central); C, super-aggregated (Waldian).

$E[Y]$, which I call population intensity, measures how much territory the average individual occupies—the area of each cell in the tessellation (Fig. 14.1). To derive the mean separation between the members of the population from the population intensity, we must make assumptions about the distribution of the areas of the tesselations, specifically the relationship between the mean and the variance of area (Miles 1970).

If the photographic process is continued at intervals of time, a trajectory of population density, and the corresponding frequency distributions, may be constructed. Both intra- and inter-generational change may be constructed by suitable choice of interval with respect to generation time. Both time-scales are of interest, but I shall restrict my discussion to inter-generational change.

The Spatial Distribution of Organisms

For a large proportion of species the shape of the frequency distribution of number of individuals per sample, or unit area or volume of habitat, has been found to vary with the population density. The most succinct expression of this phenomenon is the power law relationship between the variance, V, and mean, M, of samples estimating population density (Taylor's power law; Southwood 1966, 1978, Lincoln et al. 1982):

$$V = \alpha M^\beta \tag{2}$$

where α and β are empirically derived constants (Taylor 1961, 1984). It has been suggested that α is largely a sampling factor which may vary between data sets, while β appears to be an index of aggregation characteristic of species-stage (Southwood 1966, 1978). It has been determined empirically that the exponent lies in the range $0 < \beta < 4$, with a modal value of 2 (Taylor et al. 1983). Two is the expected value of β in two senses; it is approximately the average value across all species so far investigated, and it is expected because it states that the variance, the square of the deviations of the observations about the mean, is proportional to the square of the mean density. From a dimensional point of view this makes sense. In addition, there are a number of statistical distributions which have the variance proportional to the square of the mean (e.g., lognormal and gamma).

There are statistical distributions with other functional relations between variance and mean. The most familiar is the Poisson, for which the variance and mean are equal, and the less familiar Waldian distribution, which has variance proportional to the cube of the mean (Johnson and Kotz 1969, 1970). These three special cases I shall label the Poisson, central and Waldian cases. My reason for using "central" to describe the case where $\beta = 2$, rather than lognormal or gamma, will become apparent later.

Only a relatively few sets of data showing the distribution of density have

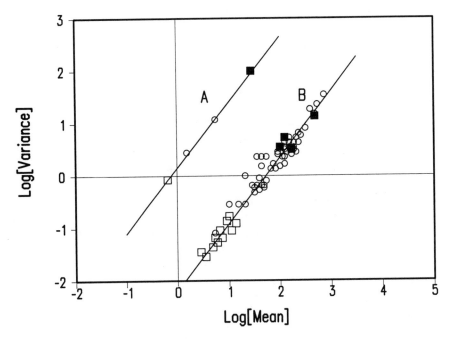

Figure 14.2   Unless the gradient of log[Variance] on log[Mean] is an integer (β = 1, 2, 3), the frequency distribution of number of organisms per sample changes from random to more aggregated as the mean number per sample (*M*) increases. Poisson distributions are represented by open squares, negative binomials by circles, and lognormals by filled squares. (Data of A, *Pyrausta nubilalis* (McGuire et al. 1957; β = 1.25) and B, *Lymantria dispar* egg masses (Brown & Cameron 1982; β = 1.27) after Taylor 1984).

an integer exponent in the relationship between variance and mean. In a study of 156 sets of data drawn from the literature, comprising 102 species from over 20 orders in seven phyla, Taylor et al. (1978) found β to average 1.45 +/− 0.39 (std. dev.), with less than one third of sets having β within +/− 0.2 of an integer number. In another sample drawn from aphids, moths and birds, different average values of β were obtained, but again the majority of sets had β significantly different from an integer (Taylor et al. 1980).

In short, a sizeable majority of such data sets have fractional exponents. The practical consequence of this is that there is no unique frequency distribution describing the population density of all but a few species: the statistical frequency distribution of numbers per sample changes as the mean number per sample changes (Fig. 14.2).

Behavior, Distribution and Density

The power law's specificity over a wide range of conditions (Taylor et al. 1988) suggests that the conditions generating it operate everywhere the same. This suggests that the power law measures an intrinsic population property, rather than extrinsic factors such as interspecific competition or predation. The way in which individuals interact with conspecifics is the intrinsic behavioral trait they all have in common. It seems probable that the specific distributional patterns that species adopt, as measured by the power law, are the consequence of those behavior patterns which contribute to the spatial organization of populations, that is, spatial behavior. Because the distributional patterns themselves are density-dependent (Fig. 14.2), it seems plausible that the underlying spatial behavior is also density-dependent (Taylor and Taylor 1977).

In addition to showing how the frequency distribution of density varies with mean density, Figure 14.2 also illustrates the flexibility of the negative binomial distribution which fits most of the data quite well provided that its parameter, k, is unconstrained. Constraining the parameter to a common value, k_c, produces much poorer fits (Perry & Taylor 1985) because k is a polynomial function of mean and variance. This can be visualized by equating the power law relationship (Eq. 3) with the moment estimator of k:

$$1/k = \frac{V}{M^2} - \frac{1}{M} = \alpha M^{\beta-2} - 1/M \tag{3}$$

Depending on the value of β, Eq. 3 takes one of four general forms, one of which is not single-valued (Taylor et al. 1979). If k is used as a generator of aggregation this could result in generated populations with high density and low variance; something which has been observed only as an artifact of sampling (Taylor et al. 1983). Consequently, the negative binomial parameter k may not be a reliable tool to generate aggregation (Taylor et al. 1979, Perry & Taylor 1986).

As an example, consider the Nicholson-Bailey model of host-parasitoid interactions which has parasitoids searching a homogeneous environment independently of each other and randomly with respect to the hosts. To describe the process, the first term in the Poisson series is used to define the proportion of hosts not attacked. The model is unstable, resulting in divergent oscillations. Several authors have shown how aggregation of the hosts and preference for the larger aggregations by the parasites could stabilize the model (e.g. Murdoch & Oaten 1975). To investigate the dynamical behavior of aggregated parasitoid models May (1978) used the first term of the negative binomial to define the proportion of hosts not attacked. With this modification, he was able to show that the aggregation of parasitoid attacks has a very similar dynamical effect to mutual interference of searching parasites (Hassell & Varley 1969), thereby demonstrating that short range interactions between individuals could have profound distributional effects at the population level.

May used his model to investigate the effect spatial aggregation might have on temporal stability, a strategic question. The answer that aggregation leads to a larger range of dynamical behavior, including stability, than randomness, would seem to be fairly robust. But had the objective been the tactical one of predicting the degree of temporal stability from an assumed degree of spatial aggregation, the result would have been much less tractable. Assuming a common k, the equilibrium parasitoid density, $P*$, as a function of host fecundity, r, is:

$$P* = k(r^{1/k} - 1) \tag{4}$$

This equation is stable for values of $k < 1$, and diverges for $k > 1$ (May 1978). Combining Eqs. 3 and 4 leads to the daunting result:

$$P* = \frac{M\{1 - r^{(\alpha M^{\beta-2} - 1/M)}\}}{1 - \alpha M^{\beta-1}} \tag{5}$$

where M is the mean number of parasitoids per patch, and α and β are the power law parameters of the variance-mean relationship describing patch size. The dynamical behavior is no longer straightforward, and depending on the values of α and β, there may be stable conditions at one or more densities, or none at all. While such rich dynamical behavior is intriguing, and may well describe real population behavior, it is sufficiently complex that visualization becomes a major obstacle to comprehension.

The main qualitative point to emerge here is that a putative behavioral parameter, k, can be shown to influence stability at one level of resolution, the population, while depending on the population density experienced by individuals within patches. Thus a complex feedback process operates across two spatial scales in this model of interspecific interaction.

Simulations

Generating Power Law Variates

Using the analytical approach, we can derive useful generalizations in the abstract, but it is frequently helpful to be able to visualize the process. Mapping the trajectories of subject populations in time we can see, for example, the transition from order to chaos in the model system.

Modeling behavior so that each individual can respond realistically to all others with which it interacts is a computational problem similar to the *n*-body problem in astronomy, but in addition we require that the simulation experiment be started with the members of the population arrayed in realistic juxtaposition, or for the model to evolve towards a realistic distribution. To model behavioral interactions which depend on the individuals' subjective densities, which vary continuously over the plane (Fig. 14.1), each individual's position relative to the others must be specified. This requires expansion of the model to include two-dimensional space explicitly; the model becomes three dimensional, presenting its own problems of visualization. The most difficult problem, however, is the assignment of coordinates, conditional on a prescribed variance-mean relationship.

To illustrate, assigning uniform random Cartesian coordinates between 0 and 1 should place each individual in the arena independently of every other. Taking random samples from the arena to estimate the mean and variance, and repeating the process with different total numbers, produces a set of mean-variance pairs. No matter what rule of sampling or total population size, the variance always grows as the square of the mean, as one would expect of a lognormal distribution (Table 14.1).

Generating a Poisson distribution ($\beta = 1$) proves to be surprisingly difficult. The arena must be divided into small cells (say 100×100) and into each cell assign a Poisson population, and then assign each member uniform coordinates within its cell. Provided the sampler is larger than the cells (say 5×5), the calculated $\beta \simeq 1$, but as the sampler approaches and becomes smaller than a cell, the calculated β exceeds 1 and approaches 2.

Algorithms to assign the coordinates directly do not exist. An indirect approach, based on box-counting (Feder 1988), uses the fact that changing the size of the box (sampler) so that the number per photograph is sX_i changes only the intercept of Eq. 2:

$$V_s = s^{2-\beta} \alpha M_s^{\beta} \qquad (6)$$

Mimicking the photographic sampling program recursively preserves the power law relationship down to the last subdivision. At the first level, we assign the number of organisms per photograph at a prescribed mean density using a random number generator. Then each photograph in turn is subdivided, and its contents are assigned at random to the sub-photographs, subject to the variance-mean criterion. This recursive subdivision is continued until only one exists in each sub-photograph. Each organism is then assigned the exact Cartesian coordinates of the center of its (sub-)photograph. If an individual-based simulation is not required, an arbitrarily large group-size may be selected, and the recursive subdivision continued to the chosen group-size.

Table 14.1. Comparison of observed to expected gradient of variance mean power law plots for different simulation strategies.

Simulation	β_{input}	β_{output}	r^2
Uniform Cartesian coordinates	—	1.935	.999
Lognormal random number per cell:			
i	0.5	0.488	.993
ii	1.0	1.031	.998
iii	1.5	1.526	.999
iv	2.0	2.009	.999
v	2.5	2.125	.855
vi	3.0	2.223	.624
vii	3.5	2.361	.413

To generate the number of organisms per photograph, we use a lognormal distribution pseudo-random number generator. The lognormal with parameters μ and σ has mean and variance given by

$$M = \exp\left(\mu + \frac{1}{2}\sigma^2\right)$$
$$V = M^2(\exp(\sigma^2) - 1) \tag{7}$$

Solving for μ and σ^2 and substituting Eq. 3 for V:

$$\mu = \log(M^2) - \frac{1}{2}\log(\alpha M^\beta + M^2)$$
$$\sigma^2 = \log(\alpha M^\beta + M^2) - \log(M^2) \tag{8}$$

Applying Eq. 8 results in variance-mean plots of high coefficient of determination for $.5 < \beta \leq 2$ (Table 14.1), but for $\beta > 2.2$ the simulated distributions do not produce the expected values of β, and the correlations are unacceptably low (Table 14.1, Figure 14.3A).

The method breaks down as β increases above 2 because the extreme values required to produce the very high variances are not generated frequently enough by the lognormal, and the distribution is not sufficiently skewed.

We work around this by using one of three schemes:

Scheme 1. Generate Poisson variates and transform twice. Healy & Taylor (1962) defined a family of exact power transformations, $z = (1 - \beta/2)$, for populations obeying the variance-mean power law which gives $z = 0$ for $\beta = 2$. They showed that the appropriate

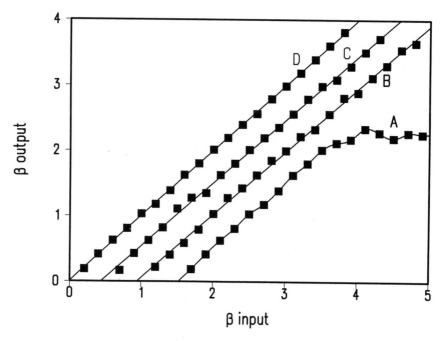

Figure 14.3 The relationship between β_{output} and β_{input} for different simulation schemes. A, direct simulation with the lognormal works well to about $\beta = 2$, but then breaks down; B-C, computed values of β using Schemes 1 and 2 are slightly lower than expected at high β, and they are slightly more variable than Scheme 3; D, the plots are staggered one half log-cycle along the abscissa for clarity.

transformation here is the logarithm, in agreement with conventional wisdom, and noted that the method selects the square root as the transformation appropriate for analysis of Poisson variates ($\beta = 1$), again as expected from other statistical arguments. Noting Healy and Taylor's result that central variates ($\beta = 2$) require logarithmic transformation and Poisson the square root transformation, we generate central values by exponentiating the square roots of Poisson variates, and generate super-aggregated variates by raising the variate to the $1 - \beta/2$ power.

Scheme 2. Generate variates with no dependence of variance on mean, using the Gaussian distribution, perhaps, and back transform using Healy & Taylor's (1962) $z = (1 - \beta/2)$ again.

Scheme 3. Generate lognormal variates with parameters μ and σ calculated from Eq. 8 for $\beta \leq 2$; to obtain the relationship between $V[Y]$ and $E[Y]$ from Eqs. 1 and 3:

$$V[Y] = \alpha M[Y]^{(4-\beta)} \qquad (9)$$

to transform the result for $\beta > 2$.

Fig. 14.3 show the results of the three schemes. These differ little in their efficiency or speed, although Schemes 1 and 2 are a little more variable than Scheme 3 at high β.

Discussion

Lognormal distributions result from the combination of a large number of stochastic multiplicative events (Kolmogorov process). As lognormal systems become more complex, the frequency functions become broader, and they coincide more with an inverse power law distribution. This is thought to be due to an amplification process which applies at frequencies greater than a threshold f_0 (West and Shlesinger 1990). Observations of many natural processes have been found to be lognormal over about 95% of their range, but in the last few percentiles the frequency function goes to $1/f$ (West & Shlesinger 1990 and references therein). The lognormal is thus only an approximate model for generating power law populations. It works best when the required $1/f$ behavior is small at $\beta \leq 2$, and performs poorly at $\beta > 2$ where $1/f$ behavior increases the variance rapidly. We see from this that as β increases above 2 the shape of the frequency distribution of numbers per sample becomes even more skewed than the lognormal. It exhibits the same transition from lognormal to $1/f$ behavior. As the distribution becomes more skewed the mean rises above 50% (in logs) and the variance accelerates. This explains why the lognormal alone cannot return power law variates for $\beta > 2$, and why I prefer to use the term "central distribution." Below 2 the transition to $1/f$ behavior does not occur in data of population density, but obviously it must in measures of population intensity.

It is also worth noting that the Waldian, lognormal and gamma distributions are continuous probability distributions, whereas the Poisson and negative binomial distributions are discrete frequency distributions. When counting organisms with $\beta < 2$, the data are expressed as integer numbers per sample (although we might rescale by dividing by the area or volume of the sampler), whereas for $\beta \geq 2$ the simulator produces continuous variates. At $\beta = 2$, there is a transition, reflected in the exact transformation, $z = (1 - \beta/2) = 0$, between positive and negative powers.

Throughout, I have assumed that Taylor's power law measures, in some sense, a component of behavior (Taylor 1981a, 1981b). This assumption is not universally held. The more cogent objections are expressed in Anderson et al. (1982) who propose that β is determined by demographic parameters, notably birth, death and migration rates, and by Downing (1986) who showed how sampling regime may influence the value of β. It is not my intention to defend the proposition that β is a behavioral parameter, but it is relevant here to inquire why it is important to determine which proposition is correct, because the consequences for population modeling differ:

1. sampling (Downing 1986): the power law is relevant only to data analysis: it provides the appropriate transformation (cf. Healy & Taylor 1962);

2. demography (Anderson et al. 1982): population distribution depends solely on demographic rates of change in time: modeling requires no spatial frame to explore population dynamics;

3. behavior (Taylor and Taylor 1977, 1978): the expression of behavior is movement: modeling population distribution requires a spatial frame in which the individuals can move, thus temporal rates of change alone are inadequate.

The power law relating spatial variance to population density says that our measurements of population density scale in a characteristic way. Regardless of the genesis of the power law, sampling, demography, or behavior, the empirical fact is disputed by few, and at our peril we fail to allow for it, because to be useful our model output must refer back to the real world (e.g., Taylor 1987).

Consider the parasitoid-host model with negative binomial aggregation. If the relative positions of organisms do not influence the outcome of behavioral interactions, the failure of a parasite to find a potential host is just a matter of probability, and the group-based approach is adequate. This is the situation analogous to the gas laws. If that probability function changes in a systematic way with density, however, as implied by the power law (Eqs 3 and 5), then for most species no single distribution is adequate (Fig. 14.2). Also, species which are found in extremely high aggregations may have not only systematically varying frequency distributions describing their spatial behavior, there may be an uncommonly large number of relevant behaviors contributing to the spatial distribution as well. Failure to account for these behaviors in our models may have profound consequences for population dynamics as a discipline, as well as applied activities like pest management which depend on a proper understanding of a species' population dynamics. If we simulate populations with the wrong density-dependent probability distribution, or none at all, how will we know that our conclusion is the correct one, or what environmental or economic damage may ensue?

Taylor's power law is one of a large number of logarithmically scaling (allometric) ecological relationships (see Peters 1983 and Schmidt-Nielsen 1984 for example). There seems to be no shortage of ecological power laws, but our knowledge of them is largely empirical and our understanding very partial. Learning to generate realistic spatial distributions may have some spin-off in the direction of other allometric relationships. Except for some special cases, realistic spatial distributions can only be generated by recursion: allometric relationships generated by recursion have been identified as belonging to the new paradigm of fractal geometry (Mandelbrot 1982, Feder 1988). The ability to simulate Taylor's power law may not only improve our understanding of that enigma and the relationship between individual behavior and population consequences, but could also shed some light on other allometric relationships in ecology.

Literature Cited

Anderson, R. M., D. M. Gordon, M. J. Crawley, and M. P. Hassell. 1982. Variability in the abundance of animal and plant species. *Nature* **296**:245–248.

Brown, M. W. and E. A. Cameron. 1982. Spatial distribution of adults of *Ooencyrtus kuvanae* (Hymenoptera: Encyrtidae), and egg parasite of *Lymantria dispar* (Lepidoptera: Lymantriidae). *Can. Entomol.* **114**:1109–1120.

Chesson, P. L. and W. W. Murdoch. 1986. Aggregation of risk: relationships among host-parasitoid models. *Am. Nat.* **127**:696–715.

Downing, J. A. 1986. Spatial heterogeneity: evolved behaviour or mathematical artifact? *Nature* **323**:255–257.

Feder, J. 1988. *Fractals*. Plenum, New York.

Gould, J. R., J. S. Elkinton, and W. E. Wallner. 1990. Density-dependent suppression of experimentally created gypsy moth, *Lymantria dispar* (Lepidoptera: Lymantriidae), populations by natural enemies. *J. Anim. Ecol.* **59**:213–233.

Hassell, M. P. and R. M. May. 1973. Stability in insect host-parasite models. *J. Anim. Ecol.* **42**:693–726.

Hassell, M. P. and R. M. May. 1974. Aggregation of predators and insect parasites and its effect on stability. *J. Anim. Ecol.* **43**:567–594.

Hassell, M. P. and G. C. Varley. 1969. New inductive population model for insect parasites and its bearing on biological control. *Nature* **223**:1113–1117.

Healy, M. J. R. and L. R. Taylor. 1962. Tables for power-law transformations. *Biometrika* **49**:557–559.

Johnson, N. L. and S. Kotz. 1969. *Distributions in Statistics. Vol. 1, Discrete Distributions*. Wiley, New York.

Johnson, N. L. and S. Kotz. 1970. *Distributions in Statistics. Vol. 2, Continuous Univariate Distributions-1*. Wiley, New York.

Kendall, M., A. Stuart, and J. K. Ord. 1983. *The Advanced Theory of Statistics, Vol. 3*. Fourth ed. Macmillan, London.

Lincoln, R. J., G. A. Boxshall, and P. F. Clark. 1982. *A Dictionary of Ecology, Evolution and Systematics.* Cambridge University Press, Cambridge, England.

Mandelbrot, B. B. 1982. *The Fractal Geometry of Nature.* W. H. Freeman, San Francisco.

May, R. M. 1974. *Stability and Complexity in Model Ecosystems.* Princeton University Press, Princeton, N. J.

May, R. M. 1978. Host-parasitoid systems in patchy environments: a phenomenological model. *J. Anim. Ecol.* **47**:833–843.

McGuire, J. U., T. A. Brindley, and T. A. Bancroft. 1957. The distribution of European corn borer larvae *Pyrausta nubilalis* (Hbn.) in field corn. *Biometrics* **13**:65–78.

Miles, R. E. 1970. The distribution of areas in random polygons. *Math. Biosci.* **6**:85–90.

Murdoch, W. W. and A. Oaten. 1975. Predation and population stability. *Adv. Ecol. Res.* **9**:2–131.

Perry, J. N. 1988. Some models for spatial variability of animal species. *Oikos* **51**:124–130.

Perry, J. N. and L. R. Taylor. 1985. Adès: new ecological families of species-specific frequency distributions that describe repeated spatial samples with an intrinsic power-law variance-mean property. *J. Anim. Ecol.* **54**:931–953.

Perry, J. N. and L. R. Taylor. 1986. Stability of real interacting populations in space and time: implications, alternatives and the negative binomial k_c. *J. Anim. Ecol.* **55**:1053–1068.

Peters, R. H. 1983. *The Ecological Implications of Body Size.* Cambridge University Press, Cambridge, England.

Schmidt-Nielsen, K. 1984. *Scaling: Why is Animal Size so Important?* Cambridge University Press, Cambridge, England.

Solow, A. R. and J. H. Steele. 1990. On sample size, statistical power, and the detection of density dependence. *J. Anim. Ecol.* **59**:1073–1076.

Southwood, T. R. E. 1966. *Ecological Methods with Particular Reference to the Study of Insect Populations,* First ed. Methuen, London.

Southwood, T. R. E. 1978. *Ecological Methods with Particular Reference to the Study of Insect Populations,* Second ed. Chapman and Hall, London.

Stewart-Oaten, A. and W. W. Murdoch. 1990. Temporal consequences of spatial density dependence. *J. Anim. Ecol.* **59**:1027–1045.

Taylor, L. R. 1961. Aggregation, variance and the mean. *Nature* **189**:732–735.

Taylor, L. R. 1984. Assessing and interpreting the spatial distribution of insect populations. *Ann. Rev. Entomol.* **29**:321–357.

Taylor, L. R., J. N. Perry, I. P. Woiwod, and R. A. J. Taylor. 1988. Specificity of the spatial power-law exponent in ecology and agriculture. *Nature* **332**:721–722.

Taylor, L. R. and R. A. J. Taylor. 1977. Aggregation, migration and population mechanics. *Nature* **265**:415–421.

Taylor, L. R. and R. A. J. Taylor. 1978. The dynamics of spatial behaviour. In *Population Control by Social Behaviour*, F. J. Ebling and D. M. Stoddart (eds.), pp. 181–212. Institute of Biology, London.

Taylor, L. R., R. A. J. Taylor, I. P. Woiwod, and J. N. Perry. 1983. Behavioural dynamics. *Nature* **303**:801–804.

Taylor, L. R., I. P. Woiwod, and J. N. Perry. 1978. The density-dependence of spatial behaviour and the rarity of randomness. *J. Anim. Ecol.* **47**:383–406.

Taylor, L. R., I. P. Woiwod, and J. N. Perry. 1979. The negative binomial as a dynamic ecological model and the density-dependence of k. *J. Anim. Ecol.* **48**:289–304.

Taylor, L. R., I. P. Woiwod, and J. N. Perry. 1980. Variance and the large scale stability of aphids, moths and birds. *J. Anim. Ecol.* **49**:831–854.

Taylor, R. A. J. 1981a. The behavioural basis of redistribution. 1 The Δ-model concept. *J. Anim. Ecol.* **50**:573–586.

Taylor, R. A. J. 1981b. The behavioural basis of redistribution. 2 Simulations of the Δ-model. *J. Anim. Ecol.* **50**:587–600.

Taylor, R. A. J. 1987. On the accuracy of insecticide efficacy trials. *Environ. Entomol.* **16**:1–8.

Taylor, R. A. J., M. L. McManus, and C. W. Pitts. 1991. The absolute efficiency of milk-carton gypsy moth traps. *Bull. Entomol. Res.* **81**:111–118.

Taylor, R. A. J. and L. R. Taylor. 1979. A behavioural model for the evolution of spatial dynamics. In *Population Dynamics* R. M. Anderson, B. D. Turner and L. R. Taylor (eds.), pp. 1–27. Blackwell Scientific Publications, Oxford, England.

Turchin, P. 1989. Population consequences of aggregative movement. *J. Anim. Ecol.* **58**:75–100.

West, B. J. and M. Shlesinger. 1990. The noise in natural phenomena. *Am. Sci.* **78**:40–45.

15

An Approach for Modeling Populations with Continuous Structured Models

Thomas G. Hallam, Ray R. Lassiter, Jia Li, and William McKinney

ABSTRACT. **This article describes an approach to modeling population dynamics that employs physiological information at the individual level. This procedure has been utilized to investigate the effects of chemicals in aquatic populations. The dynamic population protocol requires development of a mathematical model of relevant physiological processes of the individual, determination of parameters in the individual model that can lead to variation among the types of individuals in the population, and formulation of the population model. The individual model consists of ordinary differential equations, whereas the population model consists of an appropriate number of McKendrick-von Foerster partial differential equations, each of which incorporates the dynamics of the individual ecotypes present in the population. The procedure assumes clonal reproduction and is generic in that, for application, only the individual model and the diversity desired in the population need be specified. An advantage of this approach is that it focuses on the individual and utilizes a data base on the level of many biological investigations.**

Introduction

The individual organism is important for analysis of population and community dynamics. Structural variation in populations arises from differences in individuals. These individual variations can result from demographic, genetic or environmental processes. Because the demographic processes of birth, growth, and death are functions of the characteristics of the individuals that compose the population, there has been a recent surge of interest in the individual-based perspective for investigating populations (Huston et al. 1988). Several pertinent reasons sparked the renewed interest in this perspective. First, although classical approaches such as Lotka-Volterra and related state-space-aggregated approaches have been useful conceptually, a general lack of predictive capability has been exhibited by the non-mechanistic models. Second, the development of new computer architectures that provide computing power sufficient to handle the voluminous calculations necessary to follow thousands of individuals has made detailed numerical simulation of populations based on individual dynamics a feasible objective. A third reason for pursuing the development of population ecology from the individual perspective is that much data are available on properties of certain classes of organisms at the individual level and on the processes that determine growth, maturity, and death. It is generally more advantageous to utilize this information directly to estimate parameters associated with individual dynamics than to aggregate the data to estimate population level parameters.

An approach that is generally applicable and is central to issues of individual-based population theory is the mathematical model of the extended McKendrick-von Foerster type. An extensive literature is available that describes studies in which the only variables modelled in the individual are age or size; for example, the book by Metz and Diekmann (1986) and the recent paper by Vance et al. (1988) have an excellent set of references. The theory of age-dependent population dynamics is also discussed in Sharpe and Lotka (1911), Hoppensteadt (1975), Gurtin and MacCamy (1974), Cushing (1977, 1985), Nisbet and Gurney (1982), and Webb (1985).

The main purpose of this article is to provide motivation for and description of an approach that we have found to be practical in a ecotoxicological setting. We indicate the formulation of an appropriate mathematical representation of a population based on physiological attributes relevant to the individual species considered and to the problem under investigation. We also delineate the procedure used to study the dynamics of this physiologically structured population model.

The article has two main parts. The first discusses the relationship between model hypotheses and model conclusions. Problems of applicability arise from employing classical age- or size-structured models as representations of a population. Specifically, the lack of individual variation in these population models is explored. We describe certain properties of the dynamic behavior and the biological ramifications of these continuous, structured populations to demonstrate that it is often necessary to include additional physiological variables other than just age and size if one wishes to obtain information about biologically realistic, deterministic population dynamics.

The second part of the article focuses on the computation of solutions of physiologically structured models. Here we will indicate the motivation and describe the protocol for formulating a dynamic population that was employed in an investigation of effects of toxic chemicals on aquatic populations (Hallam et al. 1988, 1990a, 1990b). This protocol requires development and representation of relevant physiological processes of the individual, determination of parameters in the individual model that can generate diversity in the population, and formulation of the partial differential equation that constitutes the population model. Illustrations of the numerical solution of the population model are presented. The protocol is presented because it is generic and the approach seems to be applicable (with modification) to many environments. In Sec. II-4, we apply the method of characteristics for solving hyperbolic partial differential equations to the population model and discuss problems of interpretation. The method of characteristics is a natural procedure to utilize from a biological perspective since it corresponds to following the dynamics of cohorts that compose the population.

I. Relationships Between Model Hypotheses and Conclusions

1. The Population Model

The population model of interest here is an extension of the McKendrick-von Foerster type. In a simple form that includes variables representing age and size, it can be written as

$$\frac{\partial \rho}{\partial t} + \frac{\partial \rho}{\partial a} + \frac{\partial}{\partial m}(\rho g) = -\mu(t, a, m, \rho)\rho. \tag{1}$$

A derivation of this mathematical model may be found in Sinko and Streifer (1967) or Nisbet and Gurney (1982). In eq. (1), t is time; a is age, m is the physiological variable that measures size (biomass, length) of the organism; $\mu = \mu(t, a, m, \rho)$ is the mortality rate function ; $g = g(m; r)$ is the growth rate of an individual in the population with r denoting the vector of parameters that determine the characteristics of growth; and $\rho = \rho(t, a, m; r)$ is the population density function (usually of females in the population).

In addition to the hyperbolic partial differential Eq. (1), it is necessary to prescribe initial and boundary conditions for the population. The initial age-size distribution of the population

$$\rho(0, a, m; r) = \phi(a, m; r) \tag{2}$$

is specified. The boundary condition, often called the renewal equation, indicates the birth process of the population. In age-structured population models, the boundary condition has the form

$$\rho(t, 0; r) = \int_0^\infty \beta(t, a, \rho(t, a; r))\rho(t, a; r)da \tag{3}$$

where β is the fertility rate function.

For populations structured only by size or by both size and age, the renewal equation has a slightly more complicated form. For example, in a size-structured population in which all newborns have the same size, m_b, the equation becomes

$$\rho(t, m_b; r) = \frac{1}{g(m_b; r)} \int_{m_b}^{\infty} \beta(t, m, \rho(t, m; r))\rho(t, m; r)dm.$$

Age-size structured populations with identical newborn size have a similar boundary condition

$$\rho(0, t, m_b; r) = \frac{1}{g(m_b; r)} \int_0^{\infty} \int_{m_b}^{\infty} \beta(t, a, m, \rho(t, a, m,; r))\rho(t, a, m; r)dmda,$$

provided that g is positive. When the initial sizes of the organisms are variable, the renewal equation becomes more involved. If size is the only variable considered and the initial sizes, m_b, are in the interval $[\alpha, \gamma]$, the renewal equation is

$$\rho(t, m_b; r) = \frac{1}{g(m_b; r)} \int_{\alpha}^{m_b} B\left(t - \int_{m_0}^{m_b} \frac{dw}{g(w; r)}, m_0\right)dm_0 + B(t, m_b),$$

with

$$B(t, m_0) = \int_{\alpha}^{\infty} \beta(t, m_0, m, \rho(t, m; r))\rho(t, m; r)dm.$$

When both age and size are employed and the initial sizes, m_b, are varying from α to γ, the renewal equation is

$$\rho(t, 0, m_b; r) = \frac{1}{g(m_b; r)} \int_0^{\infty} \int_{\alpha}^{\infty} \beta(t, a, m_b, m, \rho)\rho(t, a, m; r)dmda.$$

These forms are often implemented in their discrete analogues. We found it convenient to employ a discrete renewal equation based on age.

In many analytical studies using the model (1) investigators have found it convenient to assume that the mortality rate function μ and the fertility rate function β are time independent or density independent (e.g., Sinko and Streifer 1967; Gurtin and MacCamy 1974, 1979, Cushing 1980, 1984, 1985, Wang 1980, Pruss 1983, Busenberg and Iannelli 1985, Vlad and Popa 1985, Swart 1986). There are many situations, such as when resources are scarce or when environmental stresses are present, where such assumptions are not realistic and, at least for the present, analytical studies are not generally tractable.

2. Questions about Diversity of Individuals in the Population

According to model (1), the mass of each individual in the population grows with respect to age according to the ordinary differential equation

$$\frac{dm}{da} = g(m; r). \tag{4}$$

The parameters r in Eq. (4) represent the spectrum of physiological and environmental parameters that determine the growth of an individual. With the parameters, r, fixed for the population, there is no genetic variation in the population modeled by Eq. (1); hence, the only way that individuals within the population can differ is by their initial (birth) size and their age. For example, if the organisms are born at the same size, then their growth rates are identical because the parameters that govern growth are identical. In such a situation, the population is composed of different aged clones of a single organism. The ultimate behavior of a mathematical model of an individual organism should be governed by an equilibrium value (e.g., Kooijman 1988). Certainly, any realistic model that predicts growth of an individual must ultimately saturate with respect to size as age gets suffi- ciently large. When this occurs and the approach to equilibrium or saturation is rapid, then the variation in initial birth size is quickly dissipated and this initial variation is relatively unimportant for creating structure in the pop- ulation. In such populations, an advantage of having a larger initial birth size is that a larger organism might have additional energy storage enabling it to survive under conditions of stress, such as being born in a poor feeding environment (Reed and Balchen 1982). This advantage would not be gen- erally recognized by utilizing the model (4). Even though the population represented by (1) is modeled in terms of the variables age and size, the model (1) cannot be expected in general to yield a set of individuals in the population containing much variation; indeed, (4) can only model a popu- lation of individuals that grow according to the same dynamics and are iden- tical except for age. A natural question, then, is how variation is introduced into a model of type (1) in order to have a considerable amount of structure in the population. An answer to this question is the substance of this article.

3. Minimal Structure Components in Population Dynamics

While developing a protocol to assess the effects of toxicants on a pop- ulation we found it necessary to employ physiologically structured population models that include a component for stored materials. This additional system compartment is necessary for studies in ecotoxicology, because many industrial chemicals tend to bioaccumulate in the lipid of an individual. Hence, to adequately investigate the effects of these lipophilic toxicants on populations, it is necessary to model the lipid dynamics. Lipid tends to be a very dynamic variable in many organisms with fluctuations occurring, at least on scales of approximately the reproductive period of the organisms or faster, often increasing disproportionately with size. But more importantly, lipid can vary among otherwise similar individuals as a function of available food density and quality, and other environmental stresses. From an ecological perspective, lipid is also necessary for reproduction. Certainly, for many species of animals, females cannot make eggs unless they have an adequate amount of labile lipid. Hence, for ecological reasons as well as for ecotoxicological reasons, any model that purports to mimic natural systems should have a representation for lipid in the individual organism. The existence of dynamic fluctuations in components, such as lipid, of individual organisms can lead to sustained oscillations in structured population models. These oscillations are generated by time variational processes in the coefficients of (1), that is, in the dynamics of the components in the model of the individual, and might arise through the births associated with the individual model. Nonlinearities in (1), of course, can yield oscillations; generally, nonlinearities are incorporated in (1) through either the mortality or birth processes.

There are some simple cases of the extended McKendrick-von Foerster models with just age or size structure where fluctuations do not generally occur—in particular, when the mortality function, μ, is additively separable in age and when the density dependent term factors, the solutions obtained are nonoscillatory (Busenberg and Iannelli 1985). On the other hand, with different assumptions it is possible to obtain sustained oscillations in population models with just age or size structure. Nisbet and Gurney (1982) also address the problem of lack of oscillations in structured models from the stochastic perturbation viewpoint.

4. Density Functions and Characteristics

Because the structured problem described by Eqs. (1), (2), and (3) is a well posed hyperbolic partial differential equation, it can be studied by using the method of characteristics. However, when this method is applied to standard settings where, for example, the mortality is a function of total population size, the unknown function is still involved implicitly in the expression for the solution. Hence, the solution may not be found explicitly, and the analytical study of the model is, in general, very difficult, especially when time variation in demographic parameters and density dependence are also included in the formulation. The density function ρ in (1) has dimension numbers per unit age per unit mass per volume of the environment. Along a characteristic, it describes the motion and represents the age dynamics of a cohort. The equations of the characteristics associated with the age-size structured model:

$$\begin{cases} \rho_t + \rho_a + (g\rho)_m = -\mu\rho, \\[2mm] \rho(t, 0, m_b) = \int\int \beta\rho dadm, \\[2mm] \rho(0,a, m) = \phi(a, m) \end{cases}$$

are

$$\begin{cases} \dfrac{dt}{ds} = 1, \\[3mm] \dfrac{da}{ds} = 1, \\[3mm] \dfrac{dm}{ds} = g, \\[3mm] \dfrac{dp}{ds} = -(\mu + g_m)\rho. \end{cases}$$

Along a characteristic, the death rate μ affects the survival of a cohort; however, by the last of the characteristic equations, the rate of change of the growth rate with respect to size, g_m, is also a governing factor in determining the density of the cohort (Metz and Diekmann 1986). This means that, along a characteristic, the function ρ can change not only when mortality acts but also as the growth rate of the individual changes. It is possible for the density to increase along characteristics if the growth component increases enough to compensate for mortality. For purposes of interpretation, it is convenient to transform the problem to the setting where the computations along characteristics indicate effects of mortality only.

This can be accomplished if the function representing individual size, $m(t, a)$, on each characteristic is a single valued function of t and a. In this situation, ρ is a function of t and a only, that is

$$\rho(t, a, m) = \rho(t, a, m(t, a)) \equiv \hat{\rho}\,(t, a).$$

Let $n(t, a) = \hat{\rho}(t, a)h(t, a)$ where h satisfies

$$h_t + h_a = g_m h.$$

Then along characteristics n is a decreasing density function, in the sense that only mortality is acting. The initial and boundary data for h and n need only be consistent with that of ρ. We have found it advantageous to employ these ideas in our development of a protocol to assess the risk to a model aquatic population following exposure to a toxic chemical.

II. Computation of Solutions of Physiologically Structured Models

We have developed an approach to assess the risk of a population following exposure to a toxic chemical (Hallam et al. 1990a, 1990b). In these articles we focused on the application of the procedure and provided only minor motivation for the approach. An objective of this article is to present the rationale for this approach and, at the same time, outline a procedure for studying populations from an individual perspective. We believe that the combination of ideas blends naturally in a study of population dynamics.

The rudiments of the procedure are: 1) formulation of the individual model, 2) determination of a set of parameters in the individual model that can be used to generate diversity in the population, 3) formulation of the population model incorporating the individual dynamics, and 4) numerical solution of the partial differential equations modeling the metapopulation.

1. Formulation of the individual model

We have remarked that, to determine the effects of lipophilic toxicants, it is necessary to utilize an individual model that has lipid content of the individual as a component. Thus, to keep the illustration as simple as possible, and to relate it to our previous work, we discuss the dynamics of an individual through two variables, namely, the structure, m_s, and lipid, m_l. The ordinary differential equations that are assumed to model the growth of an individual are of the form

$$\frac{dm_l}{da} = L(a, m_l, m_s; r),$$

$$\frac{dm_s}{da} = S(a, m_l, m_s; r) \tag{5}$$

We have described the equations that we employ for Daphnia dynamics in additional detail in Hallam, Lassiter, Li, and Suarez (1990) and described briefly the dynamics for fish in Hallam et al. (1990b). In these cases, the functions L and S in (5) are neither continuous nor monotone because we have chosen to allocate structure and lipid to egg production at discrete reproductive times. Other situations where reproductive allocation is treated in a continuous manner are described by Kooijman and Metz (1984) and Kooijman (1986). The Eqs. (5) represent the growth of an individual and are fundamental to an investigation of population dynamics. For reference at the population level, however, other life history information during the aging process must be obtained from the individual model. This additional life history includes the number of viable eggs produced, N_e, and their associated lipid and structure content, m_{l0} and m_{s0}, respectively. We will now indicate how these data are utilized for specification of the birth process at the population level. The procedure for the solution of the population model

$$
\begin{cases}
\rho_t + \rho_a + (\rho L)_{m_l} + (\rho S)_{m_s} = -\mu\rho, \\[2mm]
\rho(t, 0, m_{l0}, m_{s0}; r) = \iiint \beta\rho da dm_l dm_s, \\[2mm]
\rho(0, a, m_{l0}, m_{s0}; r) = \psi(a, m_{l0}, m_{s0}; r)
\end{cases}
$$

requires determination of the birth functions and the initial composition of the egg component values. This determination is accomplished by employing the individual model to govern the times of reproduction and to establish the number and size of the eggs produced. We have assumed that the times of reproduction for an individual occur discretely. In our individual models of daphnids and fish, the onset of maturity is assumed to occur at the time when a particular size is attained after which the female reproduced periodically (Kooijman and Metz 1984). The size of an egg is determined by the amount of lipid and structure that is available for reproductive allocation at the time of reproduction. Our approach assumes that control functions determine the number of eggs that can be made from each of the labile lipid and labile structure components of the individual (see Hallam, Lassiter, Li, and Suarez 1990 for the details).

Our introductory illustration above discusses a case where the number of physiological variables is two—lipid and structure. A more general setting can be developed for higher dimensional physiological problems. For example, if m is an n-vector of physiological variables for the individual, then the growth of the individual can be assumed to be described by

$$
\frac{dm}{da} = G(a, m; r), \tag{6}
$$

where $m = (m_1, m_2, \ldots m_n)$, and $G = (g_1, g_2, \ldots g_n)$. The same approach that is developed for two physiological variables can be utilized in this higher dimensional setting.

2. Determination of the parameters in the individual model that can be used to structure the population

A difficulty with employing a population model consisting of a single partial differential equation is that it cannot generate the genetic variation observed in natural populations. Variation is necessary to explain different responses of natural systems to different types of perturbation. One way to introduce variation among individuals and remain faithful to this structured approach is to vary parameters in the individual model. Variation can be obtained by employing one of several deterministic or stochastic techniques. We have chosen to discuss a deterministic method that introduces additional partial differential equations into the description of the population, one for each subpopulation consisting of clones of a single type of individual. A population composed of several different subpopulations is called a *metapopulation*.

By conducting a sensitivity study on the individual model to determine the parameters that lead to distinctly different types of individuals, we determined the parameters that were significant in structuring individuals. We also searched for parameter values that yielded the largest and smallest individuals that were consistent with known data for a specific species. As an illustration of an important parameter for the individual daphnid model, we found that the filtering rate parameter provided considerable variation among individuals and was important for structuring the population (Hallam et al. 1990a). In the individual fish model, the gut clearance rate parameter was used to introduce variation among individuals in the population (Hallam et al. 1990b) because of its sensitivity. Other parameters such as density and quality of the resource were also found to yield considerable diversity among the types of individuals that can be obtained from the individual model.

Variation can be introduced into the population by choosing a number of individual ecotypes determined by parameter values obtained from the sensitivity study. In each of the *Daphnia* and fish models, we have worked primarily with three parameters—two environmental parameters (average density and average quality of the resource to which an individual has access) and one physiological parameter for each individual type—to form the population. Choice of 3 values for each of the 3 parameters allows a diversity of structure in the population that covers 27 types of individuals. Proper selection of parameter values generates diversity and structure for the population, but there is a price to be paid. The existence of 27 ecotypes of individuals in the population requires the simultaneous solution of 27 partial differential equations to study the metapopulation.

For our illustrations, we assume that, in the individual model (6), the parameter vector r takes the values from a discrete set $R = \{r_1, r_2, \ldots r_{27}\}$.

By choosing an r_k from the set R, m is only a function of a and r_k, i.e. $m = m(a; r_k)$ and (6) becomes

$$\frac{dm}{da} = G(a, m(a; r_k); r_k).$$

As indicated above, our computations will use $p = 27$ and $n = 2$. For a fixed r_k, we denote by

$$\rho^{(k)} = \rho(t, a, m(t, a; r_k)),$$

the density function for a subpopulation of individuals that have the same r_k parameter. The dynamics of this subpopulation are described by the equation

$$\rho_t^{(k)} + \rho_a^{(k)} + \sum_{i=1}^{n} \frac{\partial}{\partial m_i} (\rho^{(k)} g_i) = - \mu^{(k)} \rho^{(k)},$$

where $\mu^{(k)} = \mu^{(k)} (a, m, \rho^{(1)}, \rho^{(2)}, \ldots \rho^{(27)})$.

Letting r_k range throughout the set R results in a system of partial differential equations that describes the metapopulation. The density function of the metapopulation is the sum of the subpopulation density functions $\rho = \Sigma_k \rho^{(k)}$.

3. *Formulation of the population model incorporating individual dynamics*

It is a routine task to incorporate individual dynamics into a population setting via the extended McKendrick-von Foerster equation. Once the individual dynamics are specified by (6) the population model can be written as

$$\begin{cases} \rho_t + \rho_a + \sum_{i=1}^{n} (g_i\rho)_{m_i} = -\mu\rho, \\[2ex] \rho(t, 0, m_b) = \int\int \beta\rho\, da\, dm, \\[2ex] \rho(0, a, m) = \psi(a, m). \end{cases} \tag{7}$$

To our knowledge, the first article that uses this n-dimensional model was published by Auslander et al. (1974).

The corresponding characteristic equations are

$$\begin{cases} \dfrac{dt}{ds} = 1, \\[2mm] \dfrac{da}{ds} = 1, \\[2mm] \dfrac{dm_1}{ds} = g_1(m_1, m_2, \ldots, m_n), \\[2mm] \vdots \\[2mm] \dfrac{dm_n}{ds} = g_n(m_1, m_2, \ldots, m_n), \\[2mm] \dfrac{d\rho}{ds} = -\left(\mu + \displaystyle\sum_{i=1}^{n} \dfrac{\partial}{\partial m_i} g_i\right)\rho. \end{cases} \tag{8}$$

The first $n + 2$ equations are the characteristic equations of a system of partial differential equations:

$$\frac{\partial}{\partial t} m_1 + \frac{\partial}{\partial a} m_1 = g_1(m_1, m_2, \ldots, m_n),$$

$$\frac{\partial}{\partial t} m_2 + \frac{\partial}{\partial a} m_2 = g_2(m_1, m_2, \ldots, m_n),$$

$$\vdots \tag{9}$$

$$\frac{\partial}{\partial t} m_n + \frac{\partial}{\partial a} m_n = g_n(m_1, m_2, \ldots, m_n).$$

Here, we are interested only in analyzing the case $a \le t$. For given boundary conditions

$$m_i(t, 0) = \phi_i(t, m_{10}, m_{20}, \ldots m_{n0}), \quad i = 1, 2, \ldots, n,$$

the system (9) represents characteristic surfaces. As discussed above, there are situations where $m_i(t, a)$ are single-valued functions of t and a. In this situation, on each characteristic surface, the function ρ is a function of t and a only. Denote by $\hat{\rho}$

$$\rho(t, a, m_1(t, a), m_2(t, a), \ldots, m_n(t, a)) \equiv \hat{\rho}(t, a)$$

and let $n(t, a) = \hat{\rho}(t, a)h(t, a)$, where h satisfies the linear partial differential equation

$$h_t + h_a = \left(\sum_{i=1}^{n} \frac{\partial}{\partial m_i} g_i + c \right) h.$$

The boundary conditions for n and h are chosen to be compatible with those of ρ. The arbitrary constant, c, is to be used as a scaling factor for computational purposes if needed. It follows that n satisfies the partial differential equation

$$n_t + n_a = -\hat{\mu}(t, a)n,$$

where

$$\hat{\mu}(t, a) = \mu(t, a, m_1(t, a), m_2(t, a), \ldots, m_n(t, a), \frac{v}{h}(t, a)) - c.$$

Since we only vary the parameter vector r_k in the set R and the rest of parameters are assumed to be fixed, the Eqs. (9) generate a set of characteristic surfaces. As r_k ranges through the set R, the family of characteristic surfaces is generated. Then, the population dynamics can be represented along this family of characteristics.

It is, in general, a nontrivial task to describe and formulate the birth process for a population model. As we indicated above, this should be formulated in connection with the individual model. To illustrate the details and concerns that one must address, we return to the approach taken in our *Daphnia* and fish model analysis.

The ith subpopulation model has the following form:

$$
\begin{cases}
\rho_t^{(i)} + \rho_a^{(i)} + (\rho^{(i)}L_i)_{m_l} + (\rho^{(i)}S_i)_{m_s} = -\mu\rho^{(i)}, \\
\rho^{(i)}(t, 0, m_{l0}, m_{s0}) = \displaystyle\int\int\int \beta(t, a, m_{l0}, m_{s0}, m_l^i, m_s^i) \\
\qquad\qquad \times \rho^{(i)}(t, a, m_l^i, m_s^i)da\,dm_l^i\,dm_s^i, \\
\rho^{(i)}(0, a, m_l^i, m_s^i) = \phi^{(i)}(0, a, m_l^i, m_s^i),
\end{cases}
\tag{10}
$$

where the L_i and S_i are the growth rates of the ith physiological variable as in (5). We assume that the initial structure content of eggs m_{s0} is constant, and that the initial lipid content of eggs m_{l0} varies in a discrete set I_0. Then, $m_l^i(t, a, m_{l0})$ and $m_s^i(t, a, m_{l0})$ represent the characteristic surfaces. On this family of characteristic surfaces, the renewal equation becomes

$$n(t, 0) = \sum_{i \in I_0} \int_0^t \beta^{(i)}(t, a, m_{l0}) \frac{h^i(t, 0)}{h^i(t, a)} n^i(t, a)da. \tag{11}$$

The fertility rate function in the *Daphnia* and fish population models illustrated in the examples below were computed by using the following techniques. Denote the onset of maturity by A_i for individuals with initial lipid content m_{l0}^i and the constant periodic time at which the female reproduces after A_i by P. Hence, an individual will reproduce first at an age A_i; this age is dependent upon the growth characteristics of that specific individual. For every individual in the population, each reproductive period after the first occurs periodically with constant frequency P. This frequency of reproduction is a population level parameter. The function $\beta^{(i)}$ is computed by the formula

$$\beta^{(i)}(t, a, m_{l0}) = \sum_{k \geq 0} N_e(x_{i,k}) \delta_{x_{i,k}}(a), \tag{12}$$

where $x_{i,k}$ denotes a discrete set of reproductive ages and, hence, is of the form $x_{i,k} = A_i + kP$ for some $i \in I_0$ and $k = 0, 1, 2, \ldots$; δ is the delta function, $a \leq t$, and $N_e(x_{i,k})$ is the number of eggs produced by an individual at age $x_{i,k}$. The renewal equation (11) becomes

$$n(t, 0) = \sum_{i \in I_0} \sum_{k \geq 0} N_e(x_{i,k}) \frac{h^i(t, 0)}{h^i(t, x_{i,k})} n^i(t, x_{i,k}).$$

Another fundamental and equally difficult problem for population ecology is the determination of the mortality representation in the population model. In our previous studies, we employed the function

$$\mu = \mu_1 + \mu_2 + \mu_3, \tag{13}$$

where the term μ_1 represents the mortality in the population due to age, μ_2 represents the mortality due to size (this is viewed as a predation term where the predator eats prey according to a size rule) and the μ_3 term is the mortality caused by all other factors lumped together and represented by density dependence. The specific forms of μ_i used in our previous application are species dependent and may be found in Hallam et al. (1990a).

The initial distribution is also relevant to population dynamics because of the deterministic approach that is utilized. If diversity in the population is desired, then the initial distributions must also contain individuals that differ in structure (parameter values).

Let n be the dimension of the parameter space used to structure the population. Then the metapopulation is composed of n hyperbolic partial differential equations of the form of (10), $i = 1, 2, \ldots, n$, where the mortality is given by (13) and the birth process is given by (12).

4. *The numerical solution of the partial differential equations modelling the metapopulation*

The approach utilized to compute the population density function, ρ, of the metapopulation is the method of characteristics. Along characteristics (e.g., (8)), the cohort dynamics are governed by ordinary differential equations; hence, the numerical techniques required are just those available to solve ordinary differential equations. The method of characteristics may be interpreted biologically as following the family tree of the clonal organisms. There is one family tree for each ecotype of individual. At branches in the family tree representing production of new individuals, a new characteristic is started in the mathematical model.

Although the individual model is highly nonlinear, the metapopulation model is linear except for the density dependence in the mortality rate term. When the effects of density-dependent mortality are assessed in the population, the numerical scheme must consider this coupling of subpopulations. This coupling process is accomplished by a computational step along each characteristic employing the density at the previous time step as the value of ρ in the mortality term μ_3.

The numerical scheme, while following the life history of cohorts, aggregates characteristics whenever it is numerically and biologically reasonable. Minimization of the number of cohorts, whose dynamics must be computed, is essential for numerical study of population and, eventually, of community dynamics.

Graphical illustrations of the computed time-dependent distributions of the physiological variables in our *Daphnia* and fish models are given in Fig. 1 and 2, respectively. The interested reader may find lists of the relevant parameter values employed in the computations in the papers by Hallam et al. (1990a, 1990b).

Techniques other than the method of characteristics have been used to solve McKendrick-von Foerster equations. We mention only the escalator boxcar train method by de Roos (1988).

Discussion and Summary

Incorporation of individual diversity into a model population can be attained conceptually in almost innumerable ways when genetic, physiological and environmental factors are considered. This natural occurrence contrasts with the classically studied single extended McKendrick-von Foerster model that allows for variation among individuals in a population only through different initial birth sizes and age. In this traditional situation all individuals have the same genetic, physiological and environmental constraints, and must grow according to the same dynamics.

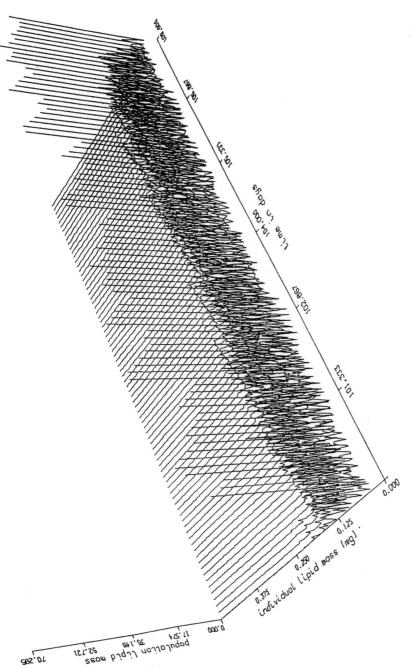

Figure 15.1 The scaled distributions of lipid (a), protein or structure (b), and age (c) in a Daphnia population. The time interval of the simulation is from day 100 to day 108 in (a), (b), and (c). Each computation during the two day period is graphed. The oscillations observed in the population on this time scale are determined by a birth process that is periodic after the first reproduction. Fig. (d) represents a simulation of the population over a period of 400 days. Observe the longer term oscillations that occur over this computation period.

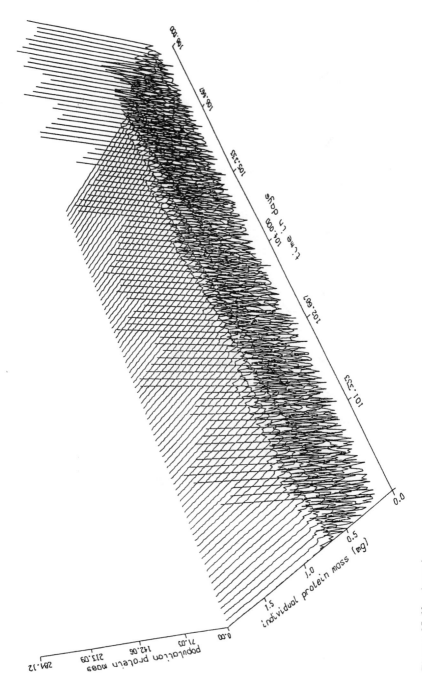

Figure 15.1b (cont.)

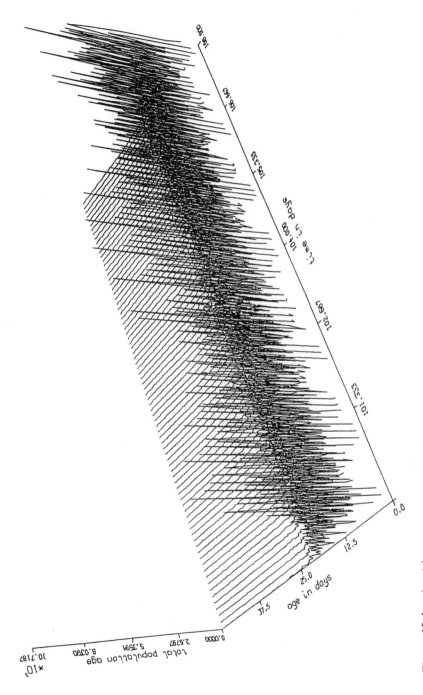

Figure 15.1c (cont.)

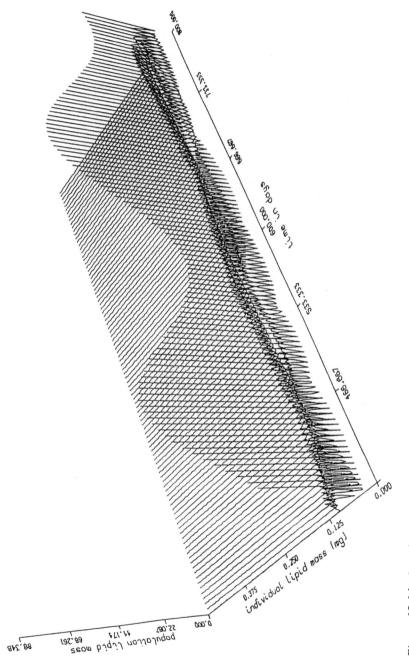

Figure 15.1d (cont.)

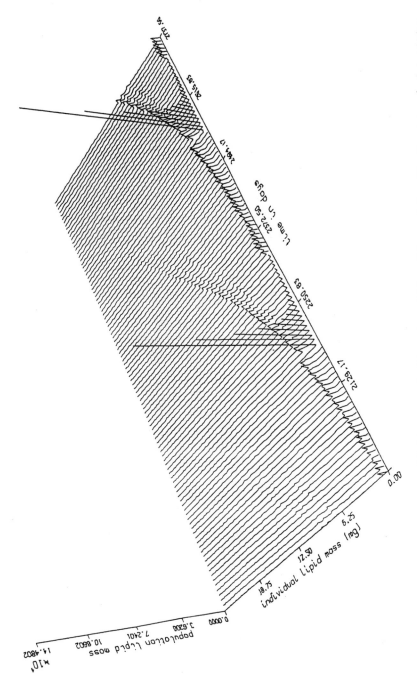

Figure 15.2 The scaled distributions of lipid (a), protein (b), and age (c) in a fish population. The time interval of simulation is from day 2000 to day 2737.

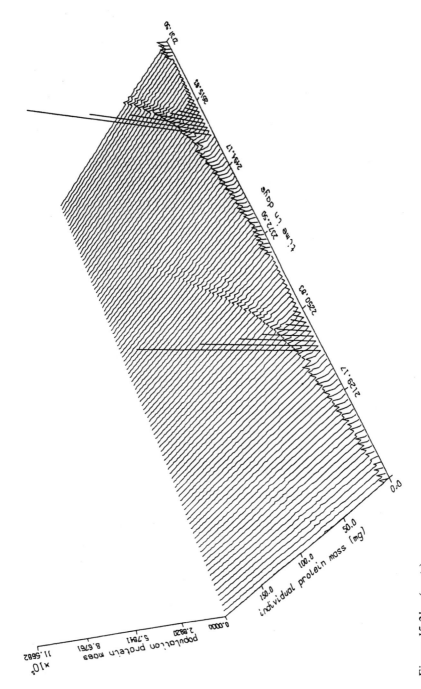

Figure 15.2b (cont.)

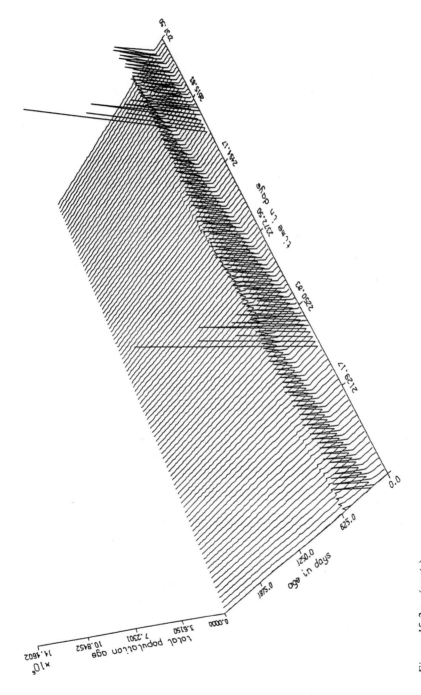

Figure 15.2c (cont.)

It often is necessary to have variation in models in order to investigate populations that have considerable diversity. If any variation other than initial size and individual age is to occur in the population, then the question arises as to what approach could be used to introduce variation into the population. The approach discussed here is a deterministic one. We couple several equations of type (1); each equation is determined by a different parameter r that is used in the description of an individual's growth. We assume individual growth is governed by the same representations of physiological processes, but the change of parameters results in different growth dynamics (4). Structure in a population model can be achieved by incorporating these additional equations modeling subpopulations that have different genetic characteristics (as reflected by different parameter values) into a larger component model. This leads to a metapopulation that exhibits variation in both initial individual size and physiological structure.

An adequate model of a natural or laboratory population or community must account for physiological variation in several (and, more probably, in many) states. Many natural populations exhibit dynamic oscillations; for example, *Daphnia* populations have fluctuations that are relatively long in comparison to the reproductive cycle (Murdoch and McCauley 1985), and a setting where fluctuations can readily occur must be used to realistically reflect these dynamic features. We have indicated that incorporation of dynamic physiological variables into the population model is one means of achieving temporal oscillations in the population.

We have presented a protocol to study the dynamics of populations. While the investigative level is the population, the foundation of the work is the individual. Fortunately, the dynamics of individuals can be incorporated directly into subpopulation models of the McKendrick-von Foerster type. By introducing a sufficient amount of genetic or environmental variation into the individual model, a metapopulation composed of a diverse group of individuals results. Our protocol requires that the individual model follow the dynamics of several physiological variables. This extra burden adds to the computation time, but introduction of the additional physiological variables does not significantly complicate the conceptual framework or the numerical code.

Our primary illustrations are from a procedure that was developed in an ecotoxicological setting to study *Daphnia* or fish populations subjected to toxic stress. In our numerical work on *Daphnia* and fish population models, the behavior of the metapopulation is consistent with evolutionary dogma. The "fittest" individual ecotypes are defined to be those that ultimately dominate the metapopulation. We found that, in the absence of stress, the fittest individual type—the fastest growing individual according to our individual model—would ultimately dominate the metapopulation. This "survival of the fittest" gives a reference level for comparison of nominal versus stress behavior of model populations. We found that, with suitable conditions on the chemical stressor, the successional winner in the model metapopulation could be any of the different individual types. For example, in the *Daphnia* population model, we were able to construct situations where the "fittest" individuals were the slowest growers and the leanest of all types in the population (Hallam et al. 1990a). Another stress by a different chemical led to a different conclusion—the fittest were organisms that grew at a level intermediate between the faster and slower growing ecotypes. Analysis of the fish model essentially led to the same general conclusions (Hallam et al. 1990b): stress can govern outcomes of population dynamics. We found that a set of individual types dominates the ultimate behavior of the population. The particular degree of fitness is determined by both biological and environmental (including exogenous and anthropogenic) factors.

This article provides the basis for modeling populations by incorporating variation among individuals. The protocol has been implemented, and we think it provides a promising numerical tool to investigate the dynamics of populations. Because the development is based on individuals, it utilizes a data base focused at the level of study of many biological investigations. Population growth parameter values are estimated from individual information whereas the mortality might contain information from the population level. In the systems terminology, this procedure is a combination of both a top-down and a bottom-up approach.

Acknowledgement

This research was supported in part by the U.S. Environmental Protection Agency with the cooperative agreement CR816081.

Literature Cited

Auslander, D. M., G. F. Oster, and C. B. Huffaker. 1974. Dynamics of interacting populations. *J. Franklin Inst.* **297**:345–374.

Busenburg, S. and M. Iannelli. 1985. Separable models in age-dependent population dynamics. *J. Math. Biol.* **22**:145–173.

Cushing, J. M. 1977. *Integrodifferential Equations and Delay Models in Population Dynamics*. Lecture Notes in Biomathematics 20. Springer-Verlag, Berlin.

Cushing, J. M. 1980. Model stability and instability in age structured populations. *J. Theor. Biol.* **86**:709–730.

Cushing, J. M. 1984. Existence and stability of equilibria in age-structured population dynamics. *J. Math. Biol.* **20**:259–276.

Cushing, J. M. 1985. Equilibria in structured populations. *J. Math. Biol.* **23**:15–39.

de Roos, A. M. 1988. Numerical methods for structurd populations: the Escalator boxcar train. *Num. Meth. Part. Diff. Eq.* **4(3):** 173–195.

Gurtin, M. E. and R. C. MacCamy. 1974. Nonlinear age-dependent population dynamics. *Arch. Rational Mech. Anal.* **54:**281–300.

Gurtin, M. E. and R. C. MacCamy. 1979. Some simple models for nonlinear age-dependent population dynamics. *Math. Biosci.* **43:**199–211.

Hallam, T. G., R. R. Lassiter, J. Li, and W. McKinney. 1988. Physiologically structured population models in risk assessment. In *Biomathematics and Related Computational Problems*, L. M. Ricciardi (ed.), pp. 197–211, Kluwer, Amsterdam, The Netherlands.

Hallam, T. G., R. R. Lassiter, J. Li, and W. McKinney. 1990a. Determination of effects of lipophilic toxicants on dynamics of *Daphnia* populations. *Environ. Toxicol. Chem.* **9:**597–621.

Hallam, T. G., R. R. Lassiter, J. Li, and W. McKinney. 1990b. Modelling the effects of toxicants on fish populations. In *Proceedings of the Symposium on Fish Toxicology, Fish Physiology, and Fisheries Management*, R. Ryans (ed.), pp. 299–320, EPA/600/9-90/011, U.S. Environmental Protection Agency, Athens, GA.

Hallam, T. G., R. R. Lassiter, J. Li, and L. A. Suarez. 1990. Modelling individuals employing an integrated energy response: Application to *Daphnia. Ecology* **71:**938–954.

Hoppensteadt, F. 1975. *Mathematical Theories of Populations: Demographics, Genetics, and Epidemics*, pp. 72, Conf Series in Applied Mathematics, SIAM, Philadelphia.

Huston, M., D. L. DeAngelis, and W. M. Post. 1988. New computer models unify ecological theory. *Bioscience* **38:**682–691.

Kooijman, S. A. L. M. 1986. Population dynamics on a basis of budgets. In *The Dynamics of Physiologically Structured Populations*, J. A. J. Metz and O. Diekmann (eds), pp. 266–297. Lecture Notes in Biomathematics 68. Springer-Verlag, Berlin.

Kooijman, S. A. L. M. 1988. The von Bertalanffy growth rate as a function of physiological parameters. In *Mathematical Ecology, Proceedings of the Autumn Course Research Seminars*, T. G. Hallam, L. J. Gross, and S. A. Levin, (eds.), pp. 3–45. World Scientific, Singapore.

Kooijman, S. A. L. M. and J. A. J. Metz. 1984. On the dynamics of chemically stressed populations: the deduction of population consequences from effects on individuals. *Ecotoxicol. Environ. Saf.* **8:**254–274.

Metz, J. A. J. and O. Diekmann. 1986. *The Dynamics of Physiologically Structured Populations*. Lecture Notes in Biomathematics 68. Springer-Verlag, Berlin.

Murdoch, W. W. and E. McCauley. 1985. Three distinct types of dynamic behavior shown by a single planktonic system. *Nature* **316:**628–630.

Nisbet, R. M. and W. S. Gurney. 1982. *Modeling Fluctuating Populations*. Wiley, Chichester, England.

Pruss, J. 1983. On the qualitative behavior of populations with age-specific interactions. *Comput. Math. Appl.* **9**:327–339.

Reed, M. and J. G. Balchen. 1982. Multidimensional continuum model of fish population dynamics and behavior: Application to the Barents Sea capelin (*Mallotus villosus*). *Model. Identif. Control* **3**:65–109.

Sharpe, F. R. and Lokta, A. J. 1911. A problem in age distributions. *Philos. Mag.* **21**:435–438.

Sinko, J. W. and W. Streifer. 1967. A new model for age-size structure of a population. *Ecology* **48**:910–918.

Swart, J. H. 1986. The stability of a nonlinear, age-dependent population model as applied to a biochemical reaction tank. *Math. Biosci.* **80**:47–56.

Vance, R. R., W. I. Newman, and D. Sulsky. 1988. The demographic meanings of the classical population growth models of ecology. *Theor. Popul. Biol.* **33**:199–225.

Vlad, M. O. and V. T. Popa. 1985. A new nonlinear model of age-dependent population growth. *Math. Biosci.* **76**:161–184.

Wang, F. S. J., 1980. Stability of an age dependent population. *SIAM J. Math. Anal.* **11**:683–689.

Webb, G. F. 1985. *Theory of Nonlinear Age-dependent Population Dynamics.* Marcel Dekker, New York.

16

Biomass Conversion at Population Level

S. A. L. M. Kooijman

ABSTRACT. The conversion efficiency of prey biomass into predator biomass (or plant-biomass into herbivore-biomass) at the individual level is not constant. Measured by the yield factor, this conversion appears to depend on size, energy reserves, and a few species-specific compound parameters involving a variety of elements from the dynamic energy budget. This efficiency is studied here on the basis of a continuous time model for the energy budget of an individual organism.

At the population level, the conversion of prey into predator biomass depends on harvesting processes. Included in harvesting is death by aging. In the absence of harvested individuals from a population of predators, a constant supply rate of prey results in a predator population of constant size. At equilibrium, this gives a conversion efficiency of zero. When the predator population is harvested, the conversion efficiency increases up to a maximum and then decreases to a threshold level, above which the population goes extinct. A practical implication is that it is possible that populations with substantial standing crops cannot stand sustained increased harvesting.

For a well tested model for energy budgets of individuals, the conversion efficiency of prey biomass into predator biomass has been compared for two modes of prey selection by the predator: random predation and fixed age predation. Both strategies gave almost identical biomass conversion efficiencies. Propagation through eggs has been found to be a bit less efficient that propagation through binary fission.

The way the efficiency depends on parameters of the individual energy budget is an important link between ecology and physiology, a fact which is highly relevant to areas of applied ecology, such as ecotoxicology. The study of this relation is complicated by population oscillations induced by the harvesting process, but computer simulations show that an analysis of the conversion efficiency on the basis of stable age distributions is still valid.

Introduction

Chronic exposure of animals to toxic substances can lead to changes in physiological characteristics, such as a reduction in life time, reproduction, feeding rate, etc. While such effects can be measured in the laboratory rather straightforwardly, the ecological relevance of such effects is not always obvious due to lack of theory relating physiological characteristics to population dynamics and subsequently to ecosystem dynamics. Such theories are needed because of experiments and theoretical considerations which show that the effect of a toxicant at a certain (fixed) concentration on a population can depend sensitively on the population growth rate (Kooijman and Metz, 1983). This rate depends on a variety of other factors, such as food availability, and consequently, effects of toxicants at population level. This is an example of one context within which it is relevant to look for the physiological basis of ecological phenomena, a procedure which is now becoming more widely recognized (see e.g., Hallam and de Luna 1984, Calow and Sibly 1990, Łomnicki 1988, Kooijman et al. 1987).

The purpose of this paper is to evaluate the conversion of food (prey) biomass into animal (predator) biomass on the basis of a dynamical energy budget (DEB) model of an individual (Kooijman 1986a). Toxic substances can change particular parameter values of this model. The change depends on the particular mode of action of the chemical. Thus the presence of a toxicant can in turn affect biomass conversion efficiencies, which should play an important role in realistic models for population dynamics. At the individual level the conversion depends on the allocation of energy to growth and reproduction, after "payment" of maintenance costs.

At the population level the situation is quite different. Think of what is known as an idealized fed batch culture in the microbiological literature, i.e. a closed population, which is supplied with a continuous, constant input of food without any outflowing food, and in which the individuals interact only through intraspecific competition. All individuals that die (for whatever reason) are instantaneously removed, an action called harvesting in this article, but those surviving remain in the population. All causes of death are lumped, because it does not make much difference for the population whether individuals become lost by aging or predation, nor does it matter for the predators whether they feed on dead individuals as scavengers or on living prey. In this way, the population actually converts food into feces and dead individuals. We could define a yield factor by dividing the total volume of dead biomass produced per unit of time by the total food volume supplied per unit of time. The standing crop and the food density are states of the system which, by definition, do not ultimately change in the mean. Feces production is assumed to be proportional to food consumption, and therefore food supply rate, when averaged over a long period. We have to multiply the yield factor with some constant to take feces production into account.

Suppose, for the moment, that we could prevent any harvesting processes from occurring, even that caused by aging. If we start the experiment with a small number of individuals, they will grow and reproduce until all the food supplied is used for maintenance. From that moment on, no production will take place, and the yield factor will be zero. If we 'admit' death by aging, the result will be that the standing crop is a bit lower compared with the former situation, but the population is actually producing dead individuals. So the yield factor is larger than zero. When additional yield processes are operational, like predation, the yield factor might increase still further.

This thought experiment shows that at the population level, conversion of food into animal biomass is controlled by harvesting processes. The energy budget at the individual level provides the constraints. To see how the constraints are imposed, I proceed using a quantitative approach. For this purpose, I will first evaluate biomass conversion at the individual level on the basis of the DEB model, and use it as a reference.

The Dynamical Energy Budget Model

A list of frequently used symbols is given in Table 16.1 Let us distinguish three isomorphic volume—defined life stages of a female animal: embryos, which do not feed or reproduce, juveniles, which feed but do not reproduce, and adults, which feed and reproduce. Volume-defined life stages differ from age-defined ones by the phenomenon that when food is scarce, growth will be retarded, resulting in a transition to the next stage that is delayed or does not occur. Suppose that ingestion, I, depends hyperbolically on food density, X, and that it is proportional to squared length, L^2. So, $I = \{I_m\}fL^2$, with $f = X/(K + X)$ and $\{I_m\}$ is the maximum surface area-specific ingestion rate. The conversion efficiency of ingested food to energy is assumed to be fixed (Evers and Kooijman 1989) at value $\{A_m\}/\{I_m\}$, where $\{A_m\}$ is the maximum surface area-specific assimilation rate. Routine metabolic costs are taken to be proportional to cubed length, ζL^3. The costs of maintaining the differentiated state are taken to be proportional to the minimum of (actual) cubed length and cubed length at the end of the juvenile stage (Thieme 1988, Zonneveld and Kooijman 1989), L_j^3. The proportionality constant is such that the individual enters a new stage at a certain volume as soon as it invested a fixed amount of energy into its development. This occurs when the costs of maintaining the differentiated state equals $[(1 - \kappa)/\kappa] \zeta \min(L^3, L_j^3)$. Energy from food is first stored. A basic difference between structural biomass and materials representing stored energy is in the way they are made temporary. Structural biomass is subjected to maintenance while stored materials are subjected to continuous supply and utilization.

Table 16.1. List of frequently used symbols

L, L_b, L_j	length, — at birth, — at end juvenile stage
$L_m = \dfrac{\kappa\{A_m\}}{\zeta}$	maximum length
$l = \dfrac{L}{L_m}$	scaled length
$l_b = \dfrac{L_b}{L_m},\ l_j = \dfrac{L_j}{L_m}$	— at birth, — at end juvenile stage
K	saturation constant
$X,\ x = X/K$	food density (volume food per volume medium), scaled —
$f = x/(1 + x)$	scaled functional response
$X_s,\ x_s = \dfrac{X_s}{mK}$	supplied volume of food per volume medium per time, scaled —
t	time
$\{I_m\}$	maximum surface area-specific ingestion rate
$I = \{I_m\}\dfrac{L_m^2}{Km}$	scaled ingestion
$\{A_m\}$	maximum surface area-specific assimilation rate
$\{A_m\}/\{I_m\}$	energy yield of a unit volume of food
ζ	volume-specific maintenance costs
η	volume-specific costs for growth
κ	fraction of utilized energy spent on growth plus maintenance
$[S],\ [S_m]$	stored energy density, maximum —
$e = [S]/[S_m]$	scaled stored energy density
$a = \dfrac{\eta}{\kappa[S_m]}$	energy investment ratio
$m = \zeta/\eta$	maintenance rate coefficient
$v = \{A_m\}/[S_m]$	energy conductance
R	reproduction rate, i.e. no. of eggs per individual per unit of time
Y	yield factor, i.e. conversion efficiency of food into biomass
$Y_i,\ Y_n,\ Y_p$	instantaneous —, non-instantaneous —, — at population level
$y = Y\dfrac{\{I_m\}[S_m]}{\{A_m\}}$	scaled yield factor
N	number of animals in the population
E	mean age of an individual
F_t	survival probability at t
$b*L_m^3$	biomass, i.e. total volume of animals in the population
$b = b*I/x_s$	scaled biomass
p	rate of random predation
q	survival probability per hatching

The dynamics of stored energy can be derived from the way growth rate and ultimate volume depend on food density when constant (Kooijman 1986b). Expressed as density, i.e. stored energy divided by cubed length, $[S] = S/W$, it follows a simple linear first order dynamics with a relaxation time proportional to length: $(d/dt)[S]/[S_m] = vL^{-1}(f - [S]/[S_m])$, where the compound parameter $v = \{A_m\}/[S_m]$ is called the energy conductance. I will write e for $[S]/[S_m]$. A fixed fraction, κ, of energy utilized from the storage is spent on growth plus maintenance, the rest on reproduction plus development (including the maintenance of the differentiated state). The volume-specific costs for growth, η, is assumed constant. This implies that the change in volume is given by $(d/dt) L^3 = (L^2 ev - L^3 am)/(e + a)$, where $m = \zeta/\eta$ is called the maintenance rate constant and $a = \eta/\kappa[S_m]$ the energy investment ratio. The cubic root of the maximum body volume is $L_m = \kappa\{A_m\}/\zeta$, which is ultimately reached for $f = e = 1$. Maintenance costs are given priority over growth or reproduction when required during starvation. The energy flow spent on development in juveniles is spent on reproduction in adults. The additional assumption that the energy storage density of the hatchling equals that of the mother at egg laying defines embryo development, and so energy investment in an egg (see Kooijman 1986c, Zonneveld and Kooijman 1991, and the appendix).

Biomass Conversion at the Individual Level

Structural body mass and reserve materials are both assumed to be of constant, but not identical, chemical composition. This poses a problem in the present consideration of the conversion of food into animal biomass: Storage density depends on food density history. So, biomass does not have a constant chemical composition. In the real world this is a well known fact, of course, and here it is an integral part of the model as well. One possibility for expressing conversions is in terms of energy, which requires a conversion factor from biomass into energy. This has to differ from η^{-1}, because growth involves a considerable overhead. Another conversion possibility is in terms of biomasses. This is also not very satisfactory because the animal itself is not able to do this conversion, so this would introduce an artificial element. One of the problems is that growth is never instantaneous and therefore intrinsically requires maintenance to occur. In this paper I consider the structural component only, which makes it much easier to deal with food chains (see the discussion).

The structural component is taken to be proportional to an appropriate cubed length measure because of the assumed isomorphism. This seems rather straightforward as far as somatic growth is concerned. Reproductive output has to be added in some way. In many species eggs are laid early in their development into young. Since no energy is added during the incubation period, I will ignore this time delay in producing a hatching individual. The delay makes the production of progeny biomass expensive compared to somatic growth, because maintenance energy is spent during the incubation period (see Zonneveld and Kooijman 1991). The production process is quantified by the change in cubed body length plus the number of produced eggs times the cubed length of a hatchling, which is a model parameter. Ingested food can be quantified in several more or less equivalent ways. I here choose volume per time.

The yield factor has a dimension of volume biomass per volume food and is defined by

$$Y_i = \left(\frac{d}{dt} L^3 + RL_b^3 \right) \Big/ I \tag{1}$$

where $(d/dt)\ L^3$ stands for change in cubed length. L_b for length at birth, R the reproduction rate, i.e. number of eggs per unit of time, and I the ingestion rate. Although cubed length is generally not equal to the volume of an animal, it does not give problems here since it differs by a fixed constant depending on the shape. We might measure length as volume$^{1/3}$, in which case the shape coefficient is 1. The reproduction rate is given in the appendix. Note that no overhead in the formation of eggs has been introduced, which makes this model unique for not having any free parameter involved in the reproduction process, other than the partitioning rule for utilized energy. Since growth and reproduction do not depend on ingestion directly, but only indirectly via stored energy, I shall concentrate on situations of constant food density, which, after a sufficiently long period, result in a constant stored energy density proportional to the scaled functional response $f = x/(1 + x)$, where $x = X/K$ is food density as a fraction of the saturation constant.

The yield factor (1) depends on length and, via R (see appendix), on stored energy density, or, at constant food density, on length and scaled functional response. Although the full model has two life stage parameters and seven energetic ones, the yield factor depends only on the partition coefficient, κ, the energy investment ratio, a, and a proportionality constant, $v/\{I_m\}$, standing for the ratio of energy yield of a unit volume of ingested food and the maximum stored energy density. The latter proportionality constant, which has dimension volume biomass per volume food, converts a dimensionless yield factor, y_i, into Y_i, i.e., the one mentioned in (1). For $l = L/L_m$ denoting length as a fraction of maximum length, the (dimensionless) scaled yield factor is given by

$$y_i(l, f) = \frac{f - l + (l > l_j)(1 - \kappa)\,(f(a + l) - (f + a)l_j^3/l^2)\,l_b^3/E}{f(f + a)} \qquad \text{for } l < f$$

(2)

$$= (l > l_j)(f - \kappa l - (1 - \kappa)l_j^3/l^2)l_b^3/f E \qquad \text{for } l > f$$

(3)

with

$$E = \left(\frac{1}{l_b(a + f)^{1/3}} - \frac{1}{3a^{4/3}} B_{a/(f+a)}\left(\frac{4}{3}, 0\right) \right)^{-3}$$

(4)

interpretated as the energy investment into an egg as a fraction of the maximum stored energy in an individual of maximum volume (see Zonneveld and Kooijman 1991 for the derivation). $B_x(a, b) \equiv \int_0^x t^{a-1}(1 - t)^{b-1} dt$ stands for the incomplete beta function. The notation $(l < l_j)$ indicates a boolean with value 1 when the expression within the brackets is true and value 0 when it is false. The expression is obtained via substitution of the equations for growth and reproduction given in the appendix into (1), for $e = f$. As described, ingestion in (1) is proportional to fl^2. This yield factor is illustrated in Fig. 16.1. The maximum scaled yield factor is

$$y_{i,\text{max}} = l_b^{-1} (1 + \sqrt{1 + a/l_b})^{-2}$$

(5)

which is reached for $l = b$, and $f = l_b + \sqrt{l_b^2 + al_b}$. For $l_b \to 0$ and $f \to 0$, $y_{i,\text{max}} \to a^{-1}$, which means that an animal of zero volume would spend no energy on development or energy storage; it just converts all energy it can obtain from food into biomass. It also means that all real-world animals, for which $l > 0$ holds, are much less efficient converters. Although Eqs. (2) and (3) look rather massive, it is surprising that they do not contain parameters like the maintenance cost, except through the scaling of length as a fraction of maximum length. The yield has a very weak local maximum for adults. The volume and functional response at this local maximum have to be obtained numerically from (2).

The yield factor, Eqs. (2) and (3), hereafter called instantaneous yield, is of limited use for studies on a longer time scale, because it is instantaneous only. It will (rapidly) change in time, as the animal changes its volume. More informative is the non-instantaneous yield factor, y_n, defined by the ratio of the cumulated biomass production from birth on, and the cumulated amount of ingested food:

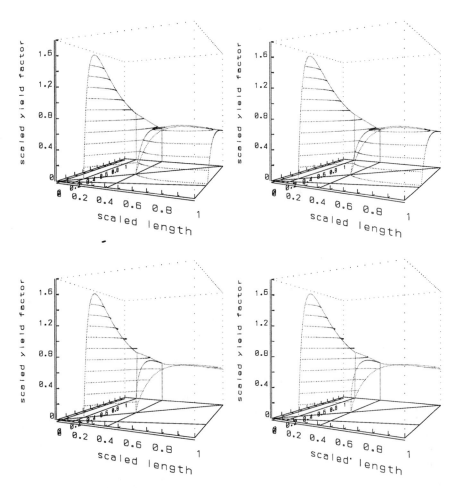

Figure 16.1.   Stereo view on the instantaneous yield factor, above, and the non-instantaneous one, below, for an individual as a function of length and, in depth, functional response (or stored energy density). Parameters: $l_b = 0.133$, $l_j = 0.417$, $a = 0.033$ and $\kappa = 0.3$. (Illuminate the picture well and homogeneously. Hold your head about 50 cm from the sheet. Do not focus at first at the picture but at an imaginary point far behind this sheet. Try to merge both middle images of the four you should see this way. Then focus on the merged image. If this fails, try stereo glasses. The use of alcohol makes the exercise much more difficult.)

$$y_n(l, f) = \frac{l^3 - l_b^3 + l_b^3 \int_0^{t(l)} R(t)dt}{amf \int_0^{t(l)} l^2 dt} \qquad (6)$$

It is identical with the conversion efficiency at population level, if yield occurs at a fixed age.

When food density is constant, the length as a fraction of maximum length is given by

$$l(t) = f - (f - l_b) \exp\left\{\frac{-tam/3}{f + a}\right\} \qquad (7)$$

or, alternatively

$$t(l) = 3\frac{f + a}{am} \ln\left\{\frac{f - l_b}{f - l}\right\} \qquad (8)$$

If we substitute these functions into (6), and for the reproduction rate R given in the appendix, with $e = f$, the integration can readily be done explicitly. The result is not a nice small formula, but a line filling one, which is illustrated in Fig 16.1. The animal, whose functional response remains constant from birth on, will obey $l < f$. This restriction was not necessary for y_i, where it is possible that an animal first grows to large volume at abundant food and is then kept at low food density for a sufficiently long time to get the amount of stored energy adapted. At $l = l_b$, the non-instantaneous yield equals the instantaneous one conceptually. For $l = f$, they are also equal, because the animal then does not change its volume for an infinitely long period. Fig. 16.1 points to the counter-intuitive result that yield at high food densities is a bit lower than at moderate ones. One would think that growth is fastest at high food densities, so that relatively little energy is lost through maintenance. The result can be explained by the increasing energy investment in storage. Is it just coincidence that laboratory cultures of many species of animals do better at 30% less than the ad libitum amount of food?

Biomass Conversion in Sandwiched Populations

In the previous section I considered an individual animal as a device that converts food into more animal. Its efficiency could be studied through the yield factor, i.e., the ratio of the biomass produced over a certain period and the cumulated food consumed over this period. I will now present some results at the population level, restricting the discussion to parthenogenetically reproducing females in spatially homogeneous equilibrium populations. It can, however, easily be extended to cover a fixed sex ratio. It will give more insight when the conversion of food into biomass is evaluated in combination with the food and biomass density. The food density will be hyperbolically transformed, giving the scaled functional response f. The biomass density, defined as the total cubed length of all individuals in the population, will be divided by the cubed maximum length of an individual. This scaled biomass, called $b = \Sigma_i l_i^3$, is thus the total cubed scaled length of all individuals in the population. As before, the yield factor will be divided by $v/\{I_m\}$, giving the scaled yield factor. To distinguish the latter from the ones discussed at the individual level, I will denote it by y_p.

Assume that, at equilibrium, food density, and so scaled functional response f, is constant and a stable age distribution exists. Then, and only then, can the DEB model be analysed as an age structured one, for which theory is well worked out. For $F(t)$ denoting the survival probability at age t, the stable age distribution is given by $F(t)/E$, where $E = \int_0^\infty F(t)dt$ is the mean age at death: see, e.g., Frauenthal (1986). The corresponding hazard function is thus $h(t) = -F(t)^{-1}(d/dt) F(t)$, I will include the case of harvesting hatchlings and animals of a specified age. This gives discontinuities in F, making $(d/dt) F$ undefined at these points. When we take each hatchling with probability $1 - q$, we have $h(0)dt = 1 - q$. Similarly, when we take an individual of age T with certainty, we have $h(T)dt = 1$ and $h(t)dt = 0$ for $t > T$. Between 0 and T, I will assume that $(d/dt)F$ exists.

The triplet scaled biomass, scaled functional response and scaled yield factor, (b, f, y_p), will be derived in four steps. The first step is to solve the scaled functional response from the characteristic equation for the equilibrium situation:

$$1 = \int_0^\infty F(t)R(t)dt \tag{9}$$

where R denotes the reproduction rate. The second step is to obtain the total number of individuals in the population, N, from the balance equation for food. For the present model, this amounts in the equilibrium to

$$x_s = IfNE^{-1} \int_0^\infty F(t)l^2(t)dt \tag{10}$$

where $x_s = X_s/mK$ relates to the food supply rate and $I = \{I_m\}L_m^2/mK$ to the maximum surface-specific ingestion rate. The scaled biomass is now

$$b* = \frac{N}{E} \int_0^\infty F(t)l^3(t)dt = \frac{x_s \int_0^\infty F(t)l^3(t)dt}{If \int_0^\infty F(t)l^2(t)dt}$$

This biomass measure still refers to the absolute size of the population, and is thus not relative to the food supply rate. A better and dimensionless measure would be $b = b*I/x_s$ or

$$b = \int_0^\infty F(t)l^3(t)dt \left(f \int_0^\infty F(t)l^2(t)dt \right)^{-1}, \qquad (11)$$

which may be interpreted as the (unscaled) biomass times $\{I_m\}/X_s L_m$: the maximum surface-specific ingestion rate divided by the food supply rate and the maximum length of an individual.

The yield factor is the ratio of the yielded biomass, $(N/E) \int_0^\infty F(t)h(t)L^3(t)dt$, and the food supply rate. Using Eq. (10), the scaled yield factor thus equals

$$y_p = I x_s^{-1} NE^{-1} \int_0^\infty F(t)h(t)l^3(t)dt$$

$$= \frac{q(1 - q)l_b^3 + F(T^-)l^3(T^{-1}) - \int_{0^+}^{T^-} \frac{dF(t)}{dt} l^3(t)dt}{amf \int_{0^+}^{T^-} F(t)l^2(t)dt}, \qquad (12)$$

where 0^+ indicates a value just a little bit larger than 0, T^- a value just a little bit smaller than T, the age at which all surviving individuals are harvested. The probability for a hatchling to survive is labeled q. The occurrence of the factor $q(1 - q)$ becomes obvious, via the small conceptual change, by harvesting just before birth, rather than just afterwards. In that case, the factor q has to be applied to the reproduction rate, instead of the survival probability. The characteristic Eq. (9), where F and R occur as a product, shows that this is totally equivalent. We add the biomass harvested as hatchling, which amounts to $(1 - q)l_b^3$, to the biomass harvested via the hazard rate. The numerator and denumerator in Eq. (12) are then both multiplied by q and this factor is subsequently taken up in the definition of F.

Before analyzing this expression for this particular model for energy budgets, note that there does not exist a stable age distribution for at least part of the appropriate range of parameter values (Kooijman et al., 1989). This model appears to generate cycles in fed batch situations, which could be partly reduced by allowing a small scatter among parameter values between different individuals. The present version differs from that used in Kooijman et al. (1989), by including a cost for maintaining the differentiated state at the expense of reproduction. The little effect this has is a further reduction of the amplitude of the cycles, which prevents death by starvation through competition between small and large individuals. The appendix gives the equations defining the dynamics, everything scaled down to dimensionless quantities. When stable age distributions do not exist, the analytical analysis of the yield factor becomes extremely complicated. Because mean yield factors over long periods of time seem the most ecologically relevant, the present approach is to assume a stable age distribution, and check the result afterwards with computer simulations.

A first useful observation in the analysis of the triplet (b, f, y_p), is that the reproduction rate affects the yield factor only through the value for the scaled functional response. At the individual level, where the latter was a free variable, it was involved more directly. At population level, the reproduction and growth processes can be viewed as extra maintenance processes compensating for the harvested individuals.

I will study three different ways of harvesting: removal of hatchlings (to simulate loss of young or a reduction of reproduction through effects of toxic substances or a specialized predator), removal of individuals upon reaching a certain age (to simulate aging), and removal randomly sampled individuals (to simulate effects of non-specialized predators or conditions of continuous cultures realized with a chemostat type of reactor design). The survivor function is thus

$$F(t; p, q, T) = (t < T)q \exp\{-pt\} \tag{13}$$

where the three parameters p, q, T are assumed to be under experimental control. The factor between the brackets denotes a boolean as before. Let us first study the three ways of harvesting separately (see Fig. 16.2).

Starting with the extreme situation of harvesting hatchlings only, so $p = 0$, $T = \infty$ and $F(t) = q$, we have the problem that the equilibrium situation depends on the starting conditions of the culture. We cope with this problem by studying the range of possible values for (b, f, y_p) in the equilibrium where no growth or non-harvested reproduction occurs. This can be done by assuming that all individuals have the same length, say l, and studying what happens if we let l range from l_b to l. In practice we would have a variety of volumes, that is, we have some weighted average over the values obtained by fixing l. When $q = 1$, so there is no harvesting at all, i.e. $y_p = 0$, the entire food input has to be spent on maintenance, i.e. on routine

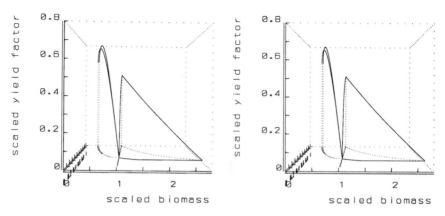

Figure 16.2. Stereo view on the yield factor for a fed batch culture as a function of biomass and, in depth, functional response. It compares three yielding strategies: yielding hatchlings (surface right), individuals of a specified age (lower curve) or randomly (upper curve). Parameters: $l_b = 0.133$, $l_j = 0.417$, $a = 0.033$ and $\kappa = 0.3$.

metabolism which is proportional to biomass plus the maintenance of the differentiated state, which is proportional to biomass as well, as long as the scaled length does not exceed l_j. This means that $f = \kappa l + (1 - \kappa)l$ min $\{l_j^3/l^3, 1\}$ and $b = (\kappa + (1 - \kappa) \min\{l_j^3/l^3, 1\})^{-1}$ for some $l \in (l_b, 1)$. When $q < 1$, we have the restriction that $l > l_j$ (otherwise no young can be produced) and $\kappa l + (1 - \kappa)l^{-2}l_j^3 < f < l$ (otherwise there will be death by starvation or growth). From the conservation law for food we obtain $b = l/f$ and $y_p = (f - \kappa l - (1 - \kappa)l_j^3/l^2) \, l_b^3/fE$, where E is given in Eq. (4). Yield is maximal for minimal biomass. i.e. $b = 1$, and maximal f, *i.e.* $f = 1$, which both occur if all individuals are of maximal length, i.e. $l = 1$. Then we have $y_p = (1 - \kappa)(1 - l_j^3)l_b^3/E$.

When we harvest at a certain age, so $p = 0$, $q = 1$ and $F(t) = (t < T)$, we do not have freedom to choose volume distributions of individuals. We have to choose $Tm > 3(1 + a) \ln(1 - l_b)/(1 - l_j)$, or the animal will not have sufficient time to reproduce, even in case of high food densities. The actual lower bound for T is still a bit higher to compensate for the lost individuals. It can be found from the characteristic Eq. (9) using $f = 1$. For increasingly large values for T, $y_p \to 0$, so the population tends to consist of individuals, all of volume l_j, at a scaled functional response of $f = l_j$ and a scaled biomass of $b = 1$. Decreasing the yielded age gives an increase in yield up to a maximum which is larger than is obtainable by harvesting hatchlings only. A further decrease in the yielded age gives a decrease in yield, up to a point where the population suddenly goes extinct. The scaled yield factor can be obtained from Eq. (12): $y_p = l^3(T)(amf \int_0^T l^2(t)dt)^{-1}$. It differs from the non-instantaneous yield factor, Eq. (6) by a value l_b^3 ($\int_0^T R(t)dt - 1$). So, when we yield at the moment the first egg is laid, both yield factors are equal. This is exactly the condition for a stable population. In other words: By setting T, the biomass and the scaled functional response evolve to a value such that the cumulative number of young per individual becomes 1.

When we yield randomly, so $q = 1$, $T = \infty$ and $F(t) = \exp\{-pt\}$, the effect on (b, f, y_p) is almost identical to that of harvesting a specified age completely. The maximum yield is only a little bit higher. This might explain the variety of harvesting strategies animals actually follow.

When we allow the three ways of harvesting simultaneously, the maximum yield is obtained by harvesting hatchlings as well as individuals of a certain age, while not harvesting randomly. The age at maximum yield with additional harvesting of hatchlings is the same as without harvesting hatchlings. The maximum yield, $y_p = 0.7646$, is obtained for $p = 0$, $q = 0.0007$ and $Tm = 1.5272$ when we choose $l_b = 0.133$, $l_j = 0.417$, $a = 0.033$ and $\kappa = 0.3$. This choice is realistic for the waterflea Daphnia magna. However, the maximum is very close to that of random harvesting. Its significance lies in the result obtained by Beddington and Taylor (1973), that maximum sustainable yield in a population of constant size on the basis of a discrete time Leslie model is obtained by complete removal of one age class, and a partial removal of a second one. Although I did not study all possible choices for the survival function, F, it seems that this result also applies to the present continuous time model with interaction between individuals.

Testing the validity of the reported results against computer simulations is not straightforward. See the appendix for the defining equations. The first problem is caused by the synchronization of individuals. When they synchronize and the survival function is of the type $F(t) = (t < T)$, competition between individuals for food can prevent growth, and subsequently the transition from juvenile to adult is affected. Such a population becomes extinct. To circumvent this problem I took instead the more realistic Weibull distribution for aging. At harvesting intensities not close to zero, i.e. when the synchronization is not very stringent, the mean scaled biomass, the mean scaled functional response and the mean yield factors were found to be very close to the predicted values. Deviations were within the accuracy set by the random events occurring in the simulations (i.e. time of death of each individual) and the subjective choice of the moment at which the population is independent from the founder population. This moment obviously depends on the harvesting intensity, and independence is obtained asymptotically only. At low harvesting intensities, details in the reproduction process tend to

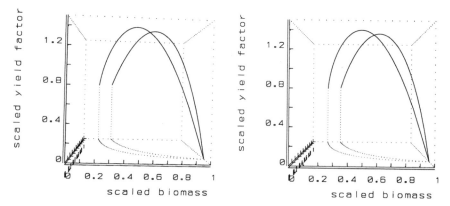

Figure 16.3. Stereo view on the yield factor for a sandwiched population as a function of biomass and, in depth, functional response. It compares propagation through eggs (lower curve, ending at the highest minimal biomass), or division upon doubling volume (upper curve). Parameters: $l_b = 0.133$, $l_j = 0.133 \; 2^{1/3}$, $a = 0.033$ and $\kappa = 0.3$.

dominate the equilibrium values. Two important details are how to translate a continuous energy flow to reproduction into a discrete number of young, and what to do with amounts of energy that are not sufficient to produce one young. In the simulations, energy drain to reproduction has been cumulated over each incubation period, after which it is instantaneously converted into eggs, while the remaining energy, not sufficient to create an egg, becomes lost. This way of reproduction is rather specific for Daphnia.

Many heterotrophs, like ciliates and some annelids, propagate through division, rather than via eggs. The DEB model also applies to this category (Kooijman *et al.* 1991). When division occurs at $l_j = l_b\sqrt[3]{2}$, the survival function for random harvesting can be written as $F(t) = (t < t_j) \exp\{-pt\}$, where $t_j = 3(a + f)/m \ln(f - l_b)/(f - l_j)$ is the age at division. So we let the animal 'die' at the moment of division, and, at the same time, adding two new ones of scaled length l_b. The stable age distribution then takes the form $p \exp\{-pt\}(1 - \exp\{-pt_j\})^{-1}$. The conservation law for numbers gives $\exp\{-pt_j\} = 1/2$, from which the scaled functional response can be solved. The conservation law for food leads to $b = \int_0^{t_j} e^{-pt}l^3(t)dt(f \int_0^{t_j} e^{-pt}l^2(t)dt)^{-1}$ and $y_p = bp/am$. In global terms, the behaviour of (b, f, y_p) closely follows that for reproducing organisms with $l_j = l_b\sqrt[3]{2}$. See Fig. 16.3. The maximum yield is a bit higher compared with organisms reproducing via eggs, due to the energy which is lost during the incubation period and the minimum biomass density is lower.

Discussion

Yield factors are not only relevant in the context of conservation biology, or economical harvesting programs, but also for a fundamental understanding of population dynamics. The factors presented in this paper differ for this purpose from the ones usually considered in these fields, by lumping all biomass leaving the closed living population, rather than taking only those individuals which were removed by commercial harvesting. For results pertinent to forestry and fisheries, see Getz and Ilaight (1989), who use age-structured discrete-time models without interaction between individuals. A survey of concepts relating to turnover rates in age-structured continuous time models without interaction is given by van Straalen (1985). A fixed monotonous volume-age relation is essential for his approach. This paper deals with the effect that increased harvesting results in a reduction of biomass, so food density is increased, which results in an enhanced production. This realistic feed-back makes the present volume- (or energy-) structured approach essentially different from an age-structured one.

The conversion efficiency of prey biomass into predator biomass is invariably treated as a constant in the extensive literature on models for non-structured populations. The population size is here measured by the number of individuals, say N. The non-structured model, which is perhaps most close to the sandwiched populations discussed in this paper, is

$$\frac{d}{dt} x = mx_s - \frac{x}{1+x} I^* N$$

$$\frac{d}{dt} N = Y_p \frac{x}{1+x} I^* N - pN$$

where I^* stands for the maximum food intake during a period m^{-1} per individual as a fraction of the saturation constant. The basic difference between this system and the one discussed in this paper is that both I^* and Y_p are treated as constants in this non-structured population model, while they depend sensitively on p in the present paper. The core of the difference lies in the dynamical behaviour of the maintenance costs: at low growth rates, relatively more energy becomes lost in maintenance. In order to take these costs into account, we have to treat Y_p not as a constant, but as a variable. This is what Pirt (1965) has done for microbial chemostat cultures, who still treats individuals as identical units. We have also accounted for the fact that the mean volume of individuals decreases for increasing harvesting rates.

The costs for maintenance is implicit in the frequently used empirical logistic growth models, where the carrying capacity can be conceived as the population size at which the inflowing food is used for routine metabolism, and for compensation for losses of individuals due to death and/or migration. The next comparison attempts to illustrate that this class of models is not an equivalent alternative for structured population dynamics.

The usual theories for harvesting non-structured populations are based on logistically growing populations:

$$\frac{d}{dt}N = rN(1 - N/K) - pN, \tag{14}$$

where r is the maximum population growth rate, K the carrying capacity and p the harvesting effort considered as a course of mortality on top of those occurring 'naturally.' At equilibrium, where the number of individuals does not change, so $(d/dt) N = 0$, we have $N = K(1 - p/r)$ and a yield of $Y = pN = pK(1 - p/r)$. The maximum yield is $Y_{max} = rK/4$, at which $N = K/2$ (see e.g., Beddington and May 1977). Food density does not follow naturally from this model, because it is not based on energetic considerations. When we try to relate this to the present results, we have to redefine the yield factor to deal with numbers rather than biomass volume and to make assumptions about natural mortality. When the latter takes the form $F(t) = (t < T)$, the complete survivor function equals $F(t) = (l < T)$ $\exp\{-pt\}$, thus the stable age distribution is then $g(l) = p \exp\{-pt\}/(1 - \exp\{-pT\})$ for $0 \le t \le T$. The number of individuals is given by $N = x_s(I f \int_0^T g(t)l^2(t)dt)^{-1}$ and the scaled yield by $Y = pN$. Length still depends on scaled food density, which as to be solved from the characteristic equation. The carrying capacity, K, corresponds with N for $p = 0$, which gives a stable age distribution of $g(t) = 1/T$ for $0 \le t \le T$. The maximum population growth rate, r, is obtained from the characteristic equation $1 = \int_0^T \exp\{-rt\}R(t)dt$ for the scaled functional response, f, is equal to 1. The way the yield depends on the number of individuals obviously depends on the choice for T. Fig. 16.4 gives the dimensionless quantity Y/rK as a function of N/K, together with the result from Eq. (14). Depending on the value for the maximum age, T, the number of individuals in the population can increase for increasing harvesting effort, because they tend to be of smaller body volume. Below a certain harvesting effort, the population size is completely controlled by food availability. The maximum of Y/rK increases with T, because r decreases. Another difference between the theory based on Eq. (14) and the present one is that in the latter the biomass does not continuously approach 0 for increasing harvesting effort, but the population suddenly becomes extinct at a biomass density well above 0 when the harvesting effort exceeds some threshold. This has strong implications for determining harvesting policies; for example, setting North Sea fishing quotas.

The results reported in this paper only apply to experimental and field

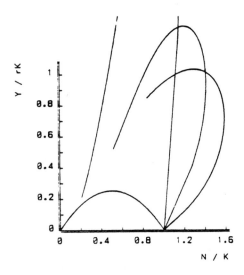

Figure 16.4. The yield, Y, of the number of individuals, N taken randomly from a population in which the individuals survive naturally with certainty up to an age T, for $Tm = 5.10.30$, $l_b = 0.133$, $l_j = 0.417$, $a = 0.033$, $\kappa = 0.3$. The yield is expressed as a fraction of the maximum population growth rate, r, times the carrying capacity, K. The lower curve is related to random yielding in a logistically growing non-structured population.

situations where equilibrium is approached. The most curious result reported here is that predictions based on stable age distributions are in sufficiently close correspondence with simulations to motivate an attempt at mathematical proof. If we would introduce time-varying environments or spatial inhomogeneities with migration, we could well arrive at situations where the present yield considerations no longer apply. We would then have to fall back on the full model, using simulations. The problem is, then, is not to become entangled by the many parameters which are necessary to describe these real world phenomena. The present approach points out that an important population characteristic is directly related to physiological properties of individuals. This allows the evaluation of effects of toxic substances at population level along the lines set out in Kooijman and Metz (1983).

It is possible to extend the theory to food chains and food webs by linking the populations directly through the triplets (b, f, y_p) (Kooi and Kooijman, in press). A given harvesting rate fixes both the biomass density as well as the functional response, and so the food density. This feature is unique for models that take care of the conservation law for energy. This gives us an important extra constraint in the study of the biomass distribution over the foodweb. The theory for the relation between parameter values with body size, as presented in Kooijman (1986b, 1988), reduces the problem for specification of the enormous amount of parameters to that of body volume distributions over the food web. A specification of the bottom population(s) of a food web or the overall turnover rate, together with the body volume distribution, would then be sufficient to determine the secondary production part of the community structure, coupled to its function, in terms of physiological properties of the species involved. Extensive computer simulations will be necessary to check whether or not the DEB model for individuals indeed produces such a structure. The next step is to study the reactions to spatial inhomogeneities and perturbations.

Acknowledgments

I would like to thank Marinus Stulp for performing the computer simulations and Bob Kooi, Cor Zonneveld, and Tom Hallam for their critical comments.

Appendix

The dynamics of the sandwiched DEB-structured population with Weibull aging and m^{-1} as unit of time based on 3 length (l)-defined life stages: embryo, $l \le l_p$, juvenile, $l_b < l \le l_j$, adult, $l > l_j$

Parameters:

	Energy budget	$a\ \kappa$	Food source, sink	$x_s\ I$
	Life stage	$l_b\ l_j$	Survival	$g\ h\ p\ q$

Dynamics of scaled food (x) in scaled time (t)
$$dx/dt = x_s - If \Sigma_i l_i^2 (l_i > l_b): \text{functional response } f = x/(1 + x)$$

Change of state variables of individual i:

scaled age, a_i scaled storage, e_i scaled length, l_i
$$da_i/dt = 1 \qquad de_i/dt = a(f - e_i)/l_i \qquad dl_i/dt = a(e_i - l_i)_+/(3e_i + 3a)$$

Boundary values of state variables at hatching

$$a_b = 3 \int_0^{a/(e_b+a)} \frac{dx}{(1 - x)x^{2/3}\left(3a\left(\dfrac{a}{e_b + a}\right)^{1/3} \Big/ l_b - B_{a/(a + e_b)}\left(\dfrac{4}{3}, 0\right) + B_x\left(\dfrac{4}{3}, 0\right)\right)}$$

e_b = scaled storage of mother at egg formation

l_b = scaled length at hatching (parameter)

Change in numbers of individuals

Reproduction rate (eggs per scaled time) for individual i:

$$R = (l_i > l_j)a(1 - \kappa)\{e_i l_i^2(a + l_i)/(a + e_i) - l_j^3\}/E \quad \text{for } e_i > l_i \text{ (growth)}$$
$$R = (l_i > l_j)a\{(e_i - \kappa l_i)l_i^2 - (1 - \kappa)l_j^3\}_+/E \quad \text{for } e_i < l_i \text{ (no-growth)}$$
$$\text{where } E = \left(\frac{1}{l_b(e_b + a)^{1/3}} - \frac{1}{3a^{4/3}}B_{a/(e_b + a)}\left(\frac{4}{3}, 0\right)\right)^{-3}$$

Survival function for individual i:
$$\text{Prob}\{a_i > a_i\} = q \exp\{-pa_i - (ha_i)^g\}$$

Initial values: x and $\{a_i, e_i, l_i\}_{i=1}^N$ at time 0

Literature Cited

Beddington, J. R. and R. M. May. 1977. Harvesting natural populations in a randomly fluctuating environment. *Science* **197**:463–465.

Beddington, J. R. and D. B. Taylor. 1973. Optimum age specific harvesting of a population. *Biometrics* **29**:801–809.

Evers, E. G. and S. A. L. M. Kooijman. 1989. Feeding, digestion and oxygen consumption in *Daphnia magna*. A study in energy budgets. *Neth. J. Zool.* **39**:56–78.

Calow, P. and R. M. Sibly. 1990. A physiological basis of population processes: ecotoxicological implications. *Funct. Ecol.* **4**:283–288.

Frauenthal, J. C. 1986. Analysis of age-structured models. In *Mathematical ecology* T. G. Hallam and S. A. Levin (eds.), pp. 117–147. Springer-Verlag, Berlin.

Getz, W. M. and R. G. Haight. 1989. *Population Harvesting*. p. 391. Monographs in population biology 27. Princeton University Press, N.J.

Hallam, T. G. and J. L. de Luna. 1984. Effects of toxicants on population: A qualitative approach III. Environmental and food chain pathways. *J. Theor. Biol.* **109**:411–429.

Kooijman, S. A. L. M. 1986a. Population dynamics on the basis of budgets. In *The dynamics of physiologically structured populations*, J. A. J. Metz and O. Diekmann (eds.), Lecture Notes in Biomathematics 68. Springer-Verlag, Berlin.

Kooijman, S. A. L. M. 1986b. Energy budgets can explain body size relations. *J. Theor. Biol.* **121**:269–282.

Kooijman, S. A. L. M. 1986c. What the hen can tell about her egg; Egg development on the basis of budgets. *J. Math. Biol.* **23**:163–185.

Kooijman, S. A. L. M. 1988. The von Bertalanffy growth rate as a function of physiological parameters: A comparative analysis. In *Mathematical ecology*, T. G. Hallam, L. J. Gross and S. A. Levin (eds.), pp. 3–45. World Scientific, Singapore.

Kooijman, S. A. L. M., A. O. Hanstveit, and N. van der Hoeven. 1987. Research on the physiological basis of population dynamics in relation to ecotoxicology. *Water Sci. Technol.* **19**:21–37.

Kooijman, S. A. L. M., N. van der Hoeven, and D. C. van der Werf. 1989. Population consequences of a physiological model for individuals. *Funct. Ecol.* **3**:325–336.

Kooijman, S. A. L. M. and J. A. J. Metz. 1983. On the dynamics of chemically stressed populations; The deduction of population consequences from effects on individuals. *Ecotoxicol. Environ. Saf.* **8**:254–274.

Kooijman, S. A. L. M., E. M. Muller and A. II. Stouthamer. 1991. Microbial growth dynamics on the basis of individual budgets, Antonie van Leeuwenhoek special issue: Quantitative Aspects of Microbial Metabolism ed. A. H. Stouthamer, to appear.

Łomnicki, A. 1988. Population ecology of individuals. *Monographs in population biology* 25. pp. 223. Princeton University Press, Princeton, N.J.

Pirt, S. J. 1965. The maintenance energy of bacteria in growing cultures. *Proc. R. Soc. Lond.* B **163**:224–231.

van Straalen, N. M. 1985. Production and biomass turnover in stationary stage-structured populations. *J. Theor. Biol.* **113**:331–352.

Thieme, H. R. 1988. Well-posedness of physiologically structured population models for *Daphnia magna*. *J. Math. Biol.* **26**:299–317.

Zonneveld, C. and S. A. L. M. Kooijman. 1989. The application of a dynamic energy budget model to *Lymnaea stagnalis*. *Funct. Ecol.* **3**:269–279.

Zonneveld, C. and S. A. L. M. Kooijman. 1991. Comparative kinetics of embryo development. (submitted.)

Part IV

Models of Plant Populations and Communities

Section Overview

Interactions among plants provide the clearest example of a situation where i-state distribution models (models that potentially have analytical solutions, such as Leslie matrices or partial differential equations) are likely to fail and i-state configuration models (computer simulations of many individuals) may be required. The critical feature of plants that complicates the application of i-state distribution models is the fact that they are rooted in a single location and cannot move. Consequently, individuals interact locally with a small subset of the total population, resulting in a violation of the mixing assumption required by i-state distribution models. The most important feature of plants from a modeling perspective is not the ability to photosynthesize, but rather the fact that plants, at least most terrestrial plants, are sessile for much of their lifespan. Free-living plants, such as planktonic algae, should be amenable to an i-state distribution modeling approach, while sessile animals, such as sponges, coral, and perhaps some molluscs, should be amenable to the same general modeling approach taken for terrestrial plants. Another potential restriction on the application of i-state distribution models to sessile organisms is the fact that the number of interacting organisms is likely to be relatively small.

On the practical issue of model parameterization and testing, sessile organisms have the major advantage of being relatively easy to capture or relocate, as well as occurring in numbers small enough to be tractable at spatial scales of biological interest. Perhaps because of the combination of readily available data, and the failure to meet the assumptions required for i-state distribution models, plant population and community modeling has a relatively long and rich history of i-state configuration models, including the "classic" individual-based computer simulation models.

The difficult issue of multispecies (>2 species) interactions has been more extensively addressed in plant community models than in animal communities. However, of the six chapters in this section, four address single species interactions, another addresses two-species interactions (one of which is a plant), and only one (a review) discusses multispecies interactions. While there is an extensive body of multispecies plant community modeling using *i*-state configuration models (see Huston, this volume), this success in modeling multispecies interactions is achieved among species at a single trophic level (see Murdoch et al., this volume). Nonetheless, primary producers are arguably the most important trophic level, providing both the energy and physical structure on which nearly all ecosystems depend, so any success in modeling these organisms is a major step toward understanding ecosystems.

A multispecies model of plant interactions is relatively simple in comparison to a multi-trophic level animal community model, because the many species of plants are much more similar in behavior, physiology, and spatiotemporal scale than are the species involved in even a two-species predator-prey interaction. All plants require the same resources (light, carbon dioxide, water, and mineral nutrients) and obtain them from the same sources. Consequently, a relative simple set of "rules of interaction" applies to all plants, and can be applied to many different combinations of species.

In spite of some success in modeling multispecies interactions among plants, most of the theoretical and modeling effort in plant ecology has been focused on populations, that is, on interactions among individuals of a single species. The review by Ford and Sorrensen documents the rich history of individual-based plant population modeling, tracing the common themes that have developed and summarizing them as a general theory of plant competition. The models they describe form a series from relatively simple, analytically tractable models (e.g., Gates, 1978) to simulation models with a detailed representation of the growth of the many branches that comprise an individual tree (e.g., Mitchell 1975, Sorrensen et al., submitted). The common feature of all models they describe is the effect of spatial structure (plant size and distance to neighbors) on the local interactions of these sessile organisms.

Many of the basic concepts developed in the field of theoretical plant population and community modeling are paralleled by the assumptions and processes that formed the basis for multi-species forest simulation models, beginning with JABOWA (Botkin et al. 1972). Unfortunately, there has been very little interaction between the field of forest simulation modeling, which developed from ecosystem ecology, and theoretical plant population and community ecology. Huston describes the history of these widely-used individual-based (*i*-state configuration) models in the context of the ecological issues that shaped their development, and thus their differences from theoretical plant population models.

Clark demonstrates how analytically tractable *i*-state distribution models can be used to understand many of the same issues addressed by computer simulation models of *i*-state configurations. Using an approach that exemplifies the method of model simplification advocated by Murdoch et al. (this volume), Clark focuses on the population dynamics of a hypothetical plant that reproduces in the "gaps" produced by treefalls in forests. While "gap dynamics" is a central feature of individual-based forest simulation models, Clark simplifies the system and uses an *i*-state distribution model to follow the changes in density, size, and mortality of a single monospecific cohort. With this approach he is able to develop a simple theoretical explanation of the temporal dynamics of self-thinning (including the 3/2 thinning phenomenon, which is also addressed in several other papers included here), net primary production, and the relative importance of density-dependent and density-independent mortality, all based on the growth of individual plants.

The remarkably detailed information that can be gathered about the growth of individual plants, and thus the *i*-state configuration of plant populations, forms the basis for two papers in this section. Benjamin and Sutherland describe an elaborate plant competition experiment using carrots, in which the treatments involved different spacing between the rows. This work exemplifies the detailed treatment of local spatial interactions that has characterized theoretical plant population modeling. Benjamin and Sutherland use their dataset to evaluate five *i*-state distribution models that differ in their representation of how local space is used and shared by plants. The five models can be classified as: 1) non-overlapping domain (e.g. Voronoi polygon), 2) overlapping domain, 3) unbounded areas of influence (with influence decreasing with distance from the plant), 4) diffuse population effect, and 5) tiers of vegetation. The authors demonstrate that the evaluation of the "best" model is dependent on subtle details of the data used for the evaluation.

Aikman also uses data from a plant competition experiment by Benjamin (1988) to evaluate *i*-state distribution models. Aikman begins by evaluating five different equations for the growth of an isolated individual plant, and having selected the equation that gave the best fit, proceeds to explore how competitive effects can be added to the model. He begins with the case of competition among evenly sized plants and proceeds to the more difficult case of competition among plants of different sizes. Aikman concludes that the theoretical models with parameter optimization that he describes do allow some mechanistic interpretations, and that more detailed mechanistic models may be sucessful in modeling the greater complexity of multispecies interactions.

The final paper in this section deals with a very interesting two-species interaction, that between a population of plants and its pollinators. Real, Marschall, and Roche conducted field experiments and developed a spatial model of a system in which the pollinators (bumblebees) transmit a fungal smut disease, as well as pollen, between plants in a population. Their work investigates the effect of the pollination behavior of individual bumblebees on the spatial distribution of infected and uninfected individual plants. Using stochastic simulations of individual bee movement and flower discrimination, Real et al. demonstrate that the preference behavior of pollinators can alter the rate of spread of disease and possibly also its spatial distribution in plant populations.

It is evident that different of types of individual-based models can provide important insights into the structure and dynamics of plants, at both the single- and multi-species levels. In spite of the limitations of small population size and inherently local interactions that are typical features of sessile organisms, simplified i-state distribution models can be applied successfully to plant populations when appropriate assumptions are met. Such simplifications can be achieved experimentally with regular plantings that guarantee that each plant has the same number of identical neighbors (e.g., Aikman, this volume, and Benjamin and Sutherland, this volume), and with the experimental or theoretical constraint that all plants are the same age and the same size (e.g., Aikman, Benjamin and Sutherland, and Clark in this volume). The full potential of both i-state distribution and i-state configuration models for understanding plant populations, communities, ecosystems, and even landscapes remains to be realized.

17

Theory and Models of Inter-Plant Competition as a Spatial Process

E. David Ford and Kristin A. Sorrensen

ABSTRACT. **A theory of competition is presented in the form of evidence for five axioms:**

   i. **Plants modify their environment as they grow and reduce the resources available for growth by other plants. This defines the existence of competition.**

  ii. **The primary mechanism of competition is spatial interaction.**

 iii. **Plant death due to competition is a delayed reaction to the mechanism of reduced growth following resource depletion.**

 iv. **Plants respond in plastic ways to environmental change, and this affects not only the result of competition, but its future outcome.**

  v. **There are species differences in the competition process.**

An almost universal approach to modeling competition between plants in single species communities has been to consider competition for space. Four categories of plant competition models of this type are reviewed; lattice models based on experiments, neighborhood models aimed at simulating processes in vegetation, forestry models designed to predict the distribution of yield between individuals, and models of the self-thinning relationship. The use these models make of the five axioms, and their contribution to them, is discussed.

In conclusion, models are discussed that simulate competition directly as a two-part process of (a) direct resource acquisition by plants, and where spatial patterns of resource depletion in the environment are calculated, and (b) individual plant growth, and where spatial pattern of the growth made then determines new patterns of resource acquisition. The requirement that (b) should include plasticity is discussed, and the suggestion is made that for models to contribute further to the advance of competition theory, they must describe differences between species and the environments in which they grow.

Introduction

Huston et al. (1988) suggest two reasons why modeling at the level of individual organisms has advantages over state space models where whole populations are represented by single variables. First, individual organism models can directly represent variation in genetic and environmental influences, and this has a controlling influence in ecological systems. Secondly, individual organism models can represent the spatial interaction between individuals. They considered that organisms are affected primarily by local interactions with other organisms. One of the most important local interactions is competition.

Almost thirty years ago, following an extensive review, Donald (1963) concluded:

> "The fuller understanding of competition among plants requires . . . a greater knowledge of the response of plants to their environment, and especially of those environmental stresses created by neighbors."

The situation has changed little. Research into competition has continued to define its effects on community structure and development, rather than make direct measurements of the processes that produce these effects (Fig. 17.1). Researchers neither make direct measurements of differences in resource uptake rate by individual plants, or the relative efficiency of resource use between different individuals. Generally they infer information about the competition process from observed effects on population structure. Most studies ask questions of the type: "Under what conditions of stand structure and resource availability does competition occur?"; "What are the consequences of competition in population regulation and community development?"

Such studies have led to substantial advances, but generally the competition models produced represent only selected aspects of the competition process. In this review, we compare advances made in competition modeling. These are (1) models of competition on regularly spaced lattice systems, (2) models of neighborhood interactions in irregularly spaced distributions, (3) models of distance-dependent processes between trees in forests, and (4) the self-thinning law. In each case, model development has contributed substantially to competition theory. In each case, however, further advance requires that a more direct approach be made to define plant properties that determine resource acquisition and the efficiency of resource use.

Before attempting a comparative assessment of competition models, we define an integrated theory of competition as a spatial process. There are two linked components; the spatial location of the plants themselves, and plant morphology that defines the pattern of spatial occupancy by an indi-

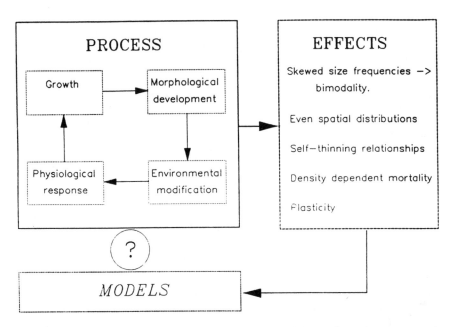

Figure 17.1. The process of competition produces continuous change in plant growth and in the environment that surrounds plants. This, in turn, has certain effects on community structure. Generally, models simulate the effects of competition without direct representation of the processes of growth and environmental modification that define how competition occurs.

vidual. The effectiveness of the model and the ability to test it are determined by the representation of these two properties.

This review is based largely on analyses of single-species, even-aged communities. This reflects the considerable research into competition conducted with such communities. In ecological research, they offer the apparent simplicity of experimentation. In agriculture and forestry, competition controls the relationship between total yield and the final size distribution of individuals that, in turn, can determine the commercial value of a crop or forest stand. Extensive experimentation has been made to investigate the effects of spacing, e.g., Verheij (1970), for a crop; Hamilton and Christie (1974), Sjolte-Jørgensen (1967) for forests; and of thinning, e.g., Assmann (1970) and Hamilton (1976) for forests, and the results used in model construction, e.g., Daniels et al. (1986).

Single species even-aged stands are ecologically important. They do occur naturally, e.g., some regenerating forests (Cooper 1961, Laessele 1965) and woodland (Hutchings 1978), or ruderal (Bradbury 1981) herbs. Recently, models originally constructed to describe individual plant interactions in single-species stands have been used as the basis for multispecies interactions, in attempts to describe vegetation dynamics (Pacala and Silander 1990).

A Theory of the Competition Process in Even-Aged Plant Communities

In a comprehensive synthesis, Donald (1963) presented experimental and observational evidence that competition between individual plants was for physical and chemical resources. He used a definition given by Clements et al. (1929): "When the immediate supply of a single necessary factor [for growth] falls below the combined demands of the plants, competition begins." To make this definition more precise, we can define a theory of competition in single-species, even-aged stands by a set of propositions. Some propositions have been considerably researched, and we call these axioms; some are less well-defined or more conjectural, and we call these postulates. As postulates are researched and more becomes known about the process of competition, then particular axioms may be qualified. This explicit acknowledgement of theory development through successive refinement is important in the current stage of competition studies. Our review of competition models shows that the competition process may be influenced by variation between species, between stages in stand development, and between the nature of the resource competed for. However, we have few precise descriptions of how plant morphology and physiology determine and interact with patterns of resource acquisition.

Axiom 1. (i) As they grow, plants change resource availability in their surrounding environment.

(ii) Larger plants obtain more resources for growth, grow more rapidly and so modify the environment of smaller plants and reduce the resources required for growth that smaller plants can obtain.

This is the fundamental axiom of competition, that plants modify their environment as they grow, and reduce the resources available for growth by other plants. It specifies that the competition process exists. Large plants will generally out-compete smaller plants, although the precise size-distance differential required to observe and model this effect may vary with species and environmental conditions. *Axiom 1* rests largely upon inference from individual plant growth rates measured in communities and not direct observations of relative resource depletions and their influence. However, few models do not include this effect of size differences in their structure (though see Slatkin and Anderson (1984)).

Koyama and Kira (1956) described a positive relationship between the size of plants and their relative growth rates. This was an important definition of the effect of the competition process. It excluded the possibility that observed size differences could simply be the effects of differences in growth rates, caused by the application of the same relative growth rate to plants of different sizes, and where these size differences had occurred by genetic or non-competition influenced environmental variation. Koyama and Kira (1956) assumed that the growth of an individual plant could be expressed* by the formula $w = w_0 e^{rt}$ where w is the weight at time t, w_0 is initial plant weight and r is relative growth rate; where competition occurs, then, w_0 is positively correlated with r.

Defining what are "large" and "small" plants, and describing the precise nature of the plant size relative growth rate correlation are the objective of many models that seek to predict the development of size frequency distributions. Westoby (1982) defined a series of "distribution modifying functions," and a range of different mathematical functions has been suggested for the development of forest stands (Moser 1974). However, size alone is an insufficient definition of the future competitive status of a plant—it depends upon relative size and distance of neighbors. Distribution modifying approaches that do depend upon size alone have been most successful when applied to stands of specific spacing and age.

Axiom 1 says nothing about the influence that smaller plants may have on larger ones. There are some observations and considerable speculation about this, and so we propose a postulate:

Postulate 1.1   The effect of small plants on large ones depends upon the measured plant parameter, species characteristics, and the resource competed for.

Diggle (1976) first defined and modeled competition as a "one-sided" process by which large plants influence small plants, but small ones do not influence large ones. This model was based on experimental analysis of competition as a canopy process (Ford 1975, Ford and Newbould 1970). A model that represents competition as one-sided can be an effective predictor of the relative status of a plant in the competitive hierarchy (Diggle 1976, Ford and Diggle 1981, Gates 1978). However, this approach may be limited where absolute plant growth increment, and not just relative status, must be predicted. For example, an almost universal result of thinning small trees from dense forest stands (Hamilton 1976) is an increase in wood increment made by bigger trees. It seems likely that at least some of this effect may be due to shading by smaller trees of the lower branches of large ones (Stiell 1969).

Thomas and Weiner (1989) fitted one- and two-sided versions of a com-

---

*Throughout the text, the use of mathematical symbols follows that given by the original authors unless otherwise stated.

petition model to experiments with different species. They concluded that competition was completely one-sided in a population of *Impatiens pallida*, a species with low root allocation and a shallow crown. Competition was less one-sided in an experimental monoculture of *Ambrosia artemisiifolia* and a natural stand of *Pinus rigida*, where Thomas and Weiner (1989) suggested competition for water and nutrients may be more important. Through an analysis of size variability in monocultures, Weiner and Thomas (1986) suggest that 14 of 16 reviewed published experiments indicated that competition was one-sided. From a study of spatial pattern changes of individual trees in a jack pine stand, Kenkle (1988) suggested that the development of a strong, regular pattern is the result of two distinct competitive phases. First, an early two-sided competition for soil resources. Second, a later one-sided competition for light. Current evidence, then, suggests that the magnitude of one- and two-sided competition requires further characterization to increase the definition of the theory of competition.

Postulate 1.2   In extreme circumstances, environmental modification by all the plants of a stand may limit the growth of all plants and not result in growth differentials.

Stand "stagnation" is described in the silvicultural literature (Oliver and Larson 1990). For example in very dense, 18-year-old, naturally regenerated lodgepole pine, Mitchell and Goudie (1980) reported a decrease in height from 4.5 to 1.0 m with increased density from 200,000 to 800,000 stems ha^{-1}. Trees at the edge of such stands grow normally, but this zone of unrestrained growth is narrow. Roots intermingle widely in such stands, but crowns do not, and there is a restriction of the competition process. Some trees are larger than others at all densities. But at the higher densities, the effect of large trees in suppressing smaller ones is limited, although when such stands are thinned, there is rapid growth of remaining trees. Mitchell and Goudie suggest stagnation may be due to a combination of shortage of water and an imbalance in the carbohydrate status of closely grown trees that results in height growth suppression. Yoda et al. (1963) present experimental evidence for an annual plant that, as site quality increases competition, seems more intense, in that it results in more rapid self-thinning, and Clark (1990) has made a theoretical analysis of this process for trees.

In order to specify the limits of operation of the competition process, the conditions under which stagnation may occur need to be defined. Stagnation may be a gradual rather than abrupt process. As stand growth takes place some changes occur to the environment that all trees experience. For example, when forest stands are thinned, greater total amounts of radiation or water may reach the forest floor and increase decomposition rates, and possibly root growth (e.g., Ford 1984). This effect may not be differentiated between neighboring plants as in a competition process, although it determines the conditions under which competition occurs.

Axiom 2.   (i)   The competition experienced by a plant depends upon the influence of its neighbors.
           (ii) The neighborhood of each individual plant is unique, but all neighborhoods are linked in their influences.

Axiom 1 is an insufficient description of the process of competition. It lacks description of the explicit process of spatial interaction and does not define the densities at which competition occurs nor the mechanisms of interaction. *Axiom 2* specifies the primary mechanism. It defines competition as being determined by spatial interactions. It is the strength of spatial dependencies, and what controls them, that must be researched. Secondly, there is an interlinked chain of dependencies. One plant may be directly influenced by its immediate neighbors, but those neighbors, in turn, may be influenced by their other neighbors. As the community grows, the size of the plants that become competing plants for any individual may not be influenced by the competition process up to that time. This is a random component in the competition process. The numbers of competitors for any individual, and the intensity of effect they will exert, cannot be predicted at the start of stand growth. These properties define competition as a spatial stochastic process; in particular, a Markov Random field.

Axiom 2 does not define the precise nature of dependencies. It does not specify when models can be effective if the influence of only nearest neighbors is incorporated, or if second nearest neighbors must also be included. Some modeling approaches shortcut the complete treatment of spatial dependency and consider the individual plant as an independent sample, e.g., neighborhood models and some tree competition models.

A corollary of this axiom is that *no two large plants can continue to exist in close proximity and both grow at high relative growth rates relative to other plants*. The consequence of competition is a trend towards spatial evenness in the distribution of large individuals or survivors. Spatial evenness occurs both in canopy forming communities, where an assumption is frequently made that competition for light is an important component, as well as in open communities such as deserts where soil resources are assumed to be key. Aggregation, associated with small plants and assumed related to dispersal, is superceded by a trend to spatial evenness of large plants (Phillips and McMahon 1981, Yeaton and Cody 1976, Beals 1968). Nobel and Franco (1986) demonstrated experimentally how the removal of individual plants of the desert bunchgrass, *Hilaria rigida*, from mono-specific communities in the Sonora Desert, results in higher shoot water potentials of neighbors. Rooting area of plants increased only 17% per year for undisturbed individuals, but increased by 125% when adjacent plants were removed.

A trend towards spatial evenness has been repeatedly observed in canopy forming communities. Kira et al. (1953) found alternate large and small plants along row crops. Kitamoto and Shidei (1972) found that large plants were spatially evenly distributed in plantations of three tree species and a natural community of *Solidago serotina*. In a natural, single-species, homogeneous stand of Solidago altissima, Kitamoto (1972) found density-dependent mortality acting over small spatial scales, and that large plants became spatially evenly distributed as growth proceeded. For a range of single-species populations, Ford (1975) also found a positive association of relative growth rate with plant size, but additionally, that plants of large size came to have an even spatial distribution throughout the population. Renshaw (1984) demonstrated that even though spatial heterogeneity may occur in community structures, the competition process has a strong effect that can be distinguished through pattern analysis. Sterner et al. (1986) demonstrated that while clumped seedling distributions of tropical trees can be found, these develop toward even spatial distributions as growth takes place. Kenkle (1988) examined the spatial distribution of live, standing dead, and live trees alone in a 65-year naturally seeded stand of jack pine. He found the combined live and dead distribution to be random, but the live trees on their own were highly regular. Cannell et al. (1977) showed that in plantations of tea bushes, which are cultivated in a closed canopy that is continuously pruned to restrict height growth, productivity per unit bush was evenly distributed. In contrast, leaf color, an attribute under genetic control, was not. Franco and Harper (1988) found negative spatial autocorrelation between plants and their nearest neighbors in experimental plantings of *Kochia scoparia*. They found a wave-like pattern that they suggested was induced by an edge effect where large individuals grow, and that alternate large and small plant sizes were propagated through the community by the competition process.

Ford and Newbould (1970), and Ford (1975), described competition in foliage canopies as a three-dimensional spatial process of crown interactions (Fig. 17.2). Generally, as described by Koyama and Kira (1956), there was a positive correlation between size and relative growth rate, but crucially, where two large plants were neighbors, the relative growth rate of one would decline. Which stand would decline would depend upon the full assessment of competition, from all of each plant's neighbors, experienced by each crown. This process defines a spatial homeostasis in development.

Postulate 2.1   (i)   In canopy-forming communities the active process that determines dominance occurs in the upper canopy. This implies a role for competition for light.

               (ii)   Once a plant is over-topped, it comes to exist in a reduced but not greatly changing environment. A distinct lower canopy of

DEVELOPMENT OF A BIMODAL DISTRIBUTION

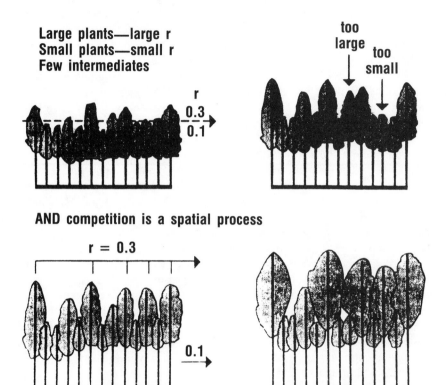

Figure 17.2. Diagrammatic representation of the process of competition for light that results in an even-spatial distribution of large trees. Two relative growth rates, 0.3 and 0.1, were applied to trees in the upper left stand. When these were distributed simply according to tree size, a stand with the structure of upper right was the result where groups of small or of large plants could occur. Distributing relative growth rates according to both size and neighbor size so that no two adjacent plants received large r, and a plant with two small neighbors always received large r, resulted in the structure of bottom center. Further application of the same rule resulted in bottom right, a bimodal height distribution and spatially even trees.

plants may form if plants are tolerant of these conditions (shade tolerant).

This three-dimensional spatial process can give rise to a two-tiered canopy structure and a bimodal frequency distribution of plant weights and heights. This bimodality may change as the lower canopy is formed and small plants die (Ford 1975). Whether or not bimodalities occur has caused some controversy. Detailed analysis of the growth process of individuals in single-species populations of natural (Kikuzawa 1988) and plantation trees (West and Burrough 1983), and *Solidago canadensis* (Bradbury 1981) have confirmed that it can occur. Certainly, processes other than competition may give rise to such a size distribution (Huston and DeAngelis 1987) but this does not exclude the competition process as a cause. Observation of bimodalities in size distributions as the result of plant competition requires four things: (1) A non-linear distribution of relative growth rates with small plants possessing smaller relative growth rates than large plants. (2) Small plants must survive long enough once they have been over-topped, i.e., there must be adequate shade tolerance to form a distinct lower canopy. Ford and Diggle (1981) showed a time delay between the expression of dominance and the occurrence of mortality in experimental plantings. It is well-known that shade tolerance varies between species, and this, too, may influence survival of plants in a lower canopy. (3) Competition should be the major process determining size distributions. As environmental or genetic variation increases in importance, bimodalities will decrease; see Ford and Diggle (1981) for simulation of this effect. (4) Adequate plants must be counted, and the actual number required to fit a bimodal distribution will depend upon the relative sizes and locations of the modes.

There can be substantial difficulties in making an appropriate statistical analysis of spatial evenness. For example, as a stand develops, the predominant influence may proceed from nearest neighbor to second nearest neighbor, and tests based on nearest neighbor alone will not indicate a pattern. Even within dense communities, processes other than competition may dominate spatial patterns. The predominant spatial pattern in naturally seeded populations is clumped (e.g., Daniels 1978) and competition may take considerable time before producing evenness. Soetono and Puckridge (1982) were unable to fit models for plant size in relation to size and distance of neighbors in dense stands of barley and wheat. They attributed this to differences in seedling emergence or the individual plant microsite. Although competition is likely to occur in such dense communities, the ability to detect it as a spatial process will depend upon how long the competition process has been dominant relative to these other processes. For this reason, it is usually more effective to analyze spatial patterns of relative growth rate rather than absolute size.

Axiom 3   Death of plants, in the lower canopy, is a time-delayed response to competition.

In canopy forming communities, competition-induced mortality is dependent upon the upper canopy assortment, but may not be completely predictable from it. A two-layer canopy stratification and bimodal size frequency distribution may not be apparent if mortality takes place quickly once a plant is suppressed. Rapid mortality following suppression might occur because the species is not shade-tolerant, or because foliage production by the large plants is rapid, and the overall intensity of competition is high.

Axiom 4   Plants undergoing competition show differences in morphology and physiology (plasticity) from those that are not.

Plasticity is a well-defined phenomenon in relation to competition (e.g., Harper 1967). It has been defined in relation to adaptations to shade, including changes in branch morphology for trees (Steingraeber 1982, Steingraeber and Waller 1986), as well as the photosynthetic response of foliage to light of different intensity (Levrenz and Jarvis 1980, Gross 1984). Plasticity is a general response of plants to variable environments (Sultan 1987; Grime et al. 1986). However, the role of plasticity in the competition process, and particularly in relation to its effects on stand structure, has not been defined. In a community where competition is occurring, even dominant plants may not have the same morphology as competition-free plants.

Axiom 5   Species differences in morphology and physiology determine differences in the rate at which competition occurs. [They will operate through Axioms 1 through 4.]

In order to specify the degree of spatial interaction, some models that describe competition explicitly as a spatial process incorporate some description of plant morphology. Under the same conditions of spacing and environmental resources, quantitative differences have been observed between species in the rate at which competition takes place, and in the resulting stand structures (Cannell et al. 1984, Norberg 1988, Weller 1987a, 1987b). The assumption that these species differences occur motivates much of the work described here.

Scope, Achievements and Limitations in Modeling Competition for Space

Four general categories of mathematical models of competition for space have been developed. Category (i), competition on regular lattice plantings, and category (ii), neighborhood models for irregular spacing; both attempt to define competition intensity in a community. The simplification of regular lattice planting enabled this to be based on effects between neighbor plants at discrete distances. The principal task in developing neighborhood models has been to define just which of the closest individuals to any subject plant influence it. Models of both of these categories have been used to define factors that control the intensity of competition and give precision to *Axioms 2 and 5*.

Category (iii) models have a practical objective of describing the relative growth of different sized individuals in forest stands. They assume competition is a spatial process and seek to define measurements of plant size, structure, or the individual area occupied for prediction. Recent advances have demonstrated that a combination of both tree size and the area potentially available to it for growth are more effective than either measurement alone. Category (iv), models of self-thinning, predict the relationship between mean plant weight and mean plant density as self-thinning occurs. They do not make an explicit representation of the spatial process, but have been used to analyze effects of plant form, function, or environmental conditions on patterns of change in community structure as competition takes place.

Regularly Spaced Lattice Systems

Mead (1967) described the intensity of competition between plants on regular triangular lattices where each plant has six equidistant nearest neighbors. He defined a competition coefficient, λ, such that the conditional expectation of the yield of a subject plant, y_i', can be represented by $\lambda \Sigma_{j(i)} y_j'$, a weighted sum of the yields of the nearest neighbors. Mead (1968) applied this model to experiments with cabbage, carrots, and radishes sown at different densities and harvested at successive times. The experiments spanned the range of comparatively wide spacings used in agricultural practice, and there was little plant mortality. Values of λ varied between spacing distances and with harvest time. For carrots at the first harvest, λ was increasingly negative for increasing plant density. In subsequent harvests λ did not show regular variation in relation to spacing density, and Mead surmised that direct competitive effects extended beyond nearest neighbors.

Mead's research (1967, 1968) demonstrated that competition could be modeled as spatial interaction, but that this interaction may occur between plants beyond nearest neighbor distances without mortality of nearest neighbors occuring. This gives an appreciation of the separation of mortality from canopy competition (*Axiom 3*). Mead's model assumed an additive effect of competition that could be expressed as a mean of all neighbors. This implies two-sided competition that may have been an appropriate model component where no distinct vertical separation in a canopy was likely, as with agricultural spacings of cabbages.

Diggle (1976) developed a one-sided model where the influence of a large plant was dependent upon its size and distance to its neighbors (*Axiom I* (ii)). Competition on a triangular lattice was considered between individuals and only their nearest six neighbors, each as pairwise encounters. Differences in the structure of size frequency distributions of plants were predicted that could be qualitatively compared with data. The experimental motivation was closely spaced annual plants of *Tagetes patula* (Ford 1975) where final height (>14 cm) was considerably greater than initial spacing (2 cm) and substantial mortality (~75%) was observed.

For each plant a competitive status was calculated, multiplicative for the surrounding neighbors. The subject plant, i, was assumed to have a circular zone of influence, r, and a neighboring competitor, j, with a circle of radius, s. The two plants were considered a distance apart of $d_{ij} = d$ where d is the distance between points of the lattice. The status, q_{ij}, of plant i relative to j was calculated in a piecewise linear way that specified the one-sidedness of large plants influencing small ones, but no reciprocal effect:

$$q_{ij}(r, s) = \begin{cases} 0, & s < \max(r, d - r) \\ (r + s - d)/2r, & \max(r, d - r) \le s \le d + r \\ 1, & s > d + r \end{cases}$$

The effect of one-sidedness was marked. As d decreased, and with other parameters constant so that competition intensity grew stronger, distributions progressed from unimodal (no competition), to bimodal with a proportionally larger number of small than large plants. With further decrease in d, distributions became bimodal with a proportionally larger number of large plants, and then, for very small d, distributions became unimodal.

Gates (1978) also considered competition as independent pairwise interactions, but represented plants as squares with their diagonals aligned, enabling an analytically tractable model that included time dependence. Bimodality of resulting size frequency distributions was predicted if, and only if, the area of overlap of squares was assigned to the larger of the two competing plants, i.e. one-sided competition. Gates (1978) found parameters for the growth rates of plants that gave similar developments in bimodality over time to the experimental data of Ford (1975). However, the pairwise calculation restricted prediction of mortality to no greater than 50 percent. Effective modeling of the tails of the frequency distributions effectively would have required a more complex function for plant growth that, in turn, would have made analysis intractable.

McMurtie (1981) defined a differential equation model for individual plant growth and interaction with neighbors, and which partitioned the available resources for growth (assumed to be radiation) according to plant size (a condition of one-sidedness). He showed that the effectiveness of the model used for the growth suppression-dominance phenomenon was not dependent upon the precise model used for growth.

There have been two developments of these lattice based models representing competition as *one-sided* encounters. Gates et al. (1979) made an extensive theoretical analysis of resource partitioning in overlapping areas between plants. They investigated possible effects of crown shape on the extent to which competition is one-sided. The other development was to fit a model to data that included an explicit growth function for individual plants, a separate mortality function, and made use of Monte Carlo procedures (Ford and Diggle 1981).

Gates et al. (1979) started with a two-dimensional disk model, considered as the horizontal projection of the three-dimensional crown, and used the following assumptions:

(i) **Optimal space.** There exists an optimal individual space that plants with no neighbors use. A similar construction is used in a number of distance-dependent tree competition models.

(ii) **Single individual space.** The three-dimensional space within which a plant can be influenced is equal to the space within which it can influence others. This assumption may not hold when light comes from an angle causing shadows to extend the crown space of a tree.

(iii) **All points equal.** All points within the optimal space potentially contribute equally to plant growth. This is assumed when trees are represented as identically proportioned cones with external surfaces that render interior foliage insignificant in its contribution to growth. Mitchell (1975) also used this reasoning to project crown surfaces to two dimensions.

Gates et al. (1979) defined a domination index, η, the proportion of overlapping area of two plants allocated to the larger plant (Fig. 17.3). When $\eta = 1$, large plants totally dominated, i.e., complete one-sidedness as used by Diggle (1976) and Gates (1978). For $\eta = 1/2$, overlap area is partitioned by the common chord, and this favors the larger plant, while $\eta = 0$ gives unbiased (symmetric or two-sided) competition, and the overlap area is partitioned by an hyperbola. For two circles, x and y, with radii R_x and R_y, the boundary between them, $\Gamma_{x,y}(R_x, R_y)$ has an equation of the form

$$R_x^\alpha - |r - x|^\alpha = R_y^\alpha - |r - y|^\alpha,$$

where x and y are the centers of the discs and r is a point of the boundary of the respective circle (Fig. 17.3).

Gates et al. (1979) made an additional assumption that the crown morphologies between competitors are similar (non-plasticity, i.e., that Axiom 4 does not apply). This implies that the boundry between two disks is the projection of the three-dimensional curve created by the intersection of the outer surfaces of two overlapping three-dimensional zones of influence (pic-

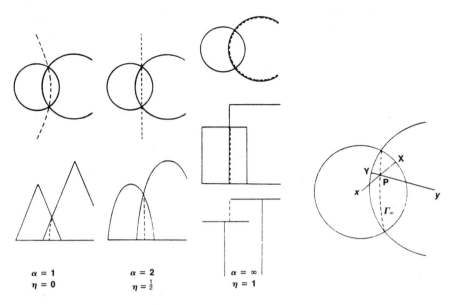

$\alpha = 1$
$\eta = 0$

$\alpha = 2$
$\eta = \frac{1}{2}$

$\alpha = \infty$
$\eta = 1$

Figure 17.3. Methods of partitioning the overlap between two discs that represent competing plants, and their equivalence to different crown shapes. η is the domination index. Graphical representation of the nomenclature in the formula (see text) is shown at the right. [Redrawn from Gates et al. (1979)]

ture the "horse shoe" shaped curve created by the intersection of two cone surfaces). The shape of the three-dimensional zone, from flat-topped to round-topped to conical, is specified by α. The special cases, Fig. 17.3, are $\eta = 0$ for cones, $\eta = 1/2$ for ellipsoids of revolution, and $\eta = 1$ for cylinders. The relative dominance for a species increases as the flatness of its crown increases.

The detailed rules of Gates et al. (1979) shed light on assumptions that are implicit in many two-dimensional competition indices. Many equations can produce a relative resource division term (e.g., the ratio of the subject stem diameter to neighbor stem diameter, possibly taken to some power). However, none of those used in other competition indices have coefficients with such specific interpretation as α. This essential advantage comes from the explicit statement of hypotheses used to define the model.

The approach taken by Gates and co-workers suggests that different crown morphologies may result in very different outcomes in the competition process. However, a further analysis by Gates (1982), to consider multiple pairwise interactions between plants on a square lattice, illustrated difficulties associated with extending rigorous, formalized mathematical constructions.

If the part of the model that considers variation in crown shape is retained, and (a necessary realism to account for plant growth functions (not just space occupancy) is added,) then the model requires numerical simulation.

Ford and Diggle (1981) extended an overlapping disc representation of competition to a form that could be fitted to data. This model was suited to the situation where plant height is large relative to inter-plant distance, and competition is occurring in an elevated canopy. A four-parameter model was optimized for a replicated series of harvests from 2 cm triangular lattice plantations of *Tagetes patula*. The competitive status of each plant was calculated from plant height at the start of a growth period relative to the heights of neighbors. This calculation was weighted relative to a crown apex angle, Θ, that estimated the intensity of competition within the whole stand (Fig. 17.4). Taken together, Θ and plant height define a circle, but competition was only deemed to take place (i) if an individual plant location fell within this circle, rather than if plant circles intersected at all, and (ii) if a plant, within the circle of another, had a height less than the surface of the cone over its point of location. This use of crown apex angle is different from that of Gates et al. (1979), although they discuss the extension of their model to the condition of plant survival if it is within the estimated circle of a dominant plant. That would involve values of $\alpha < 1$ in their model, and additional rules would be required to weight the area of circle that was totally enclosed within the circle of the dominating plant.

The estimation of status of each plant was then applied, as a modifier, to a relative height growth rate selected at random from a distribution with mean, μ, that represented a mean relative height growth rate for plants experiencing no overtopping by neighboring plants, and variance σ. This variance represents non-competition influences on growth rate. Plants were considered to have died if their relative growth rate did not achieve a value ϕ. An optimization routine, which minimized the residual between data and model for a statistic incorporating both individual plant size and size of neighboring plants, was used to solve for the four parameters of the model.

Ford and Diggle's model incorporated direct, multi-way interaction, and, quantitatively illustrated the stochastic nature of the competition process. Future success depends not only on how big a plant is, but also how big the neighbors are that will influence competition in the next time period, and that is a matter of chance. It depends upon their previous sets of interactions as well as the operation of σ possible non-competition effects. Furthermore, the complete set of mutual influences can be felt across the lattice; competition was not limited to nearest neighbors only, and at latter stages in the growth of the experimental plants, significant interaction was found with next but one neighbors (second order). Θ gives some indication of the relative lateral extension of plants during the growth period, and is a partially interpretable parameter, though it also requires an appreciation of absolute

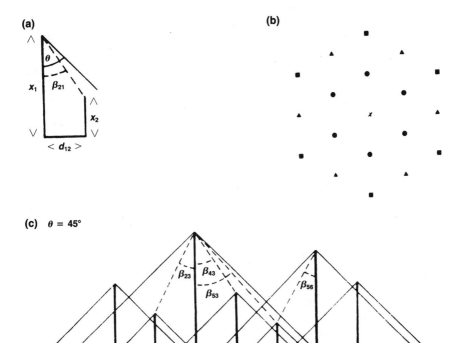

Figure 17.4.   (a) The relationship between two plants of heights x_1 and x_2, and planted d_{12} apart. Plant 1 reduces the relative growth rate of Plant 2 by a factor of β_{21}/θ. (b) Part of a triangular lattice showing first-order neighbors at 20 mm (●) and second-order neighbors at 34.64 (▲) and 40 mm (■) from the subject plant x. The total influence of all these first- and second-order neighbors can be calculated through the multiplication of their values of β/θ if this value is <1. (c) Diagrammatic representation of plant heights along a single row to demonstrate which plants may influence others when $\Theta = 45°$. [Redrawn from Ford and Diggle (1981)]

height. This model quantified distinct time differences in the intensity of competition in the upper canopy, and the occurrence of mortality.

The model had some important limitations. First, Θ has a limited morphological interpretation as a crown angle approximation. The leaves of *Tagetes patula* grow slightly upward and laterally out from the plant apex. For large plants, a few leaves tend to extend laterally over a much wider distance than predicted by Θ. Thus the geometric approximation implied by Θ for the experimental data was not a close approximation of crown form. Second, the complete possibilities of the stochastic nature of competition were not explored. Variation could be applied with equal logic to Θ and ϕ, as it was to μ, although the biological rationale for that would itself require study.

Cannell et al. (1984) measured relative growth rates of successive annual height increments in closely spaced *Picea sitchensis* and *Pinus contorta* planted on a 0.14 m hexagonal lattice. Competition developed as the stands grew to a height range of 0.5 m to 1.15 m in *Picea sitchensis*, and 0.30 m to 0.90 m in *Pinus contorta*. Under this close spacing, the two species showed very marked differences in crown morphology. Branches of *P. contorta* responded by bending close to the main stem and growing almost vertically upwards. Branches of *P. sitchensis*, on the contrary, maintained a horizontal posture. Marked differences were found between the species in a model for competition that calculated neighbor influence on height increment, and weighted the effects of neighbors themselves of different heights. For *P. contorta*, all neighbors with an apex greater in height than the subject plant were found to have an effect. For *P. sitchensis* neighbors had an effect only when their apex was 50° above that of the subject plant. It seems possible that the greater competitive effect between *P. contorta* trees could be the result of the close to cylindrical crown shape they developed as a plastic response to competition, whereas the conical shape of *P. sitchensis* crowns produced a different effect per unit of height difference.

The study of inter-plant competition on lattices has been useful. The mathematical simplification provided has permitted an analysis of the control of competition intensity between plants. The problems posed by Mead (1968) of estimating effects over variable numbers of neighbors were largely resolved by Ford and Diggle (1981), where estimates could be made for first and second order competition, and by Cannell et al. (1984), where competition effects were seen to increase over successive years in an ordered way. A precise analysis of the effects of crown structure on competition has not been made, although there is experimental evidence that plant form may influence neighbor effects (Cannell et al. 1984 and Gates et al. 1979), illustrated with theoretical analysis that differences in crown structure may result in different effects. None of these models attempts to account for plasticity within stands.

Neighborhood Models

A general objective of ecologists has been to define the population dynamics of species. What controls plant numbers? How many seeds are set, and what is their distribution? How do these processes influence vegetation structure and development? The principal task has been to characterize the competitive influence that different species may have. Central to this has been the development of the concept of a neighborhood. To research this, experiments have been made by ecologists in which seeds of annual plants have been broadcast in irregular spacings of one, two, and multiple species. In some instances such planting mixtures have been allowed to set seed and have been followed through a number of generations. Pacala and Silander (1990) describe a one- and two-species system, and reference others.

Mack and Harper (1977) proposed a neighborhood model of interference whereby individual plant performance could be described by a regression equation that accounted for the size and proximity of neighbors. They experimented with five autumn germinating annual plant species of stabilized sand dunes. Seeds were hand broadcast in all possible pairings at 3700 seeds m^{-2}. Biomass at harvest, seed production, and spatial coordinates of each plant were recorded. Mack and Harper made an *a priori* categorization that each plant's neighbors be considered in three concentric annuli around it. For the species with the largest individuals, *Vulpia fasciculata*, the successive annuli were 0–0.5, 0.51–1.0, and 1.01–2.0 cm; for the other species, they were 0–0.5, 0.51–1.0, and 1.01–1.5 cm.

Mack, and Harper described neighborhood relationships in terms of size of a plant's neighbors, their distance, and, for the two largest species, *V. fasciculata* and *Phleum arenarium*, there were also directional effects. In pairwise mixtures, directional effects were accounted for by weighting plant effect according to the proportion of the four quadrants of the two outer annuli that were occupied.

This weighting, δ, varied from 1.0 to 0 as the neighbors varied from tightly clumped to regularly dispersed, and 1/δ was used as a weighting factor. Mack and Harper found that the effects of neighbors in the two outer annuli were influenced by their relative mean directions. A further weighting factor, z, was computed as 1.0, 0.75, 0.5 when mean angles of neighbors from the second and third annuli came from opposite, adjacent, or the same quadrants, respectively. The mean direction is measured as the mean angle, $\bar{a}$, of the *Phleum* and *Vulpia* neighbors in each annulus: $\mathrm{Tan}\,\bar{a} = y/x$.

Taking into account these weightings for directional effects, and an additional weighting to reduce the relative effect of plants in the outermost annulus, but that varied according to species, Mack and Harper found the biomass of any *V. fasciculata* individual in a five-species mixture to be proportional to an interference quotient:

$$-\log\{\Sigma_{\text{annulus 1}}\text{biomass} + \delta\Sigma_{\text{annulus 2}}\text{biomass} + \delta z\Sigma_{\text{annulus 3}}\text{biomass}\}.$$

In two-species mixtures, regressions of individual plant biomass against interference quotients of this form had r^2 values between 0.32 and 0.69. In five-species mixtures, r^2 was 0.37 to 0.43. They noted that all species had simple leaf morphology and considered it probable that the neighborhood effects were produced in relation to soil nutrient demands.

There are two categories of problem associated with neighborhood models of this type, model identification and model fitting, that should include sensitivity analysis.

Mack and Harper (1977) suggested spatial effects were unlikely to occur in the discrete way represented by annuli, and this poses a problem of model identification. Lattice experiments' indicate that neighborhood effects cannot be defined as distinct distances, or combination of size and distance, but vary between species and over time. Subsequently, attempts have been made to search for an optimum neighborhood size (Silander and Pacala 1985). This still requires an *a priori* decision of the relative weighting to apply to changes in size of a neighbor per unit of increasing distance from the subject plant. This problem in model identification is a symptom of selecting any sample of pattern, e.g. annuli, from a larger spatial matrix, where the extent of the dependencies is actually what must be researched.

Spatial dependencies also pose problems in model fitting. The regression procedures of Mack and Harper (1977) make multiple use of individual plants, both as subjects and then as competitors. This transgresses a fundamental requirement of regression analysis; that the individuals should be independent and identically distributed. Of particular concern is that plants in dense patches contribute to more interference quotients than plants in sparse patches, and so have a biased influence. Mead (1968) overcame this problem of dependence between neighbors on lattice systems by using sample units where plants were counted only once, though at the cost of having to use very large experimental plots. Ford and Diggle (1981) overcame the problem by solving models for the complete community, and then each community is the experimental unit.

Waller (1981) suggested a more appropriate procedure in model formulation is to develop several relatively simple interference indices for each plant, and included them in stepwise linear regression to determine which of the indices contributes significantly to individual plant performance. "However, this approach requires that the data be transformed to linear and provides no systematic way to determine the functional forms of the competition index." The problem of model identification and fitting remain, however. Hamilton (1969) used a similar procedure in developing functionally based models for tree competition.

Weiner (1982), working with a single species, developed a simpler expression than that used by Mack and Harper (1977), based on the inverse-yield relationship between plant mass, V, and plant density, N

$$\frac{1}{V} = a + bN.$$

He assumed that a maximum value for the seed yield of an individual, R_{max}, could be found so that the seed yield of any individual in a community, R, could be expressed

$$R = \frac{R_{max}}{1 + W}.$$

W is the total competitive effect of all neighbors

$$W = \sum_{i=1}^{n} \frac{2}{d_i^z} \sum_{j=1}^{m} C_j N_{ij}$$

where d_i is the mean distance to the ith neighborhood and where neighborhoods are concentric circles around the subject plant. C_j is the mean effect of an individual of the jth species and N, the number of individuals of species j in neighborhood i. The one-species version of this model was more successful than the multispecies models of Mack and Harper (1977). However, it cannot be determined if this is because of the nature of the competition in the natural patches of open grown *Polygonum minimum* and *Polygonum cascadense* examined by Weiner (1982), or to the particular form of the model.

Subsequently, Thomas and Weiner (1989) used a modified version of this model to estimate varying degrees of asymmetry between effects of large-on-small and small-on-large plants. Different species showed different degrees of asymmetry. Incremental developments of this type of model, applied to suitably simplified experiments, can result in progress in understanding the competition process.

Silander and Pacala (1985) had the objective of systematically developing an interference index so as to improve the fit of a given non-linear neighborhood model. Their objective was to assess individual plant seed set through an index of interference. The size of the annulus within which neighbors were considered as competitors was increased incrementally and solved for $S = M/(1 + CW)$, where S is the number of seeds per subject plant, M is the maximum number of seeds produced (i.e., by plants with no competitors, a constant), C is a decay constant, and W is the index of crowding. The crowding index weighted each neighbor by its distance from the subject plant:

$$W = \sum_{i=1}^{N} \left(1 - \frac{d_i}{r}\right)^{\Theta}$$

where d_i = the distance of neighbor i from the focal plant, r = the neighborhood radius, and Θ = a weighting constant. The effect of distance is minimized as d approaches r and is maximized as d approaches 0. For Θ = 0, the index is the same as the number of neighbors; for $\Theta < 1$, the effect of neighbor distance is convex, for $\Theta = 1$, the effect is linear, and for $\Theta > 1$, the effect is concave. For the data they examined, Pacala and Silander (1985) found a minimum of residual sums of squares/total sums of squares, (RSS/TSS) at $r = 5$ cm. Only when larger neighborhoods were examined did Θ have an effect; under the condition of $r = 7$ there was a 20% reduction in the residual sums of squares, with Θ between 2 and 3, i.e., nearest individuals had a greater effect as the neighborhood considered was enlarged. They did not include this in the model they used for further analysis, but used the 5-cm radius. This is an interesting indication of lack of uniqueness in model identification. A weighted 7 cm radius model is very close to an unweighted 5 cm radius model.

Silander and Pacala (1985) also looked for an effect of angular dispersion. Their premise was that within a 5-cm zone, neighbors should be spatially regularly distributed if they are to have equal effect on the subject plant. The angular dispersion index they used was

$$z = 1 - \sqrt{(\Sigma_{i-1}^{N} \sin \alpha_i)^2 \times (\Sigma_{i=1}^{N} \cos \alpha_i)^2}/N,$$

and a weighted crowding coefficient $W = N \cdot z^\theta$, where α_i is the angle to the neighbor i, N is the number of plants in the neighborhood, and Θ a constant used to weight W. They found a 13 percent reduction in RSS/TSS when $\Theta = 1$.

Pacala and Silander (1990) used this basic formulation of a neighborhood, and defined a predictor of fecundity comprised of functions for adult biomass and for seed set. A biomass predictor was a function of neighborhood crowding. They experimented with two annuals, *Abutilon theophrasti* (velvet leaf) and *Amaranthus retroflexus* (pig-weed). Spatial locations were recorded for two annual broadcast experiments of single species and mixtures. Some plots were permitted to regenerate over four successive years and, on those, plant numbers and approximate locations were recorded.

Velvet leaf seedling survival was density independent, while pigweed seedling survival was density dependent. Survivorship neighborhoods were small, inconsistent across plots and years, and interference affecting seed production was greater than that affecting survival. They made four conclusions from fitting an adapted version of the model of Pacala and Silander (1985) to biomass prediction:

(1) Inter-plant interactions occur over short distances, estimated radii ≤ 20 cm, and velvet leaf neighbors interfered with neighbors over greater distances.

(2) Interspecific interference was asymmetrical; velvet leaf affected pig-weed more than the reverse.

(3) Growth of plants with "no neighbors" varied between the plots, and between years.

(4) Where local crowding is proportional to $1/\alpha$, as α decreases residuals became increasingly skewed. The variance of residuals at any one level of local crowding is proportional to $1/\alpha$.

Pacala and Silander's (1990) objective was to use parameter estimates obtained from the repeated annual experiments to predict the extended course of development on the permanent plots. However, to do this they also derived a "simple" version of the model, by setting the crowding adjustment factor to 1 so that full spatial interactions were not included.

Pacala and Silander (1990) suggested that the number of seeds predicted to be set by the two models, with and without full spatial interaction, showed a good approximation, differing by 4 percent for velvet leaf and 8 percent for pig-weed. However, there were consistent differences in the mismatch between predictions made by the full and reduced spatial models. Over 9 plots the reduced model estimated smaller seed set on all 9 in the first year, but estimated greater seed set on 7 out of 9 in the third year. Consistent trends in residuals can be more important than mean differences in assessing model fit.

Populations in the four-year plots became weakly spatially aggregated. Pacala and Silander (1990) suggested this might be due to soil heterogeneity, which was not accounted for in their model. They concluded that simple non-spatial mean models may be adequate for predicting population development in systems such as velvet leaf and pig-weed, where spatial distributions are highly non-random, being caused by short seed dispersion distances or heterogeneous soils. However, this conclusion, to reduce the scope of model expression of interactions, does not seem justified, either in this case, or for neighborhood models in general. An important challenge is to make neighborhood models more representative of the processes of growth and interaction involved in competition.

Pacala and Silander (1990) point to a difficulty in translating neighborhood models to describe competition for resources. "Neighborhood models are based on phenomenological descriptions of performance and density dependence. Unfortunately, there is simply no obvious way to determine how the parameters of a phenomenological model of plant competition will change in response to changes in the physical environment." This implies that such models cannot be developed to analyze the fundamentals of the process of competition, where resource acquisition is dependent on plant morphology. Neighborhood models are still being used in an exploratory way. Until a more tractable theory of the interaction process can be put forth, and the structure of the neighborhood defined effectively *a priori*, problems over the fitting of such models cannot be solved.

Distance-Dependent Models of Tree Status and Growth

Prediction of forest yield has two components: (1) projection of total stand timber increment and (2) estimation of the distribution of that increment between trees of different sizes. Distance-independent methods for (2) usually comprise mathematical descriptions of stem size frequency distributions and how parameters vary as the stand grows (Moser 1974). This latter approach is similar to that used by Westoby (1982) in competition studies. Distance-dependent models are most frequently solved for individual trees measured on sample plots. Some are similar to neighborhood models, and seek effective definitions for numbers of competitive plants and their influence in terms of sizes and distance to include in competition indices.

Kent and Dress (1979) considered that "the progression of trees from competition class to competition class might be adequately modeled for many purposes by a stochastic process that does not require or assume measurements on neighboring trees." They developed a model of transition probabilities between five crown classes for successive periods over a forest plantation's life that considered only tree size at the start of a period. Despite using a non-spatial model, they made predictions of the effect of competition on spatial structure. Not surprisingly, they calculated that an initially random population of trees would continue to have a random spatial distribution (Kent and Dress 1979), that an initial lattice pattern would develop to a random distribution, and that a clumped distribution would remain clumped (Kent and Dress 1980). These results are contradicted by the field observations cited previously in this review. Kent and Dress (1980) compared frequency distributions of individuals between the five tree classes for simulated and actual data, and considered it adequate for forest mensuration purposes, though they attempted no test procedure. The justification for distance-independent methods lies in the argument of utility for a management purpose, relative to collecting spatial data and developing more complex models without scientific precision.

Daniels et al. (1986) reviewed four types of distance-dependent loblolly pine growth models and compared their effectiveness against each other, and against a distance-independent model (Fig. 17.5).

For area overlap indices, an estimate is made of the crown size that a tree would have if it were open-grown, and thus free from competitive influences. Then the overlap onto this crown from neighboring tree crowns, estimated in the same way, is calculated. In Ek and Monserud's (1974) index the overlap terms are weighted by the relative size of competitors to the subject tree (Fig. 17.5). Arney's (1973) version has a minimum value of 100 when there is no overlap.

Hegyi's (1974) size ratio-distance index is the sum, for all competitors, of the diameter ratios between competitor and subject tree, divided by dis-

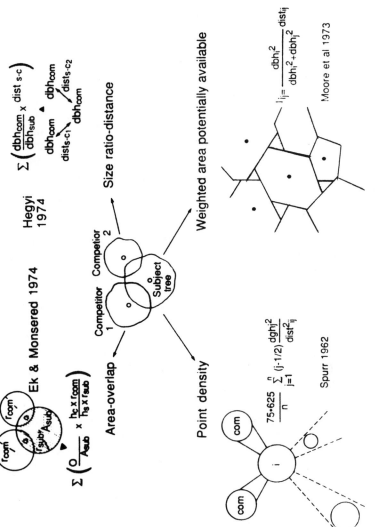

Figure 17.5. Procedures for calculating four categories of tree status or competition indices with an example of each. *dbh* = diameter at breast height. *A* = crown area estimated from open grown tree of the same *dbh*. *sub* indicates subject free, *com* indicates a competitor

tance between them. Hegyi defined the number of competitors by a fixed radius around the tree. Daniels (1976) used a fixed angle gauge sweep around the subject tree so that trees with diameters that subtended a greater angle than set for the gauge were defined as competitors.

Spurr's (1962) point density estimate also used an angle sweep to define inclusion for competitors. The competition index is a weighted average estimate of basal area of these competitors.

The area potentially available index requires construction of Voronoi polygons (via the Dirichlet tessellation) around each tree (Fig. 17.5). The distance between each tree is bisected at right angles, and these successive right angle lines are joined to form a polygon. The area of the polygon is the presumed area potentially available for growth (APA). Moore et al. (1973) weighted division of the distance between trees according to relative tree sizes; Daniels et al. (1986) weighted on diameter^2. The resulting polygons leave some open areas on the plot not "potentially available" to trees. A three dimensional index can be calculated by multiplying the polygon areas by tree height (Pelz 1978). Nance et al. (1988) present an empirical analysis of the weighted APA approach.

Daniels et al. (1986) calculated these competition indices for individual loblolly pine (Pinus taeda) trees in sample plots of a thinning experiment measured annually between years 5 and 10 and again at year 13. Each tree was considered an independent sample. Correlation coefficients between basal area growth and all distance-dependent competition indices increased from year 5, while that between basal area growth and a distance-independent measure, the ratio of a trees basal area to the mean tree basal area (Glover and Hool 1979), decreased (Fig. 17.6). The most effective index was the APA index weighted by basal area, P_2. Clearly, in the early years of growth, tree size itself is required as a predictor, and whereas P_2 calculates the modification of this by neighbors, G_B does not. It is interesting that the unweighted calculation of APA, P_1, gives a lower value of correlation than P_2 for all ages. Among the remaining indices there is no consistent ranking.

These comparisons illustrate that in these forest plantations competition is a spatial process. The comparative success of P_2, not just relative to other distant-dependent models from year 10, but also to the early phases of stand growth, illustrates the utility of including plant size as well as neighbor relationships.

Alemdag (1978) compared a similar set of competition indices on three plots of white spruce over three five-year periods of growth from age 32 for one and 33 for two plots. These plots were thinned at the start and again after 10 years. Not surprisingly, where the greatest thinning occurred and density was reduced from 3600 to 1025, and then to 400 trees.ha^{-1}, correlations between competition indices and growth were low, even where initial tree diameter was included in the predictive equation. For two less

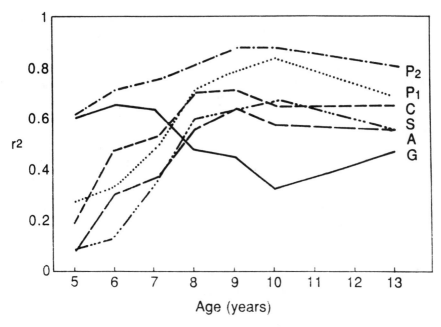

Figure 17.6. r^2 values for different competition indices fitted to loblolly pine stands as they grow to age 13. Area potentially available, P_1 unweighted, P_2 weighted. Area overlap, A. Point density, S. Size ratio distance, C. Distant-independent model, G. For exact models used, see Daniels et al. (1986). [Redrawn from Daniels et al. (1986)]

heavily thinned stands, Bella's (1969) influence zone overlap model Hegyi's (1974) size-ratio distance index, and two versions of a weighted APA model were consistently the best ranked, though not in the same relative positions for all time periods and plots. The least successful models were those of Arney (1973), Newnham (1964), Staebler (1951) (each of which are overlap indices), and Brown (1965), an unweighted APA model.

Again, a weighted APA model seems most effective. The lack of distinction between other model types may be because within each, some major choices must be made, either of an angle sweep factor to decide inclusion of neighbors, or of the formula and parameters to define apparent open grown crown occupancy. These models are not simple, logical procedures to define model structure would require substantial additional amounts of data. A possible approach would be to follow procedures of Pacala and Silander (1990), but with independence maintained between subject trees (Mead 1967, 1968).

Larocque and Marshall (1988) described four major factors that limit the effectiveness of single-tree, distant-dependent models. First, consistent estimates of competition intensity cannot be made across a variety of stand conditions. Many studies have concentrated on competition for light, but differing water and nutrient regimes may also alter the mode of competition. Further, the concentration of most models on predicting growth, rather than relative growth rate, places emphasis on the effect of tree size in the competition process. Second, most models are developed using narrow ranges of data on age, site index, and stand density, and repeated measurements of spacing or thinning experiments are short term. Limited data can give an incomplete view of the competition process. For example, it would be of interest to obtain an extension of the type of information presented in Fig. 17.5 through the complete life of the plantations. Third, Larocque and Marshall (1988) point to the generally inadequate treatment of variation in growth not due to competition, and suggest that the structure of competition models should change to facilitate this. Fourth, multiple linear regression techniques are inadequate (as discussed here for neighborhood models). Larocque and Marshall suggest development of process-based models, and that "the next generation of single tree-distant dependent growth models could focus on modeling inter-tree mechanistic relationships."

It is perhaps surprising that information on the relationships between crown structure and yield, e.g., Krammer (1966), has not been used more extensively in distance-dependent models. Hamilton (1969) examined the effectiveness of crown projection, crown volume and crown surface area in growth prediction of trees released from competition by thinning and unreleased trees. With a stepwise multiple regression technique, he accounted for up to 92% of variation in tree volume increment. This approach requires more information in model construction, i.e., measures of crown structure, but offers the possibility of developing a more functional approach to growth itself.

Mitchell (1975) developed the most comprehensive distance-dependent model. Increments in height and branch extension produce successive crowns represented as shells of foliage. Relative productivity of each shell changes with age. Empirically determined relationships are used between crown foliage amount, tree volume, and height increment; and between timber volume increment, bole shape and bole increment. Trees on a given plot are grown in crown dimensions. They interact by over-topping each other, causing reduction in a neighbor's crown volume. First this affects its timber volume and then its height growth. Mitchell (1975) simulated silvicultural operations such as thinning by tree removal, or fertilizer application, by manipulating the growth equations. Mitchell's model requires more detailed assumptions, and more parameters to be estimated, than other tree competition models. Perhaps it is this that has, unfortunately, excluded this model in the comparison exercises that have been made.

Mean Plant Weight and Mean Space Occupied During Self-Thinning

Morphological and physiological differences between species may influence both the rate of resource acquisition and growth rates, and so may influence intensity of inter-specific competition. Many investigations have been made of the relationship between mean plant size and plant density—the inverse of mean space occupied—as self-thinning due to competition occurs (Weller 1987a, Zeide 1987). As mortality occurs, mean plant mass, $\overline{m}$ (grams), and plant density, N (plants.m^{-2}), are related:

$$\log \overline{m} = \gamma \log N + \log k$$

where k is a constant. Following the original proposition by Yoda et al. (1963), many investigators have concurred with a constant value for $\gamma = -3/2$. However, substantial evidence (Weller 1987a, Zeide 1987) has now accumulated that both γ and k may vary between species and within species under different conditions. Theoretical analyses (Norberg 1988) suggest that this may be due to differences in the patterns of spatial occupancy and growth.

Biometrically, it is more appropriate to investigate the relationship

$$\log B = \beta \log N + \log k$$

where $\beta = \gamma + 1$ and so $\beta = -1/2$ if $\gamma = -3/2$ (Weller 1987a). In a comprehensive review, Weller (1987a) showed that of 75 reported estimates, 35 had values of β not significantly different from $-1/2$, 14 had β greater, and 27 had β significantly less than $-1/2$. Zeide (1987) suggested that slopes steeper than $-3/2$ seem typical for dense stands in optimal conditions, while flatter slopes are common in suboptimal conditions of high latitude, altitude, low light, arid conditions, or restricted soil environment.

Weller (1987a) proposed that systematic biological differences between species, particularly differences in shade tolerance, cause differences in the thinning relationship. For angiosperms, β was significantly and negatively correlated with shade tolerance, i.e., more shade tolerant trees have steeper, more negative thinning slopes, and total biomass increases more rapidly per unit decrease in density. For gymnosperms, Weller (1987a) found the opposite, but also that α (the projected value of $\log B$ when $\log N = 0$, i.e., at 1 plant/m^2) increased with increasing shade tolerance. However, this estimation of α lies outside the measurement ranges of the self-thinning relationship for most investigations.

Zeide (1985) suggested that intuitive classifications of a species as tolerant, based on its survival in mixed stands, does not necessarily imply that it will show less mortality than a lesser tolerant species when grown in monocultures. Intra- and inter-specific competition may involve different plant attributes. For pine species of the southern USA, he reported a negative correlation between tolerance and self-tolerance, as assessed by the rate of self-thinning in numbers relative to average tree diameter in pines of the southern USA. For some angiosperm species, he reported the reverse.

Zeide (1987) developed a critique of the simple isometric geometric model proposed by Yoda et al. (1963). They had suggested that a value of $\gamma = -3/2$ could be explained as a result of plants having constant proportionality in size and space occupancy as they grew. Zeide noted that this was based on two assumptions:

1. The combined action of crown growth and self-thinning maintains complete canopy closure.

2. Plants of the same species are always geometrically similar in shape, irrespective of the growth stage and habitat condition.

If one assumption does not hold, then for a constancy in γ, compensation must occur in the other process. Zeide reasoned that crown growth and mortality are events that happen to individual trees, and that gaps occur in the canopy of single-species stands. Species tolerance and site quality can determine the rate of canopy closure. Zeide (1987) suggested that if plant shape were fixed, β would increase from $-1/2$ at pole stage when crown closure is maximal. From mechanical considerations, the total above-ground mass of the tree must increase more rapidly than crown mass. Because of this, Sprugel (1984) concluded that a thinning slope of $-3/2$ would be the exception rather than the rule.

Variation in β can be correlated with variation in plant allometry (Weller 1987b). The thinning exponent depends upon the proportionality between plant mass and ground area covered. Area covered by a plant can be considered proportional to $m^{2\theta}$ where Θ can vary from $1/3$ to reflect changes in slope with size. The thinning equation is then

$$\overline{m} = KN^{-1/2\theta},$$

and the exponent $\gamma = -1/(2\theta)$ equals $-3/2$ only if plant shape does not vary as growth takes place. A plant may increase in mass in three ways: with height increment, $h \propto m^\theta$, with increase in the area occupied, $R \propto m^\theta$ (where R = radius of the occupied area), and through biomass increase in the area already occupied, i.e., the density, $d \propto m^\delta$. Weller (1987b) assumed that volume occupied (υ) is approximately cylindrical, i.e., that $\upsilon = \pi R^2 h$, and plant mass, m, follows the relationship $m = \upsilon d = \pi R^2 h d$. Given the previously defined proportionalities, $m \propto m^{2\theta} m^\theta m^\delta$ so that $2\phi + \theta + \delta = 1$ and $\phi = 0.5 - 0.5\theta - 0.5\delta$. Weller commented that this expression is a clear representation of the biological compromises inherent in plant growth: ". . . allocation of resources to height growth [higher Θ] or to packing more biomass in the space already occupied [higher δ] leaves fewer resources for radial expansion [lower ϕ]. Less radial expansion means less conflict with neighbors, so a given amount of biomass can be added with less attendant mortality." This assumes radial increment is the cause of competitive influence.

Weller (1987b) used a modified equation in order to analyze empirical data

$$\phi' = \frac{-1}{(2\gamma)} = 0.5 - 0.5\theta.$$

For trees, $\overline{h} \propto \overline{DBH}^\Lambda$, an increment of height requires increment in *DBH*, and λ must be positive. Similarly, since *DBH* and *BA* are related, $\overline{h} \propto \overline{BA}^\psi$ then ψ should equal 2λ. Weller suggested that plants "can be more massive at a given density and can grow with less attendant mortality [i.e., have more negative thinning exponent] if they allocated more resources to height growth than to radial growth."

Weller (1987b) solved Θ, λ, and ψ for a range of experimental and natural populations. ϕ' was significantly and negatively correlated with λ, Θ and ψ for forestry yield table data, and similarly with Θ and ψ for experimental and natural populations, but there were few observations to examine the $\theta - \lambda$ relationship. He concluded that the self-thinning exponent is related to the allometry of plant growth.

Norberg (1988) developed theoretical arguments of why values of γ should deviate from -1.5. For isometric growth height to breadth ratio of a plant remains constant

$$H = h2R$$

where H is height and $2R$ is the diameter of ground occupied. With V_1, the volume of plant material, and k_1 is a conversion coefficient incorporating h, then

$$V_1 = kR^3.$$

Plant volume is a cubic function of any linear plant dimension. When plants are densely packed, their number per unit ground area, N_1, is inversely related to the average ground area

$$N_1 = k_2R^{-2}$$

where k_2 is a conversion coefficient that varies with the units of density. These equations can be combined by substituting k_3 for $k_1 k_2^{3/2}$, so that

$$V_1 = k_3 N_1^{-3/2},$$

which is equivalent to the self-thinning rule if the density of plant material remains constant.

Norberg (1988) points out that differences in height-to-breadth ratio of space occupied by different species will give different elevations of the thinning line (values of k_3), but the slope will not be affected by this. He suggested that observed differences in k, with higher values for grasses and conifers compared with dicotyledonous and deciduous trees (Lonsdale and Watkinson 1982), may be explained by differences in height-to-breadth ratio rather than by differences in canopy efficiency of light interception, as Lonsdale and Watkinson (1982) suggested.

Norberg (1988) notes that geometric similarity implies that a constant number of branches be maintained, and that their spatial separation is only subject to isometric enlargement. Actual branch growth does not follow this pattern. Violation of the geometric-similarity constraint on branch number, increasing rather than remaining constant, might help explain why self-thinning stands of some plants show values of γ between 1.5 and 2.0

Norberg (1988) presented a formal argument concerning why trees may show values of $\gamma > -1.5$, based upon the reasoning that their growth allometry follows elastic, rather than geometric, similarity. Under elastic similarity longer branches and trunks are proportionately thicker than short ones. There is a taper ratio such that diameter increases as the 3/2 power of length. Consequently, as trees grow in height, wood mass increases more rapidly per unit of horizontal space occupied.

Where t = trunk diameter, assuming that branch diameter is proportional to trunk diameter, the volume of the trunk, $V_2 \propto t^2 H$, can be expressed entirely in terms of t by assuming that $H = k_8 t^{2/3}$, so that $V_2 = k_8 t^{8/3}$. Norberg described the radius of a plant's exclusive space, as determined by trunk radius at the level of the longest branches, and the horizontal projection of those branches, so that:

$$R = 0.5t + k_3 t^{2/3}.$$

The population density, N_2, of plants maintaining elastic similarity is:

$$N_2 \equiv k_2 R^{-2} = k_2 (0.5t + k_7 t^{2/3})^{-2}.$$

Norberg (1988) showed that as t, trunk radius, approaches 0 the slope of the thinning curve, $d \ln V_2 / d \ln N_2$, approaches -2; and as t approaches

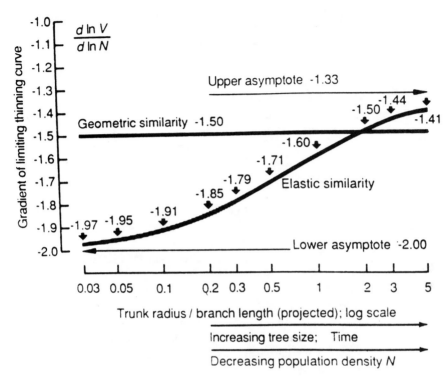

Figure 17.7. Variation in the gradient of the self-thinning curve, γ, with mass estimated as plant volume, as the ratio of trunk radius to branch length changes. For geometric similarity, the gradient is constant at −1.5. For trees that have elastic similarity, the gradient of the self-thinning line increases as trees grow, and self-thinning occurs (after Norberg 1988).

infinity, $d \ln V_2 / d \ln N_2$ approaches $-4/3$. Thus, as trunk radius/branch length varies, so would the expected gradient of the thinning curve (Fig. 17.7). Norberg (1988) stated that "when elastic similarity prevails throughout individual growth, self-thinning of populations proceeds along a curve, rather than along a line, in a log-log diagram of volume versus density. When trunk diameter t is small and population density N is high, the gradient of the limiting thinning curve is close to -2, with a theoretical minimum at -2. . . . As trunk diameter increases and population density decreases, the gradient increases monotonically, with a theoretical maximum at -1.33 . . . .".

The applicability of this model depends upon the branch diameter:branch length relationship holding under the conditions of crown competition, and requires further investigation. Norberg concluded that the elastic similarity model fits tree allometry data much better than the geometric similarity model. Populations of many trees exhibit thinning gradients close to -1.9, though some tree species come close to the geometric similarity exponent of -1.5. These differences might be due to differences in allometry, and particularly to the relationship between aerial parts and roots.

These analyses of the self-thinning line provide an analytical background to *Axiom 5,* that species differences influence the rate at which competition occurs. Here the rate is defined as change in plant number per unit plant biomass increment. Experimentation may prove valuable in defining the influence of form or physiology on the parameters of the self-thinning relationship. However, the self-thinning relationship does not include time as a parameter, and so may not be easy to extend to a direct analysis of resource acquisition and utilization.

Spatial Variation in Resource Depletion and Growth Plasticity

In each category of competition modeling, further advance requires a new conceptualization of the competition process. Models of canopy-forming plants on lattices illustrated that species differences may determine the course of the competition process, but that simple geometric models of crown structure may not provide suitable descriptions of these differences. Neighborhood models, unless used for discrete experiments, e.g., those of Weiner and Thomas (1986), have an imprecise distinction between exploratory analysis necessary in model formulation and rigorous procedures required in model fitting. As with the lattice models, there is an indication that species differences dictate major differences in model structure. Distance-dependent models of tree competition also suffer from underspecification of the mechanisms of how trees grow and compete. Self-thinning models show how individual plant characteristics may determine the progress of competition, but the form of the model, based on mean plant parameters, limits its effectiveness for further analysis.

We have reached the stage where models must be written in terms of competition for specific resources, and where plant properties that affect resource acquisition must be incorporated in model structure. Clearly, morphology will be an important component, but morphology is not constant, showing great plasticity under competitive stress. Models that do not account for this may be underspecified.

Wu et al. (1985) presented a theoretical analysis of plant growth in response to resource distribution. They defined an ecological interference potential, whereby all influences within a community combine at a point to contribute to the intensity of interference. They proposed a series of mathematical functions for the gradient of resources around a plant, i.e., root effects on soil surface environment, theoretical distributions for light for crown effects on soil surface, and stem effects on the soil surface, particularly for stem flow. Distributions all had the general property of defining a finite gradient of influence, but did not have the same mathematical form. Influences on plant growth could then be scaled over the range from 0 (beneficial) to 1 (detrimental). Multiple individuals can be placed within a field, and the potential calculated for any species at any point. This appears to be a definitive statement of model structure for resources. However, it lacks the important component that plant response to a particular environmental condition may change. In that sense, the model is unspecified. In another respect, the model may be overspecified because there is no indication that all the influences suggested by Wu et al. (1985) need to be incorporated. Theoretical models of this type need to be tested under particular experimental conditions where both resource changes and growth can be measured.

Sorrensen et al. (in press) constructed a process-based simulation of competition for light to model the range of competitive effects in a stand with heterogeneous spacing. The model simulates competition as a sequential series of:

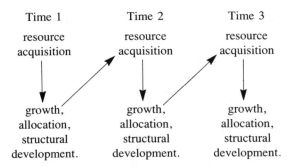

This required two features. First, that each individual plant resource acquisition be modeled directly, according to the amount of foliage held by the individual and its position. Second, that variation between individuals in the use of resources that had been acquired be incorporated. Taken together, these features enabled plasticity in growth to be simulated over, time and its effects on stand development could be assessed.

Each tree is represented as a series of whorls stacked along the tree trunk, and irradiation is considered vertically from above in a contiguous series of columns, each divided into cells (Fig. 17.8). Interaction between individuals was simulated at the level of individual branch and foliage units which overlap horizontally, or in vertical projection, in the columns. Each whorl is considered in radially divided segments which represent branches. Successive whorls intercept a fractional part of light as it passes down a vertical

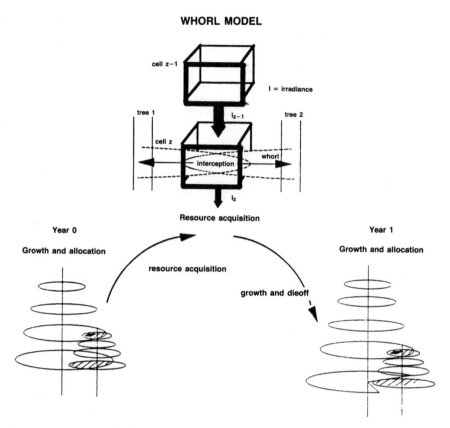

Figure 17.8. Representation of the competition process between individual trees, with their crowns represented as whorls of foliage, and competition calculated for light received in vertical columns. Growth depends upon the amount of radiation intercepted (after Sorrensen et al. in press).

column. This permits the evolution of a structured canopy where both horizontal and vertical variation can influence the future growth of individuals. Empirical relationships were used to determine the reduction in foliage density as light levels decrease with increasing interception.

Radiation interception is translated to growth production according to a conversion efficiency, E, and is assigned to each branch, and branch and tree height increments are calculated dependent upon the total production made by the branch and whole tree, respectively. Foliage death was estimated as a function of irradiance level in a cell.

The model was used to simulate growth and structural development over a 30 year period of naturally regenerated A. amabalis, assuming typical conditions of seedling distribution and structure at the time of harvest of the preceding generation of over-story trees. The effectiveness of the model was judged by assessing model outputs of tree height, height frequency distribution, and mortality against those from remeasurement of permanent plots.

Variation in light interception efficiency and conversion efficiency between trees of different sizes was required to simulate the observed differences in crown structure between large and small trees, i.e., to simulate the type of plasticity in the canopy. Acquisition plasticity increased the growth and survival of suppressed branches on suppressed trees, and was important to enable prediction of adequate tree survival. Permitting differential branch death, as opposed to calculating a mean value for all branches on the same whorl, was important in simulating crown depth correctly.

Sorrensen-Cothern et al. (in press) conducted a sensitivity analysis of the components of shade tolerance, i.e., decreased mortality and increased growth, as related to the foliage-irradiation regime. In a pure stand, shade tolerance, as judged by the survival of suppressed trees, was enhanced where foliage densities were low and radiation interception was reduced. Increased foliage survival at low light intensities did not enhance shade tolerance in a pure stand. However, if increase in foliage density was differentially incurred by suppressed branches and trees, i.e., there is plasticity in the acquisition of resources, then shade tolerance in a pure stand was enhanced.

Independence of growth and death between branches within a whorl and between whorls was an essential component of the model. As increasingly clumped stands were simulated, this asymmetric growth became more advantageous for survival, and increased mean tree height in the stand. Where whorl growth was simulated as symmetric, total foliage amount supported decreased with increasing aggregation of stems. Where branch growth was permitted to be asymmetric, each sector of growth varying according to its resource capture, total foliage amount remained the same with increased stem aggregation.

Construction of process-based models has been urged by those who have developed or reviewed other types of competition models. The process-based model constructed by Sorrensen-Cothern et al. (in press) certainly provides some insights, but also brings with it some challenges. First, such models are integrators of different types of information, e.g., the relation between light intercepted and growth attained, and how this varies between trees of different sizes. Development of such models becomes a comprehensive ecophysiological research task. Secondly, the model predicts a range of outputs, e.g., the spatial structure of individuals, total foliage amount, and frequency distribution of tree heights. The very variety of these outputs means that there is no simple, single statistic with which to judge the model's effectiveness. If we are to use complex process-based models, our approach to statistical inference will have to become a component of, but not a surrogate for, the logical appraisal of theory.

Conclusions

Two distinct approaches to modeling competition have been reviewed, those that make an equivalence between the occupancy of space and the acquisition of resources, and the far smaller group that attempt direct simulation of resource depletion itself. All four categories of models that use space as a surrogate, (i.e., lattice, neighborhood, distance-dependent tree, and self-thinning models), have *Axiom 1* as their central assumption. They define a functional relationship between plant size or space occupied and distance to neighbors that describes the magnitude of a competitive effect (*Axiom 2*). Few models make an explicit description of mortality (*Axiom 3*). However, models of the self-thinning relationship assume a direct relationship between the increasing space occupancy of larger plants and a mortality rate of smaller ones, but only the analysis of West and Burrough (1983) considers the consequences of separation of these effects for self-thinning.

No clear body of theory has emerged to define differences between species in morphology and/or physiology (*Axiom 5*) may generate differences in the competition process. The analytical models of Gates et al. (1979) provided a theoretical examination of the effects of differences in crown shape on competition for space. Unfortunately, development of that type of mathematical model seems unlikely to provide a further advance because i) the form of the equations limits their solution to very restricted cases (Gates 1982), and ii) the important attribute of crown structure may be how it changes as density stress occurs, and this makes the required analysis yet more complicated (Assmann 1970, Hamilton 1969). Analyses of the self-thinning relationship provide empirical evidence of the extent of variation in the competition process. Differences between shade tolerant and intolerant species occur (Weller 1987a, 1987b; Zeide 1985, 1987). Norberg's (1988) proposed model to incorporate species differences into the self-thinning rule does not separate effects that may be due to crown expansion, and the rate of weight increment per unit volume of space occupied.

We suggest that competition models have reached the stage where species-specific differences in form and physiology must be specified directly in relation to resource acquisition and utilization. Models of this structure are necessary if we are to use species or genotype comparisons to advance the understanding of the general parts of competition theory, i.e., *Axioms 1 through 4*, and to produce specific propositions for *Axiom 5*. However, experience to date with such models (Wu et al. 1985, Sorrensen et al. submitted) indicates three important challenges. First, incorporation of more detailed information on plant structure and growth requires the development of process-based models for competiton between plants, where the unit of interaction is the branch or twig, e.g., Franco (1986) and Jones and Harper (1987). These models must be based upon morphology (Ford and Ford 1990, Ford et al. 1990) and account for plasticity. Although theories of growth have been advanced that emphasize control through modular development (Watson and Casper 1984, Hardwick 1986), the crucial unanswered question is of balance between independent branch components and their integration in the complete plant body, and what theory of control is most effective for this.

Although there is a clear requirement for process-based individual plant growth models, there is no unequivocal theory of plant growth on which to base this. This leads to the second problem of bounding, what should and should not be included in the model. Sorensen-Cothern et al. (in press) made a decision to simulate competition as a process of light interception by foliage; soil and root-mediated effects were not considered, yet under some conditions, this would be essential. It will be important to make such choices in relation to a theory for competition.

The third problem is to define a procedure for inference for process-based models. The strength of process-based models is that they can make a range of predictions. For example, Sorrensen-Cothern et al. (in press) predicted absolute growth, size frequency distributions, crown structures for trees of different sizes, total foliage amounts, and spatial distributions of surviving trees. Such a range of outputs can be particularly valuable when conducting a sensitivity analysis, but beyond that, statistical descriptions of these outputs are required so that comparisons can be made with real data. This is a particularly challenging task for outputs that describe spatial structure for which there are dependencies in the data. When adequate statistics have been developed, an overall problem remains of how such statistics, each of which gives different information, should be incorporated in a general model assessment.

These conclusions are particularly challenging for those who wish to predict vegetation change and development. For example, Smith and Huston (1989) used an individual-based competition model to examine species interaction. Their questions of concern were how vegetation may develop over time or under different site conditions. They modeled competition for light and predicted how the balance of species competition would change with the basic resource levels. An important and as yet unanswered question is how sensitive such predictions are to variation in the form of the competition model.

Acknowledgements

We wish to thank Dr. Ross Kiester for valuable suggestions and Ms. Cindy Helfrich for preparation of the manuscript. This research is sponsored by the Synthesis and Integration Project of the Joint U.S. Environmental Protection Agency-U.S.D.A. Forest Service Forest Response program, a part of the National Acid Precipitation Assessment Program. Although the research reported in this article has been wholly funded by the U.S. Environmental Protection Agency under the cooperative agreement CR814640 with the University of Washington, it has not been subject to Agency review and therefore does not necessarily reflect the views of the Agency, and no official endorsement should be inferred.

Literature Cited

Alemdag, I. S. 1978. Evaluation of some competition indexes for the prediction of diameter increment in planted white spruce. *Can. For. Serv. For. Manage. Inst. Rep.* FMR-X-108.

Arney, J. D. 1973. Tables for quantifying competitive stress on individual trees. *Can. Dep. Environ. Pac. For. Res. Cent. Inf. Rep.* BC-X-78.

Assmann, E. 1970. *The Principles of Forest Yield Study.* Pergamon, Oxford England.

Beals, E. W. 1968. Spatial pattern of shrubs on a desert plain in Ethiopia. *Ecology* **49:**744–746.

Bella, I. E. 1969. Competitive influence-zone overlap: A competition model for individual trees. *Bi-M. Res. Notes Canada Dep. For.* **25(3):**24–25.

Bradbury, I. K. 1981. Dynamics, structure and performance of shoot populations of the rhizomatous herb *Solidago canadensis* L. in abandoned pastures. Oecologia **48:**271–276.

Brown, G. S. 1965. Point density in stems per acre. *New Zealand For. Res. Notes* 38.

Cannell, M. G. R., C. K. Njunga, E. D. Ford, and R. Smith. 1977. Variation in yield among competing individuals within mixed genotype stands of tea: A selection problem. *J. Appl. Ecol.* **14:**969–985.

Cannell, M. G. R., P. Rothery, and E. D. Ford. 1984. Analysis of the dynamics of competition within stands of *Picea sitchensis* and *Pinus contorta. Ann. Bot.* **53:**349–362.

Clark, J. S. 1990. Integration of ecological levels: Individual plant growth, population mortality and ecosystem processes. *J. Ecol.* **78:**275–299.

Clements, F. E., J. E. Weaver, and H. Hanson 1929. Publication 398. Carnegie Institute, Washington, D.C.

Cooper, C. F. 1961. Pattern in ponderosa pine forests. *Ecology* **42:**493–499.

Daniels, R. F. 1976. Simple competition indices and their correlation with annual loblolly pine tree growth. *For. Sci.* **22:**454–456.

Daniels, R. F. 1978. Spatial patterns and distance distributions in young seeded loblolly pine stands. *For. Sci.* **24:**260–266.

Daniels, R. F., H. E. Burkhart, and T. R. Clason. 1986. A comparison of competition measures for predicting growth of loblolly pine trees. *Can. J. For. Res.* **16**:1230–1237.

Diggle, P. J. 1976. A spatial stochastic model of inter-plant competition. *J. Appl. Probab.* **13**:6–671.

Donald, C. M. 1963. Competition among crop and pasture plants. *Adv. Agron.* **15**:1–118.

Ek, A. R. and R. A. Monserud. 1974. Trials with program FOREST: Growth and reproduction simulation for mixed species even- or uneven-aged forest stands. In *Growth Models for Tree and Stand Simulation*, J. Fries (ed.), pp. 56–73. Research Notes 30. Department of Forest Yield Research, Royal College of Forestry, Stockholm, Sweden.

Ford, E. D. 1975. Competition and stand structure in some even-aged plant monocultures. *J. Ecol.* **63**:311–333.

Ford, E. D. 1984. The dynamics of plantation growth. In *Nutrition of Forest Trees in Plantations*, G. D. Bowen and S. Nambiar (eds.), pp. 17–52. Academic Press, London.

Ford, E. D., A. Avery, and R. Ford. 1990. Simulation of branch growth in the *Pinacea*: Interactions of morphology, phenology, foliage productivity and the requirement for structural support on the export of carbon. *J. Theor. Biol.* **146**:15–36.

Ford, E. D. and P. J. Diggle. 1981. Competition for light in a plant monoculture modelled as a spatial stochastic process. *Ann. Bot.* **48**:481–500.

Ford, E. D. and P. J. Newbould. 1970. Stand structure and dry weight production through the sweet chestnut (*Castanea sativa* Mill.) coppice cycle. *J. Ecol.* **58**:275–296.

Ford, R. and E. D. Ford. 1990. Structure and basic equations of a simulator for branch growth in the *Pinaceae*. *J. Theor. Biol.* **146**:1–13.

Franco, M. 1986. The influence of neighbours on the growth of modular organisms with an example from trees. *Phil. Trans. R. Soc. Lond.* B **313**:209–225.

Franco, M. and J. L. Harper. 1988. Competition and the formation of spatial pattern in spacing gradients: An example using *Kochia scoparia. J. Ecol.* **76**:959–974.

Gates, D. J. 1978. bimodality in even-aged plant monocultures. *J. Theor. Biol.* **71**:525–540.

Gates, D. J. 1982. Analysis of some equations of growth and competition in plantations. *Math. Biosci.* **59**:17–32.

Gates, D. J., A. J. O'Connor, and M. Westcott. 1979. Partitioning the union of disks in plant competition models. *Proc. R. Soc. Lond.* A. **367**:59–79.

Glover, G. R. and J. N. Hool. 1979. A basal area ratio predictor of loblolly pine plantation mortality. *For. Sci* **25**:275–282.

Grime, J. P., J. C. Crick, and J. E. Rincon. 1986. The ecological significance of plasticity. In *Plasticity in Plants*, D. H. Jennings and A. Trewaras (eds.), pp. 5–

29. Symposium of the Society for Experimental Biology 40. Cambridge University Press, Cambridge, England.

Gross, L. J. 1984. On the phenotypic plasticity of leaf photosynthetic capacity. In *Mathematics in Ecology*, S. Levin (ed.), pp. 2–14. Lecture Notes in Biomathematics 54. Springer-Verlag, Berlin.

Hamilton, G. J. 1969. The dependence of volume increment of individual trees on dominance, crown dimensions and competition. *Forestry* **42**:133–144.

Hamilton, G. J. (ed.) 1976. Aspects of thinning. *Bull. For. Comm. (Lond.)* **55**:138 pages.

Hamilton, G. J. and J. M. Christie. 1974. Influence of spacing on crop characteristics and yield. *Bull. For. Comm. (Lond.)* **52**:1–91.

Hardwick, R. C. 1986. Physiological consequences of modular growth in plants. *Philos. Trans. R. Soc. Lond. B* **313**:161–173.

Harper, J. L. 1967. A Darwinian approach to plant ecology. *J. Ecol.* **55**:247–270.

Hegyi, F. 1974. A simulation model for managing jack-pine stands. In *Growth Models for Tree and Stand Simulation*, J. Fries (ed.), pp. 74–90. Research Notes 30. Department of Forest Yield Research, Royal College of Forestry, Stockholm, Sweden.

Huston, M. A. and D. L. DeAngelis. 1987. Size bimodality in monospecific populations: a critical review of potential mechanisms. *Am. Nat.* **129**:678–707.

Huston, M. A., D. J. DeAngelis, and W. Post. 1988. New computer models unify ecological theory. *BioScience* **38**:682–691.

Hutchings, M. J. 1978. Standing crop and pattern in pure stands of *Mercurialis perennis* and *Rubus fruticosus* in mixed deciduous woodland. *Oikos* **31**:351–357.

Jones, M. and J. L. Harper. 1987. The influence of neighbours on the growth of trees I. The demography of buds in *Betula pendula*. *Proc. R. Soc. Lond. B* **232**:1–18.

Kenkle, N. C. 1988. Pattern of self-thinning in jack pine: Testing the random mortality hypothesis. *Ecology* **69**:1017–1024.

Kent, B. M. and P. Dress. 1979. On the convergence of forest stand spatial pattern over time: The case of random initial spatial pattern. *For. Sci.* **25**:445–451.

Kent, B. M. and P. Dress. 1980. On the convergence of forest stand spatial pattern over time: The case of regular and aggregated initial spatial patterns. *For. Sci.* **26**:10–22.

Kikuzawa, K. 1988. Intraspecific competition in a natural stand of *Betula ermanii*. *Ann. Bot.* **61**:727–734.

Kira, T., H. Ogawa, and N. Sakazaki. 1953. Intraspecific competition among higher plants. I. Competition-yield-density interrelationship in regularly dispersed populations. *J. Inst. Polytech. Osaka City Univ. Ser. D*, **4**:1–16.

Kitamoto, T. 1972. The spatial pattern in a natural population of goldenrod (*Solidago altissima* L.), with particular reference to its change during the shoot growth. *Res. Popul. Ecol. (Kyoto)* **14**:129–136.

Kitamoto, T. and T. Shidei. 1972. Studies on the spatial pattern in forest trees. (I) Distribution of dominant and supressed trees in even-aged forest. *Bull. Kyoto Univ. For.* **43**:152–161.

Koyama, H. and T. Kira. 1956. Intraspecific competition among higher plants. VIII Frequency distribution of individual plant weight as affected by the interaction between plants. *J. Inst. Polytech. Osaka City Univ. Ser. D,* **7**:73–94.

Krammer, H. 1966. Crown development in conifer stands in Scotland as influenced by initial spacing and subsequent thinning treatment. *Forestry* **39**:40–58.

Laessele, A. M. 1965. Spacing and competition in natural stands of sand pine. *Ecology* **46**:65–72.

Larocque, G. and P. L. Marshall. 1988. Improving single-tree distant dependent growth models. In *Forest Growth, Modelling and Prediction,* pp. 94–101. Proceedings, IUFRO Conference, Minneapolis, MN. General Technical Report Ne-120 U.S. Forest Service,

Levrenz, J. W. and P. G. Jarvis. 1980. Photosynthesis in Sitka spruce (*Picea sitchensis* (Bong.) Carr.) X. Acclimation to quantum flux density within and between trees. *J. Appl. Ecol.* **17**:697–708.

Lonsdale, W. M. and A. R. Watkinson. 1982. Light and self-thinning. *New Phytol.* **90**:431–445.

Mack, R. N. and J. L. Harper. 1977. Interference in dune annuals: spatial pattern and neghborhood effects. *J. Ecol.* **65**:345–363.

McMurtrie, R. 1981. Suppression and dominance of trees with overlapping crowns. *J. Theor. Biol.* **89**:151–174.

Mead, R. 1967. A mathematical model for the estimation of inter-plant competition. *Biometrics* **23**:189–205.

Mead, R. 1968. Measurement of competition between individual plants in a population. *J. Ecol.* **56**:35–45.

Mitchell, K. J. 1975. Dynamics and simulated yield of Douglas fir. *For. Sci. Monogr.* 17, pp. 1–39.

Mitchell, K. J. and J. W. Goudie. 1980. *Stagnant Lodgepole Pine.* Progress Report E.P. 850.02. British Columbia Ministry of Forestry, Victoria, B. C.

Moore, J. A., C. A. Budelsky, and R. C. Schlesinger. 1973. A new index representing individual tree competitive status. *Can. J. For. Res.* **3**:495–500.

Moser, J. W. 1974. A system of equations for the components of forest growth. In *Growth Models for Tree and Stand Simulation,* J. Fries (ed.), pp. 260–287. Research Notes 30. Department of Forest Yield Research, Royal College of Forestry, Stockholm, Sweden.

Nance, W. L., J. E. Grissom, and W. H. Smith. 1988. A new competition index based on weighted and constrained area potentially available. In *Forest Growth, Modelling and Prediction,* pp. 134–142. Proceedings, IUFRO Conference, Minneapolis, Minnesota, Gen. Tech. Rep. Ne-120 U.S. Dept. Ag. For. Ser.

Newnham, R. M. 1964. The development of a stand model for Douglas-fir. Ph.D. Dissertation. University of British Columbia.

Nobel, P. S. and A. C. Franco. 1986. Annual root growth and intraspecific competition for a desert bunchgrass. *J. Ecol.* **74:**1119–1126.

Norberg, R. A. 1988. Theory of growth geometry of plants and self-thinning of plant populations: Geometric similarity, elastic similarity, and different growth modes of plant parts. *Am. Nat.* **131:**220–250.

Oliver, C. D. and B. C. Larson. 1990. *Forest Stand Dynamics.* McGraw-Hill, New York.

Pacala, S. W. and J. A. Silander. 1985. Neighborhood models of plant population dynamics. I. Single-species models of annuals. *Am. Nat.* **125:**385–411.

Pacala, S. W. and J. A. Silander. 1990. Field tests of neighborhood population dynamic models of two annual weed species. *Ecol. Monogr.* **60:**113–134.

Pelz, D. R. 1978. Estimating individual tree growth with tree polygons. In *Growth Models for Long Term Forecasting of Timber Yields,* J. Fries, H. E. Burkhart, and T. A. Max (ed.), pp. 172–178. Virginia Polytechnical Institute State University, School for Wildlife Resources. FWS-1-78.

Phillips, D. L. and J. A. McMahon. 1981. Competition and spacing patterns in desert shrubs. *J. Ecol.* **69:**97–115.

Renshaw, E. 1984. Competition experiments for light in a plant monoculture: An analysis based on two-dimensional spectra. *Biometrics* **40:**717–728.

Silander, J. A. and S. W. Pacala. 1985. Neighborhood predictors of plant performance. *Oecologia* **66:**256–263.

Sjolte-Jørgensen, J. 1967. Influence of spacing on coniferous plantations. *Int. Rev. For. Res.* **2:**43–94.

Slatkin, M. and D. J. Anderson. 1984. A model of competition for space. *Ecology* **65:**1840–1845.

Smith, T. and M. Huston. 1989. A theory of spatial and temporal dynamics of plant communities. *Vegetatio* **83:**49–69.

Soetono and D. W. Puckridge. 1982. The effect of density and plant arrangement on the performance of individual plants in barley and wheat crops. *Aust. J. Ag. Res.* **33:**171–177.

Sorrensen-Cothern, K. A., E. D. Ford, and D. G. Sprugel. A process based competition model incorporating plasticity through modular foliage and crown development. Ecological Monographs, in press.

Sprugel, D. G. 1984. Density, biomass, productivity, and nutrient-cycling changes during stand development in wave-regenerated balsam fir forests. *Ecol. Monogr.* **54:**165–186.

Spurr, S. H. 1962. A measure of point density. *For. Sci.* **8:**85–96.

Staebler, G. R. 1951. Growth and spacing in an even-aged stand of Douglas-fir. Master's thesis, University of Michigan.

Steingraeber, D. A. 1982. Phenotypic plasticity of branching pattern in sugar maple (*Acer saccharum*). *Am. J. Bot.* **69:**638–640.

Steingraeber, D. A. and D. M. Waller. 1986. Non-stationarity of tree branching patterns and bifurcation ratios. *Proc. R. Soc. Lond. B* **228:**187–194.

Sterner, R. W., C. A. Ribic. and G. E. Schatz. 1986. Testing for life historical changes in spatial patterns of four tropical tree species. *J. Ecol.* **74**:621–633.

Stiell, W. M. 1969. Stem growth reaction in young red pine to the removal of single branch whorls. *Can. J. Bot.*, **47**:1251–1256.

Sultan, S. E. 1987. Evolutionary implication of phenotypic plasticity in plants. *Evol. Biol.* **21**:127–178.

Thomas, S. C. and J. Weiner. 1989. Including competitive asymmetry in measures of local interference in plant populations. *Oecologia* **80**:349–355.

Verheij, E. W. M. 1970. Spacing experiments with Brussels sprouts grown for single-pick harvests. *Neth. J. Agric. Sci.* **18**:89–104.

Waller, D. M. 1981. Neighborhood competition in several violet populations. *Oecologia* **51**:116–122.

Watson, M. A. and B. B. Casper. 1984. Morphogenetic constraints on patterns of carbon distribution in plants. *Ann. Rev. Ecol. Syst.* **15**:233–258.

Weiner, J. 1982. A neighborhood model of annual-plant interference. *Ecology* **63**:1237–1241.

Weiner, J. and S. C. Thomas. 1986. Size variability and competition in plant monocultures. *Oikos* **47**:211–222.

Weller, D. E. 1987a. A reevaluation of the $-3/2$ power rule of plant self-thinning. *Ecol. Monogr.* **57**:23–43.

Weller, D. E. 1987b. Self-thinning experiment correlated with allometric measures of plant geometry. *Ecology* **68**:813–821.

West, P. W. and C. J. Burrough. 1983. Tree suppression and the self-thinning rule in a monoculture of *Pinus radiata* D. Don. *Ann. Bot.* **52**:149–158.

Westoby, M. 1982. Frequency distributions of plant size during competitive growth of stands: The operation of distribution modifying functions. *Ann. Bot.* **50**:733–735.

Wu, H., P. J. H. Sharpe, J. Walker, and L. K. Penridge. 1985. Ecological field theory: A spatial analysis of resource inference among plants. *Ecol. Modell.* **29**:215–243.

Yeaton, R. I. and M. L. Cody. 1976. Competition and spacing in plant communities: The northern Mohave desert. *J. Ecol.* **64**:689–696.

Yoda, K., T. Kira, H. Ogawa, and K. Hozumi. 1963. Self thinning in overcrowded pure stands under cultivated and natural conditions. *J. Inst. Polytech. Osaka City Univ. Ser. D,* **14**:107–129.

Zeide, B. 1985. Tolerance and self-tolerance of trees. *For. Ecol. Manag.* **13**:149–166.

Zeide, B. 1987. Analysis of the 3/2 power law of self-thinning. *For. Sci.* **33**:517–537.

18

Individual-Based Forest Succession Models and the Theory of Plant Competition

Michael Huston

ABSTRACT. **One of the few types of individual-based models that address multi-species interactions and community- and ecosystem-level phenomena is the class of forest succession models known as JABOWA or FORET models. This paper briefly reviews the history and major features of these models, which are probably the most widely used individual-based computer simulation models. The assumptions and mechanisms underlying these models, which originated over 20 years ago, are virtually identical to those that have been developed in the context of theoretical plant competition and plant population models. Interest among theoretical ecologists in mechanistic models of plant growth and competition has led to the convergence of recently developed models with the assumptions, mechanisms, and structure of a model that has been in use by ecosystem scientists for over 20 years.**

Introduction

In 1972, at the height of the ecosystem research efforts of the International Biological Program (IBP), a fundamentally new type of ecological model was described in a major journal and subsequently ignored by most ecologists. Perhaps is it not surprising that one new model did not attract much attention and was lost in the flood of ecosystem models produced by the developing field of ecosystem ecology. The unique features that distinguished the new "forest growth simulator" from its IBP siblings were difficult to decipher from the two publications that announced its development. Nonetheless, it differed radically from the pool and flux ("box and arrow") models for which the IBP is remembered. The new "ecosystem" model was actually an individual-based population model (specifically, an i-state configuration model, Metz and Diekmann 1986) of a multi-species forest, based on the growth and competitive interactions of many individual trees.

The model in question was developed by computer scientists at IBM's Thomas J. Watson Research Center, in collaboration with ecologists from Yale University who were part of IBP's Hubbard Brook Ecosystem Study. It was first published in the *IBM Journal of Research and Development* and is known by the acronym "JABOWA" (Botkin et al. 1972a, and 1972b.) Although the JABOWA model played little further role in the Hubbard Brook study, its basic concepts (and much of the original FORTRAN code) have proliferated into a host of closely-related "forest succession" or "gap" models which have been applied to many different forest types (see Shugart 1984, 1990) and used to address many different issues of population, community, and ecosystem ecology.

The JABOWA model was one of the first individual-based computer simulation models and, through its offspring such as the FORET model (Shugart and West 1977), has become one of the most widely used and best known models of this type. Although a growing group of researchers interested in the mechanisms of plant competition, succession, nutrient cycling, and other ecosystem processes has been using and publishing the results of these individual-based forest succession models for over 20 years, the models remain poorly understood by the majority of ecologists. The basic hypotheses about plant growth and competitive interactions that were implemented in the original JABOWA forest simulation model are still controversial ideas that are undergoing conceptual development in the fields of plant population and community ecology. If the JABOWA model were reinvented in 1991, it might be hailed as a major new contribution to plant ecology.

The goal of this brief review is to examine the assumptions and mechanisms of JABOWA-type models in relation to the major issues and modeling approaches in the field of theoretical plant population modeling. Hopefully, this exercise will contribute to improved communication and understanding between scientists who are pursuing the same goals from different modeling and philosophical backgrounds.

Individual-based Forest Succession Models: JABOWA and FORET

Two primary constraints shaped the structure of the JABOWA model. First, its objective was to reproduce the behavior of the *mixed-species, mixed-age* forest of the Hubbard Brook Experimental Station. This objective can be contrasted with that of most theoretical plant competition and plant population models, which were designed to be applied to and tested in even-aged monocultures. A second constraint on the JABOWA model was to produce output that could be compared directly with data collected from the forest, which were the diameters and growth increments of many individual trees in forest inventory plots. Consequently, the model was designed to grow individual trees and produce size distributions of the populations of many different species. Although the model output was often summed and presented as the total biomass per species, it was also possible to follow the simulated growth of a single individual, and to observe how it responded to such changes in its environment as the death of a neighboring tree (Botkin et al. 1972a).

The ability to account for variation due to local spatial interactions is one of the primary rationales behind the individual-based modeling approach (Huston et al. 1988). However, the JABOWA model takes a very different approach to spatial interactions than that developed in most theoretical plant competition models. On first inspection, it appears that spatial location is not even considered in the JABOWA model, which does not keep track of the locations of any of the hundreds of individual trees in a single simulation. This contrasts with the experimental and modeling approach in plant population ecology, where positions on a regular lattice or explicit interplant distances are used to model variation in the intensity of interplant interactions (see Ford and Sorrensen, this volume).

Local spatial interactions are handled in the JABOWA model simply by limiting the size of the simulated plot to an area small enough to be considered a "neighborhood." The original JABOWA model was designed to model trees growing on a 10 × 10 m plot. However, further work with the model demonstrated that the size of the plot had a critical effect on the model's behavior (Shugart and West 1977). Specifically, when the plot size was scaled to the canopy size of a full-grown individual of the largest species, the model reproduced the phenomenon known as "gap dynamics," that is, the establishment and growth of large numbers of saplings in the high light levels in the "gap" created by the death of a large tree. This practical definition of the competitive neighborhood as the crown area of a single large tree allowed the model to ignore the exact spatial locations of individual trees (most of which died as a result of simulated shading, anyway), while preserving the essential features of forest dynamics.

The most important aspect of the way in which spatial structure is modeled in JABOWA is the vertical dimension. Most theoretical plant competition and population models have focused on the "nearest-neighbor" distance, which is used as a two-dimensional index of the intensity of competition between neighboring plants. Analytical models of this type (e.g., Mead 1968; Diggle 1976; Gates 1978; Aikman and Watkinson, 1980) included the assumption that competition between plants was "one-sided," that is, large plants had a negative effect on small plants, but small plants had little or no effect on large plants. Ford and Diggle (1981) developed an index based on the angle between the tops of neighboring trees that was used to determine whether one tree fell within the "cone of influence" of a larger tree. More recently, "neighborhood" models have attempted to parameterize the effect of competition on plant growth and reproduction in detailed spacing models (Pacala and Silander 1985, 1990).

The approach taken in the JABOWA model was not to calculate an index of competition, but to model the mechanism of competition directly. On the assumption that competition for light was the primary interaction between trees, the creators of JABOWA explicitly modeled the height growth of each individual tree, summed the leaf areas of all individuals of a given height class, and used a simple Beer-Lambert extinction equation to calculate the amount of light available at each height in the forest stand. Thus, the growth of each individual was determined by the light available at the top of its canopy, and each individual reduced the light available to shorter individuals. Although light availability was indexed in the original model in proportion to full sunlight at the top of the canopy, Botkin et al. (1972a) clearly intended that the model could be modified to use quantitative calculations of quantum flux.

The JABOWA model ignored most of the details of crown structure and growth plasticity that have played a major role in some plant competition models, in which interactions between trees have been modeled at the level of branches and foliage (Hamilton 1969, Mitchell 1975, Sorrenson et al., submitted). Much of the theoretical interest in the growth form of competing trees relates to the attempt to explain the phenomenon known as the 3/2 thinning rule (Yoda et al. 1963) which describes the relation between mean plant size and plant density as the size of plants increases through time. This "rule" is a consequence of the geometric relationship between plant mass or volume (a cubic function of linear size) and the area on which the mass is located (a square function of linear size) and has turned out to be more flexible than originally supposed (cf. Weller 1987a, b, Zeide 1987, Norberg 1988). The 3/2 thinning rule is a phenomenon of even-aged, monospecific stands rather than the complex forests for which JABOWA was designed, and, not surprisingly, the JABOWA/FORET model does not produce this phenomenon even when monospecific stands are simulated (D. Weller, personal communication). Some versions of JABOWA/FORET have been developed that do include horizontal spatial locations as well as height (e.g., Doyle 1990, Busing 1990), and these models could be expected to approximate the 3/2 thinning phenomenon.

One population-level phenomenon that has been of continuing interest in plant competition theory and modeling is size bimodality, which develops from initially unimodal size distributions under some conditions (Ford 1975, Gates 1978, Hara 1984, Huston 1986, Huston and DeAngelis 1987, Weiner and Thomas 1986). While bimodality can result from many different mechanisms (Huston and DeAngelis 1987), one important mechanism in plant populations is dominance and suppression resulting from competition for light. JABOWA/FORET models can produce bimodal size distributions in monospecific populations.

A key feature of any individual-based model is the representation of the individual organism. Given the current interest in the development of "mechanistic" and "process-based" models of plant populations, communities, and ecosystems (Schoener 1986, Tilman 1987), it is interesting to examine the mechanisms of plant growth incorporated in the original JABOWA model. "The model consists of a basic growth-rate equation that may be taken to represent the growth of a tree with optimal site quality and no competition from other trees. . . . this growth rate is decreased by factors that take into account shading and shade tolerance, soil quality, and average climate as measured by the number of growing degree-days . . . The equation states that the change in volume (D^2H) of a tree over a period of one year is proportional to the amount of sunlight the tree receives, derated by a factor $(1 - DH/D_{max}H_{max})$, which takes some account of the energy required to maintain the living tissue" (Botkin et al. 1972a). Thus the model incorporates a simple description of net carbon uptake proportional to leaf area, as well as allocation of carbon to aboveground growth versus respiration.

The authors note that "all growth curves tend to be sigmoid in shape and our final growth equation exhibits this overall property. We realize that some readers may feel that the equations . . . are occasionally based on rather arbitrary assumptions, but we expect that they will concur with us that there is no unique model of forest growth and that many equations based on different assumptions could yield quite similar results" (Botkin et al. 1972a). The essential features of the sigmoid growth curve, that is, low growth rates at small sizes, maximum growth rates at intermediate sizes, and a "leveling off" to low growth rates as the maximum size is approached, are common to nearly all growth models in biology, whether they are applied to cells, organs, individual organisms, or entire populations. The well-known logistic curve of the Lotka-Volterra competition and predation models is of this general form, as are the growth equations discussed by Aikman (this volume) and Clark (this volume).

The leaf area of a tree of a given size is calculated as a species-specific function of stem basal area, an assumption supported by later work on tree ecophysiology (Waring and Schlesinger 1985). The photosynthetic light response of trees is generalized into two types, representing either shade tolerant or shade intolerant species. Each type has an appropriate light compensation level and a light response curve that matches the classic generalizations of ecophysiology (cf. Larcher 1980, Ledig 1969). Growth suppression resulting from shading (or other factors) produces delayed mortality based on a mortality algorithm which states that a tree whose diameter increments fall below a specified minimum has a 0.368 probability of mortality in a given year (which results in a one percent probability of such a tree surviving for ten successive years).

Competition for belowground resources was modeled as a logistic decrease in the growth of all species as total basal area on the plot approaches a maximum that is related to site quality. ". . . the function S is a crude expression of the competition for soil moisture and nutrients on the plot" (Botkin et al. 1972a). Although there were no species-specific differences in competition for belowground resources in the original JABOWA, subsequent versions of the model did incorporate species-specific differences in nutrient uptake and response to soil nutrients (Aber and Melillo 1982, Pastor and Post 1985, 1986, Weinstein et al. 1982, Bonan 1989).

Discussion

The recent interest in "mechanistic" ecological models as the next step toward a better theoretical understanding of communities and ecosystems (Schoener 1986, Tilman 1987) was presaged by the individual-based forest succession model developed over 20 years ago. The JABOWA model and its descendants have been modified to apply to many different forest ecosystems (e.g., Doyle, 1981; Shugart 1984), as well as grasslands (Coffin and Lauenroth 1990), mixed herbaceous and woody vegetation (Prentice et al. 1989), and even phytoplankton (Lehman et al. 1975). The models have been modified to incorporate ecosystem processes such as decomposition (Aber and Melillo 1982, Weinstein et al. 1982, Pastor and Post 1985, 1986), hydrology (Mann and Post 1980, Pastor and Post 1985, Bonan 1989) and soil thermodynamics (Bonan 1989, 1990). Notwithstanding the evolution of the model to incorporate improved understanding of plant growth and ecosystem processes, the original model incorporated the basic concepts and assumptions that have been redeveloped over the past twenty years in the field of theoretical plant population and community modeling.

Perhaps the ecosystem orientation of the model, and its use to address "applied" problems such as forest yield (Aber et al. 1979, 1982, Shugart et al. 1980, Busing and Clebsch, 1987), response to fire (Mielke et al. 1977, 1978, Shugart and Noble 1981, Overpeck et al. 1990), and climate change (Solomon et al. 1981, Davis and Botkin 1985, Solomon 1986, Pastor and Post 1988, Dale and Franklin 1989, Urban and Shugart 1989, Overpeck et. al. 1990, Shugart 1990, Bonan et al. 1990), have served to isolate it from those ecologists interested primarily in theory and theoretical models. Nonetheless, there has been some use of JABOWA/FORET models to address such theoretical issues as size bimodality (Huston and DeAngelis 1987), life history strategies and plant succession (Huston and Smith 1987, Smith and Huston 1989) and spatial pattern in forests and landscapes (Smith and Urban 1988, Smith and Huston 1989, Urban et al. in press). It is interesting that some recent models developed in theoretical plant community ecology (e.g., ALLOCATE, Tilman 1988) show a remarkable convergence with the JABOWA/FORET models.

Significant recent developments in JABOWA/FORET models include 1) a more physiologically explicit representation of plant growth, and 2) a more realistic representation of the light regime of trees in large heterogeneous areas. While the species-specific parameters used in the original JABOWA model implicitly incorporated tradeoffs in resource allocation for life history processes and physiological responses, recent efforts have been made to make both the consequences and the representation of these tradeoffs more explicit (Smith and Huston 1989, Luxmoore et al. 1990, Weinstein and Yanai, 1991).

Huston and Smith (1987) demonstrated that the single mechanism of competition for light was able to produce a wide variety of successional patterns, depending on the correlation between life history and physiological traits such as maximum size and growth rate, reproductive rate, maximum age, and shade tolerance. While the model was able to reproduce the anomalous patterns of species replacement and coexistence that are often cited as exceptions to general models of succession, the "classical" successional sequence of one species replacing another was produced only by a specific set of inverse correlations among the traits. These inverse correlations, between traits such as maximum size and maximum growth rate, were hypothesized to be the result of tradeoffs resulting from basic energetic and physiological constraints that apply to all plants, and consequently explain the remarkable similarity in successional patterns found in all plant communities.

The effect of such basic physiological and energetic constraints as the linkage between CO_2 uptake and water loss, and the allocation of resources to the capture of aboveground versus belowground resources, have been incorporated into the description of individual tree growth (Smith and Huston 1989). This increase in physiological detail allows the model to predict changes in species composition and canopy leaf area along moisture gradients, as well as to predict how successional patterns will differ at different positions along the gradient. The leaf area supported by a particular tree is a function of light available to that tree and its physiological status with regard to water, along with its size and species. This allows the model to predict not only the distribution of different species along resource gradients, but also the distribution of such aggregate ecosystem properties as leaf area. The connection between shade tolerance, drought tolerance, and leaf chemistry has major implications for decomposition and the nitrogen cycle (Post and Pastor 1990).

The treatment of light penetration through the forest canopy has become more sophisticated in versions of the model that consider diffuse as well as direct light, and calculate leaf area along paths parallel to the angle of direct beam penetration (Leemans and Prentice 1987, Bonan 1989, Urban 1990). A model version known as ZELIG extends the basic concept of the individual-based gap model to a grid of interacting cells, each of which can potentially interact with neighboring cells (Smith and Urban 1988, Urban 1990, Urban et al., in press). Shading interactions are calculated based on tree height and solar angle, while seed dispersal is calculated as a function of the size and number of mature trees, with an inverse-square decrease with distance from seed source (Urban 1990). This model, which is spatially explicit at the scale of large trees, reproduces large-scale spatial phenomena, such as gradients, variation in gap size, and wave phenomena. Versions of the model that are spatially explicit at the scale of individual trees within a plot reproduce smaller-scale phenomena that result from interactions between close neighbors.

Further improvement in the reliability of individual-based forest succession models depends not on an improvement of model structure or the representation of specific mechanisms, but on an improved understanding of the growth of individual trees. The regulation of the growth of large woody plants remains one of the major challenges in plant physiology, with major unknowns such as the regulation of stomatal opening and the control of resource allocation. "It is our conclusion that the population dynamics of forest trees can be simulated, that such simulation can be done from first principles, and that at present the improvement of simulation is prevented primarily by a lack of data that accurately describe the relation between tree growth and the environment" (Botkin et al. 1972a).

In summary, individual-based forest succession models, beginning with JABOWA (Botkin et al. 1972a, 1972b), are based on the same five theorems that have been developed as a theory of plant competition (see Ford and Sorrenson, this volume). The parallel development of theoretical plant population biology and forest succession modeling has been driven by different motivations. Theoretical plant population and competition models have been designed to address experimentally tractable systems, and have consequently focused on monospecific populations, either herbs or even-aged forest stands. Forest succession models have been designed to address complex problems in the dynamics of natural and managed forests, and the prediction of ecosystem response to disturbance and environmental change. The evolution of theoretical plant competition models toward a more mechanistic, process-based approach represents a convergence with the assumptions and structure of individual-based ecosystem models. From a theoretical perspective, the forest simulation models can be seen as tools to investigate the natural selection processes that determine which life history and resource allocation strategies are most successful under different environmental conditions of resource availability and disturbance (Huston and Smith 1987, Smith and Huston 1989).

Acknowledgments

I want to thank Dean Urban for access to an up-to-date bibliography of FORET descendents, and Don DeAngelis and Mac Post for helpful suggestions during the development of this paper. This is Publication Number (0) of the Environmental Sciences Division, Oak Ridge National Laboratory. Research sponsored by Walker Branch Watershed Project, Ecological Research Division, Office of Health and Environmental Research, U.S. Department of Energy, under Contract No. DE-AC05-840R21400 with Martin Marietta Energy System, Inc.

Literature Cited

Aber, J. D., D. B. Botkin, and J. M. Melillo. 1979. Predicting the effects of different harvesting regimes on productivity and yield in northern hardwoods. *Can. J. For. Res.* **9**:10–14.

Aber, J. D. and J. M. Melillo. 1982. FORTNITE: a computer model of organic matter and nitrogen dynamics in forest ecosystems. *Univ. Wisc. Res. Bull* R3130.

Aber, J. D., J. M. Melillo, and C. A. Federer. 1982. Predicting the effects of rotation length, harvest intensity, and fertilization on fiber yield from northern hardwood forests in New England. *For. Sci.* **23**:31–45.

Aikman, D. P. and A. R. Watkinson. 1980. A model for growth and self-thinning in even-aged monocultures of plants. *Ann. Bot.* (Lond.) **45**:419–427.

Bonan, G. B. 1989. A computer model of the solar radiation, soil moisture, and soil thermal regimes in boreal forests. *Ecol. Modell.* **45**:275–306.

Bonan, G. B. 1990. Carbon and nitrogen cycling in North American boreal forests. I. Litter quality and soil thermal effects in interior Alaska. *Biogeochem.* **10**:1–28.

Bonan, C. G., H. H. Shugart, and D. L. Urban. 1990. The sensitivity of some high-latitude boreal forests to climatic parameters. *Clim. Change* **16**:9–29.

Botkin, D. B., J. F. Janak, and J. R. Wallis. 1972a. Rationale, limitations, and assumptions of a northeastern forest growth simulator. *IBM J. Res. Dev.* **10**:101–116.

Botkin, D. B., J. F. Janak, and J. R. Wallis. 1972b. Some ecological consequences of a computer model of forest growth. *J. Ecol.* **60**:849–872.

Busing, R. T. 1990. A spatial model of forest dynamics: *Bull. Ecol. Soc. Am.* 71 (suppl.):110.

Busing, R. T. and E. E. C. Clebsch. 1987. Application of a spruce-fir forest canopy gap model. *For. Ecol. Manage.* **20**:151–169.

Coffin, D. P. and W. K. Lauenroth. 1990. A gap dynamics simulation model of succession in a semiarid grassland. *Ecol. Modell.* **49**:229–236.

Dale, V. H. and J. F. Franklin. 1989. Potential effects of climate change on stand development in the Pacific Northwest. *Can. J. For. Res.* **19**:1581–1590.

Davis, M. B. and D. B. Botkin. 1985. Sensitivity of cool-temperate forests and their fossil pollen record to rapid temperature change. *Quat. Res.* **23**:327–340.

Diggle, P. J. 1976. A spatial stochastic model of inter-plant competition. *J. Appl. Probab.* **13**:662–671.

Doyle, T. 1981. The role of disturbance in the gap dynamics of a montane rain forest: an application of a tropical forest succession model. In Forest Succession Concepts and Applications, D. C. West, H. H. Shugart, and D. B. Botkin (eds.), pp. 56–73. Springer-Verlag, New York.

Doyle, T. 1990. An evaluation of competition models for investigating tree and stand growth processes. In *Process Modeling of Forest Growth Responses to Environmental Stress*, R. K. Dixon et al. (eds.), pp. 271–277. Timber Press, Portland, OR.

Ford, E. D. 1975. Competition and stand structure in some even-aged plant monocultures. *J. Ecol.* **63**:311–333.

Ford, E. D. and P. J. Diggle. 1981. Competition for light in a plant monoculture modelled as a spatial stochastic process. *Ann. Bot.* **48**:481–500.

Gates, D. J. 1978. Bimodality in even-aged plant monocultures. *J. Theor. Biol.* **71**:525–540.

Hamilton, G. J. 1969. The dependence of volume increment of individual trees on dominance, crown dimensions and competition. *Forestry* **42**:133–144.

Hara, T. 1984. Dynamics of stand structure in plant monocultures. *J. Theor. Biol.* **110**:223–239.

Huston, M. A. 1986. Size bimodality in plant populations: an alternative hypothesis. *Ecology* **67**:265–269.

Huston, M. A. and D. L. DeAngelis. 1987. Size bimodality in monospecific populations: a critical review of potential mechanisms. *Am. Nat.* **129**:678–707.

Huston, M. A. and T. M. Smith. 1987. Plant succession: life history and competition. *Am. Nat.* **130**:168–198.

Huston, M. A., D. L. DeAngelis, and W. M. Post. 1988. New computer models unify ecological theory. *BioSci.* **38**:682–691.

Larcher, W. 1980. *Physiological Plant Ecology*. Springer-Verlag, Berlin.

Ledig, F. T. 1969. A growth model for tree seedlings based upon the rate of photosynthesis and the distribution of photosynthate. *Photosynthetica* **3**:263–275.

Leemans, R. and Prentice, I. C. 1987. Description and simulation of tree-layer composition and size distributions in a primaeval *Picea-Pinus* forest. *Vegetatio* **69**:147–56.

Lehman, J. T., D. B. Botkin, and G. E. Likens. 1975. The assumptions and rationales of a computer model of phytoplankton population dynamics. *Limnol. Oceanogr.* **20**:343–364.

Luxmoore, R. J., M. L. Tharp, and D. C. West. 1990. Simulating the physiological basis of tree-ring responses to environmental change. In *Process Modeling of Forest Growth Responses to Environmental Stress*, R. K. Dixon et al. (eds.), pp. 393–401. Timber Press, Portland, OR.

Mann, L. K. and W. M. Post. 1980. Modeling the effect of drought on forest growth and composition. *Bull. Ecol. Soc. Am.* **61**:80.

Mead, R. 1968. Measurement of competition between individual plants in a population. *J. Ecol.* **56**:35–45.

Metz, J. A. J. and O. Diekmann. 1986. *The Dynamics of Physiologically Structured Populations.* Springer-Verlag, Berlin.

Mielke, D. L., H. H. Shugart, and D. C. West. 1977. *User's manual for FORAR, a Stand Model for Upland Forests of Southern Arkansas.* ORNL/TM-5767. Oak Ridge National Laboratory, Oak Ridge, TN.

Mielke, D. L., H. H. Shugart, and D. C. West. 1978. *A Stand Model for Upland Forests of Southern Arkansas.* ORNL/TM-6225. Oak Ridge National Laboratory, Oak Ridge, TN.

Mitchell, K. J. 1975. Dynamics and simulated yield of Douglas fir. *For. Sci. Monogr.* **17**:1–39.

Norberg, R. A. 1988. Theory of growth geometry of plants and self-thinning of plant populations: geometric similarity, elastic similarity, and different growth modes of pure plant parts. *Am. Nat.* **131**:220–250.

Overpeck, J. T., Rind, D., and Goldberg, R. 1990. Climate-induced changes in forest disturbance and vegetation. *Nature* **343**:51–53.

Pacala, S. W. and J. A. Silander. 1985. Neighborhood models of plant populations dynamics. I. Single-species models of annuals. *Am. Nat.* **125**:385–411.

Pacala, S. W. and J. A. Silander. 1990. Field tests of neighborhood population dynamics models of two annual weed species. *Ecol. Monogr.* **60**:113–134.

Pastor, J. and W. M. Post. 1985. *Development of Linked Forest Productivity-Soil Process Model.* ORNL/TM-9519. Oak Ridge National Lab, Oak Ridge, TN.

Pastor, J. and W. M. Post. 1986. Influence of climate, soil moisture, and succession on forest carbon and nitrogen cycles. *Biogeochem.* **2**:3–17.

Pastor, J. and W. M. Post. 1988. Response of northern forests to CO_2-induced climatic change: Dependence on soil water and nitrogen availabilities. *Nature* **334**:55–58.

Post, W. M. and J. Pastor. 1990. An individual-based forest ecosystem model for projecting forest response to nutrient cycling and climate change. In *Forest Simulation Systems,* L. C. Wensel and G. S. Biging, (eds.), pp. 61–74. Bulletin 1927, Division of Agriculture and Natural Resources, University of California, Berkeley.

Prentice, I. C. et al. 1989. *Developing a Global Vegetation Dynamics Model: Results of an IIASA Summer Workshop.* International Institute for Applied Systems Analysis, Laxenburg, Austria.

Schoener, T. W. 1986. Mechanistic approaches to community ecology: new reductionism? *Am. Zool.* **26**:81–106.

Shugart, H. H. 1984. *A Theory of Forest Dynamics.* Springer-Verlag, New York.

Shugart, H. H. 1990. Using ecosystem models to assess potential consequences of global climatic change. *Trends Ecol. Evol.* **5**:303–307.

Shugart, H. H., M. S. Hopkins, I. P. Burgess, and A. T. Mortlock. 1980. The development of a succession model for subtropical rain forest and its application to assess the effects of timber harvest at Wiangaree State Forest, New South Wales. *J. Environ. Manage.* **11**:243–265.

Shugart, H. H. and I. R. Noble. 1981. A computer model of succession and fire response of the high altitude Eucalyptus forest of the Brindabella Range, Australian Capital Territory. *Aust. J. Ecol.* **6**:149–164.

Shugart, H. H. and D. C. West. 1977. Development of an Appalachian deciduous forest succession model and its application to assessment of the impact of the chestnut blight. *J. Environ. Manage.* **5**:161–179.

Smith, T. M. and M. A. Huston. 1989. A theory of the spatial and temporal dynamics of plant communities. *Vegetatio* **83**:49–69.

Smith, T. M. and D. L. Urban. 1988. Scale and resolution of forest structural pattern. *Vegetatio* **74**:143–150.

Solomon, A. M. 1986. Transient response of forests to CO2-induced climate change: simulation modeling experiments in eastern North America. *Oecologia* **68**:567–579.

Solomon, A. M., D. C. West, and J. A. Solomon. 1981. Simulating the role of climate change and species immigration in forest succession. In *Forest Succession: Concepts and Applications*, D. C. West, H. H. Shugart, and D. B. Botkin (eds.), pp. 154–177. Springer-Verlag, New York.

Sorrensen, K. A., E. D. Ford, and D. G. Sprugel. A process based competition model incorporating plasticity in foliage and crown development. Submitted.

Tilman, D. 1987. The importance of the mechanisms of interspecific competition. *Am. Nat.* **129**:769–774.

Tilman, D. 1988. *Dynamics and Structure of Plant Communities*. Princeton University Press, Princeton, NJ.

Urban, D. L. 1990. *A versatile Model to Stimulate Forest Pattern: A User's Guide to ZELIG Version 1.0.* Environmental Sciences, University of Virginai, Charlottesville.

Urban, D. L., G. B. Bonan, T. M. Smith, and H. H. Shugart. Spatial applications of gap models. *For. Ecol. Manage.*, in press.

Urban, D. L. and H. H. Shugart. 1989. Forest response to climatic change: a simulation study for southeastern forests. In *The Potential Effects of Global Climate Change on the U.S.: Appendix D—Forests*, J. B. Smith and D. A. Tirpak, (eds.), Office of Policy, Planning, and Evaluation, U.S. EPA, Washington, DC.

Waring, R. H. and W. H. Schlesinger. 1985. *Forest Ecosystems: Concepts and Management*. Academic Press, Orlando, FL.

Weiner, J. and S. C. Thomas. 1986. Size variability and competition in plant monocultures. *Oikos* **47**:211–222.

Weinstein, D. A., H. H. Shugart, and D. C. West. 1982. *The Long-term Nutrient Retention Properties of Forest Ecosystems: a Simulation Investigation*. ORNL/TM-8472. Oak Ridge National Laboratory, Oak Ridge, TN.

Weinstein, D. A. and R. Yanai. 1991. Models to assess the response of vegetation to global climate change. Submitted, Topical Report, Electric Power Research Institute EPRI EN-7366 Project 2799-1, Palo Alto, CA.

Weller, D. E. 1987a. A reevaluation of the $-3/2$ power rule of plant self-thinning. *Ecol. Monogr.* **57**:23–43.

Weller, D. E. 1987b. Self-thinning experiments correlated with allometric measures of plant geometry. *Ecology* **68**:813–821.

Yoda, K., T. Kira, H. Ogawa, and K. Hozumi. 1963. Self thinning in overcrowded pure stands under cultivated and natural conditions. *J. Polytech. Osaka City Univ. Ser. D.* **14**:107–129.

Zeide, B. 1987. Analysis of the 3/2 power law of self-thinning. *For. Sci.* **33**:517–537.

19

Relationships Among Individual Plant Growth and the Dynamics of Populations and Ecosystems

James S. Clark

ABSTRACT. **Population- and ecosystem-level processes in forests have been increasingly studied with complex simulation models in recent years. These models have been advocated in part because analytical models have not been developed that accommodate several important features of population dynamics, and because many existing models do not make reasonable predictions. Here I argue i) that analyzable models are needed to facilitate understanding of results from large simulation models, and that ii) models that link population and ecosystem dynamics to the growth of individual plants can address many of the important processes associated with the dynamics of individual plants. I summarize an analytical approach that permits analysis of population thinning, density-independent mortality, gap area, net primary production, nutrient turnover, and the self-thinning rule in terms of parameters that describe growth of individual plants within the population. Analysis of these models here and elsewhere demonstrates relationships among these processes by exploiting the relationship of each to individual plant growth. The "stages" in stand development identified by forest ecologists arise as a natural consequence of individual plant growth. This approach is complementary to the more complex numerical models that have been a primary focus of much research on forest dynamics.**

Introduction

Understanding the dynamics of plant populations requires knowledge of processes that operate at vastly different scales. Plant growth depends on the local physical environment, including light, water, temperature, and nutrient availability. These factors depend, in turn, on "ecosystem-level" variables such as stand-leaf area, leaching and runoff, rainfall interception, evapotranspiration, accumulation of organic matter in forest understories, and microbial mineralization and immobilization of nitrogen. The population falls somewhere between these levels of complexity and scale, describing a space–time pattern of numbers, age classes, and demographic variables. The complexity that arises as a consequence of this mixture of scales as different levels makes the "population level" rather difficult to explain in terms of factors that affect it. Moreover, populations also have a feedback effect on ecosystem-level processes by way of their influence on stand characteristics. Thus, models containing these different levels of organization tend to be intricate and, therefore, difficult to analyze.

Because of the complexity required to adequately link processes at different scales, and because analytical models have not described well the qualitative behavior of such systems, population dynamics have increasingly been studied in recent years with complex simulation models. This potential to address the many assumptions contained in simple analytical models has been achieved at a cost of reduced tractability. Experiments with large simulation models have produced many valuable insights concerning succession (Botkin et al. 1972, Huston and Smith 1987), resource competition (Tilman 1988, Smith and Huston 1989), net primary production (Borman and Likens 1979, Aber et al. 1979, 1982, Shugart 1984), and nutrient cycling (Pastor and Post 1986). Each approach has its limitations, however, and the behaviors of complex numerical models are consistently difficult to comprehend. Moreover, they have not addressed many of the recurring questions in plant ecology that concern interrelationships across these scales. Several examples follow:

- What governs the rate of plant death in even-aged cohorts? Mortality rates are "remarkably constant" in even-aged populations (Harper 1977, Peet and Christensen 1980, Waring and Schlesinger 1985), presumably because growth influences crowding, and crowding governs self-thinning (Aikman and Watkinson 1980). In some aspect of individual plant growth responsible for constant per-capita mortality rate? Perhaps a different mortality schedule is more realistic (e.g., Sprugel 1984).

- What is the relationship between density-dependent (DD) mortality, which *results* from crowding, and density-independent (DI) mortality, which *reduces* crowding?

- How should standing crop change with stand age (Peet 1981, Shugart 1984, Sprugel 1985)? Sometimes this pattern may be best described by a simple saturation function, while at other times it might be more sigmoid. Presumably, standing crop integrates individual plant growth and stand mortality.

- Foresters manage several "ecosystem" properties (e.g., yield per unit area) through the manipulation of a population variable, density. The link between these quantities is achieved with the aid of a "yield table," which contains entries for density and tree volume at different stand ages. A separate yield table exists for each species on each of several "site indexes" (an ecosystem variable), which are defined by the performance of the average individual plant (expressed as height at a standard age). How are the empirical patterns contained within yield tables related across these different scales of complexity?

- The "self-thinning rule" is a power relationship that seems robust to some (Yoda et al. 1963, Harper 1977, White 1981, Westoby 1984) and not so to others (Sprugel 1984, Zeide 1987, Weller 1987). Beyond the established knowledge that the relationship only applies in crowded stands, when should we expect this relationship?

Each of these topics involves more than one "level" of complexity (i.e., individuals, populations, and/or ecosystems). The purposes of this paper are (i) to suggest that simpler analyzable models can be constructed to describe and link these processes, provided they take cognizance of several important features of plant populations, and (ii) to summarize aspects of these simpler models that relate to dynamics within a cohort of plants that become established at roughly the same time. The principal assumptions of the approach are (i) that plant recruitment is episodic, (ii) that mortality has a DD component, which results from individual plant growth, and a DI component that influences density both directly and indirectly through its effect on DD mortality rate, and (iii) that plants require a "balanced diet" of resources (e.g., Tilman 1982, Vitousek et al. 1988) that results in a particular ecosystem stochiometry (Reiners 1986, Vitousek et al. 1988). Assumptions (i) and (ii) are necessary components of a "shifting-mosaic" landscape, where recruitment is limited locally to short events (Fig. 19.1) separated by longer periods when population dynamics are dominated by mortality processes (Watt 1947).

It is worth mentioning that Forest *gap* models contain essentially all of these attributes, at least implicitly, while analytical models generally do not. It is not surprising, therefore, that numerical models have seemed more useful for understanding some properties of plant populations (Botkin 1981, Huston et al. 1989). Limitations associated with existing theory that derives from simpler models are not entirely problems of complexity, however, but rather of approach. By incorporating the factors mentioned above, it is possible to analyze many of the population and ecosystem processes that occur during forest succession, for example, using rather simple theory. In developing the theory, I define and stress the importance of "density compensation" (DC), a dimensionless sensitivity coefficient describing the response of DD mortality rate to DI mortality. DC plays a central role in

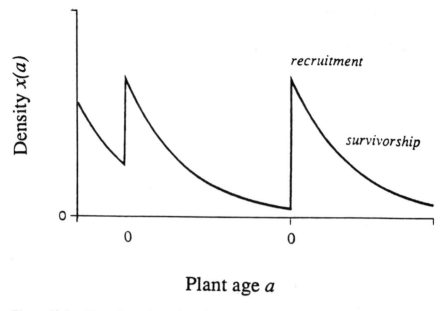

Plant age *a*

Figure 19.1. Two-phase dynamics of small cohorts, with episodic recruitment episodes separated by times when dynamics are dominated by mortality processes (from Clark, 1991a).

explaining sources of mortality, gap area, self-thinning, and net primary production. This summary of theory that applies to a single cohort has been extended to accommodate metapopulation structure across a shifting mosaic landscape and life-history strategies in separate papers (Clark 1991a, 1991b).

Model Summary and Implications

Beginning with an expression for individual plant growth, I outline how cohort density, gap area, net primary production, the self-thinning coefficient, and some aspects of resource competition change with stand age. I assume that a cohort of plants becomes established at age $a = 0$ at density $x(0)$, which can depend, for example, on availability of seed. We will subsequently track several attributes of this cohort of plants, each of which occupies exclusive crown area $A(a)$ described by weight $W(a)$, and subject to mortality forces that result from crowding (DD mortality) and from DI factors. We will acknowledge an "interaction" between DD and DI mortality forces, and we will relate the intensity of this interaction to such variables as canopy coverage, thinning rate, net primary production, and the "self-thinning rule." Some simple analyses of the model will be used to explore implications of individual plant growth for population- and ecosystem-level processes. In order to simplify the process, I assume that individual plant growth does not respond to density. Including this effect will quantitatively modify some of the conclusions that arise from this model, but plastic growth does not change the general relationships among processes explored here (Clark 1990).

Plant Growth and Local Population Dynamics

I begin with a simple population model that describes how density $x(a)$ changes with stand age a on a small patch following a recruitment event, i.e., the establishment of an even-aged cohort:

$$\frac{1}{x(a)} \frac{dx(a)}{da} = \phi(a) + \lambda(a).$$

$$\begin{array}{ccc} \text{per-capita} \\ \text{mortality rate} \end{array} = \begin{array}{c} \text{density-dependent} \\ \text{mortality} \end{array} + \begin{array}{c} \text{density-independent} \\ \text{mortality} \end{array}$$

The "thinning rate", $dx(a)/da$, is non-increasing because recruitment only occurs (by definition) when $a = 0$, and both $\phi(a)$ and $\lambda(a)$ are non-increasing. The way in which this model is used to link individual growth to population- and ecosystem-level processes is summarized in several steps. I first discuss how growth of plants within the cohort influences the thinning rate. The two principal effects here are a growth effect and a crowding effect. I then consider DI mortality rate $\lambda(a)$, which is independent of all other processes that will be included in the model. The final component is a "density-compensation" effect, which describes the dependence of $\phi(a)$ on $\lambda(a)$.

Plant growth and cohort thinning rate

DD thinning depends on growth of plants and consequent crowding. In order to solve for this dependency, I rely on two relationships. The first relationship is an expression for individual plant growth. Let $A(a)$ be the crown-area projection of a plant at age a (Norberg 1988, Hara 1988), and assume that increase in crown-area projection can be represented by a sigmoid (logistic) expression (Assmann 1970, Aikman & Watkinson 1980, Charles-Edwards et al. 1986),

$$\frac{dA(a)}{da} = A(a)r\left[1 - \frac{A(a)}{A_m}\right], \tag{2}$$

where A_m is the maximum crown-area projection that an individual plant can attain, and r is a rate constant. This growth equation establishes a relationship between size and age. Clark (1990) discusses sensitivity of several stand characteristics on the assumption of sigmoid crown-area increase. Relaxing this assumption does not have important influences on results presented here.

The second relationship is one between size and density. In a fully stocked stand (i.e., one in which canopy coverage is complete), the average area occupied per individual plant is proportional to the inverse of density x,

$$\kappa = xA \tag{3}$$

(Assmann 1970, Norbert 1988, Tait 1988, Valentine 1988), where κ depends on plant form and the units that are used to measure plant area and density.

Together, these relationships among age, crown area, and density provide an expression for change in density, i.e., a thinning equation.

$$\frac{dx(a)}{da} = \frac{\partial x}{\partial A}\frac{dA}{da} = \frac{-x}{A}\frac{dA}{da}$$
$$= -x(a)r\left[1 - \frac{A(a)}{A_m}\right]. \tag{4}$$

Equation (4) implies that the rate of thinning within the cohort is nearly constant until plants approach maximum size. Moreover, this near constant per-capita growth rate approaches the maximum per-area rate of crown-rate increase, r (Fig. 19.2b).

The crowding effect

Although these expressions relate density to crowding, they assume that canopy coverage is complete. This assumption provides a good fit to data in Fig. 19.2a, because the stands used to construct yield tables are chosen for their complete site occupancy. This condition does not hold when such processes as initial seed density and episodic reductions in density at other times in the life of a stand result in a noncontinuous canopy. Equation (4) must be modified in such a way that thinning sets in gradually as full canopy coverage is approached. Thus, Eq. (4) must contain a dependency on the proportionate canopy coverage,

$$F(x, A) = \frac{Ax}{\kappa}. \tag{5}$$

F is zero if no plants are present. Complete canopy coverage is described by $F = 1$. This second crowding effect is incorporated by replacing the maximum per-capita mortality rate r with rF,

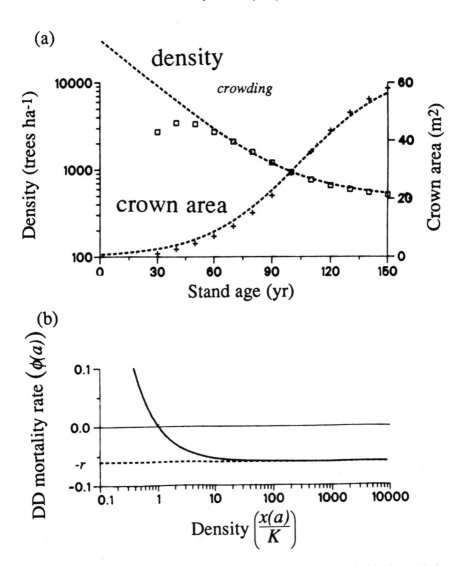

Figure 19.2   Relationship between individual plant growth and thinning. (a) Assuming proportionality between stem diameter and crown area, crown area $A(a)$ increases sigmoidally according to Eq. (2). This increase in individual size results in decreasing density according to Eq. (4). Model predictions are compared with data from Husch et al. (1972). (b) The decrease in density proceeds at near constant rate, which is equivalent to the maximum per-area rate of crown-area increase r, until plants approach maximum size. This DD thinning rate declines to zero as plants approach maximum size and density approaches K, the density that obtains when all plants assume crown area A_m (from Clark 1990). From Eq. (3), $K = \kappa/A_m$.

$$\frac{dx}{da} = -xrF\left[1 - \frac{A}{A_m}\right] \tag{6}$$

(Fig. 19.3a). This substitution results in a thinning rate that approaches r when the canopy is closed and when $A \ll A_m$ (Fig. 19.3c). When the canopy is open (Fig. 19.3b), however, the thinning rate is less than this value. As a result of these crowding effects on the thinning rate, stand densities converge to Eq. (4) from a range of initial densities in the fashion described by Harper (1977) and Westoby (1984) (Fig. 19.2a, 19.3a).

Density-independent mortality and density compensation

Density-independent (DI) mortality tends to displace the thinning curve from the trajectory described by Eq. (6). The extent to which DI mortality actually does decrease density, however, depends on (i) the relative importances of DD and DI mortality, and (ii) the response of DD mortality to DI mortality. DI mortality is generally most important when plants are young and susceptible to a fluctuating physical environment (reviewed by Oliver and Larsen 1990) and late in life with senescence (Loehle 1988). A process I term "density compensation" offsets this tendency to a degree that depends on individual plant growth and, thus, stand age. The DI mortality rate $\lambda(a)$ can be viewed as the sum of mortality forces u that affect juveniles and v associated with senescence

$$\lambda(a) = \lambda_u(a) + \lambda_v(a),$$

where $d\lambda_u(a)/da \leq 0$, and $d\lambda_v(a)/da \geq 0$. The conditions stipulate that juvenile mortality forces diminish with age, while those associated with senescence eventually increase in importance. DI mortality has a direct effect on thinning, because it is independent of stand processes (by definition), and an indirect effect, because DI mortality reduces density and therefore alleviates crowding. This density compensation (DC) effect can be incorporated in the thinning Eq. (6) by reducing the maximum per-capita thinning rate r to $(r - \lambda(a))$. The form of this DC effect implies that DI mortality "substitutes" for DD mortality. Equation (6) can now be written as

$$\frac{1}{x(a)}\frac{dx(a)}{da} = -[r - \lambda(a)]F\left[1 - \frac{A(a)}{A_m}\right] - \lambda(a), \tag{7}$$

(Fig. 19.4). The thinning Eq. (7) now contains DD mortality, a direct effect of DI mortality, and the indirect effect of DI mortality on DD mortality. We can use this model to explore the role of density compensation at the level of a plant cohort.

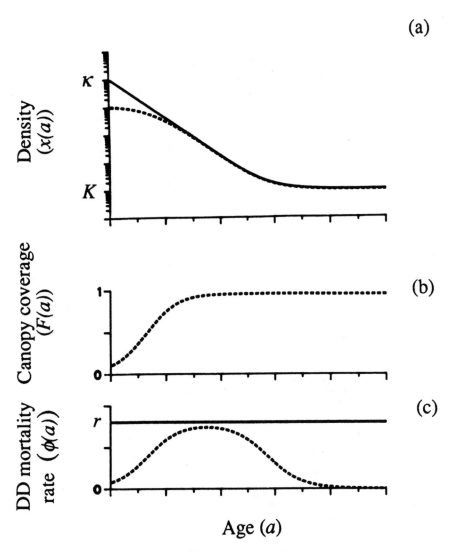

Figure 19.3. The effect of canopy coverage on DD thinning rate. The thinning Eq. (4) is approached from different initial densities (a) as canopy coverage increases (b) according to Eq. (6). The result is a mortality rate that is near the maximum rate of crown-area increase (r) when canopy coverage is high and when plants are small (c).

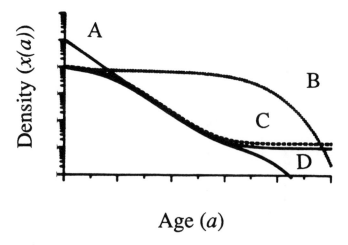

Figure 19.4. The combined effects of growth, canopy coverage, and DI factors on density. (A) Isolated effect of growth [Eq. (4)]. (B) Isolated direct effect of DI factors. (C) Combined effects of growth and canopy coverage. (D) Composite effects of DD and DI processes.

Implications for local population dynamics

The analysis of Eq. (7) and a related model by Clark (1992) provides insights concerning how the thinning process unfolds with stand age and the changing contributions of growth and DI factors. One useful result is the role of density compensation, or the link between DD processes, which owe their existence to the growth of individual plants, and DI processes, the changing importance of which determine the phases through which the cohort passes over time. The effects of DI mortality are made somewhat complex as a result of this linkage. In the absence of DI mortality ($\lambda(a) = 0$ for all a), DD mortality eventually drives density to $K = \kappa/A_m$ rather than to zero (Fig. 19.3a). Plants do not live forever, however, and local extinction can occur only if we allow for DI mortality, which insures that density eventually tends to zero (Fig. 19.4 curves B and D). This conclusion is important, because it implies that DD interactions are essentially nonexistent when plants become large and old. Thus, the relative importances of DD and DI mortality must change with age.

I have defined density compensation to be the direct effect of DI mortality rate λ on DD mortality rate

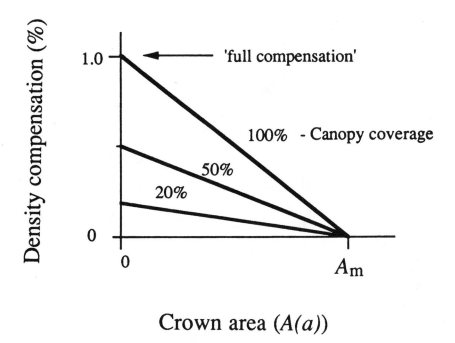

Crown area $(A(a))$

Figure 19.5. Effects of plant size and canopy coverage on density compensation.

$$DC = -F\left[1 - \frac{A}{A_m}\right] \tag{8}$$

(Clark, 1992). This DC response describes how DD mortality rate compensates for the DI mortality that alleviates crowding within cohorts. Any displacement from the thinning Eq. (7) will be compensated by an attendant decrease in DD mortality rate. Eq. (8) shows that DC tends to zero as plants approach maximum size or as canopy coverage declines to low values (Fig. 19.5). Thus, as plant growth slows with age, DD mortality can no longer compensate for DI mortality. As a result, DI mortality progressively drives canopy area below full coverage, and canopy gaps begin to appear. Because DC is also weak when the canopy is open ($F \ll 1$), DD mortality is less responsive to DI mortality in open stands. Together, these effects of stocking and plant size in Eq. (8) cause DC to be maximized at young to intermediate stand ages (Fig. 19.6d) These predictions are borne out by evidence from studies of stand development (Oliver and Larson 1990).

The degree of crowding, given by F in Eq. (5), also changes with stand age as

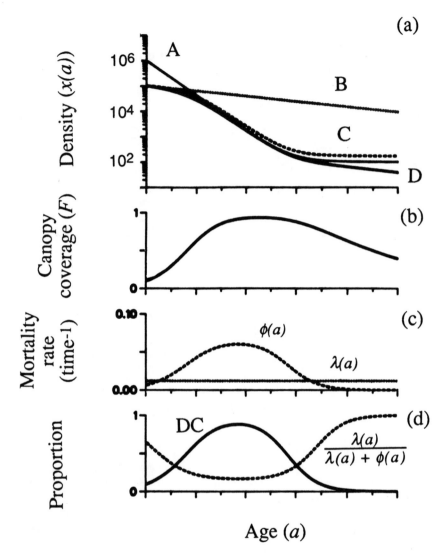

Figure 19.6. Time development of stand processes with constant DI mortality rate. (a) Density (labels as in Fig. 19.5). (b) Canopy coverage *F*. (c) Per-capital DD and DI mortality rates. (d) Density compensation (DC) and the fraction of total mortality accounted for by DI factors.

$$\frac{dF(x, A)}{da} = \frac{\partial F}{\partial x}\frac{dx}{da} + \frac{\partial F}{\partial A}\frac{dA}{da},$$

where the first term on the right-hand side is the rate at which canopy coverage is lost due to mortality, and the second is the rate of new canopy-area production due to growth. In the absence of DI mortality, this rate of change is positive and decreasing, tending to zero as the canopy approaches full occupancy ($F \to 1$) (Fig. 19.3b) and/or plants approach full size ($A \to A_m$) (Clark, 1992). Thus, canopy coverage tends to full occupancy with time from a range of initial values. Only with DI mortality will canopy coverage eventually decline. Moreover, the sensitivity of canopy coverage to DI mortality depends on plant growth rate, which decreases with age. The sensitivity to a constant DI mortality rate λ,

$$\frac{\partial(dF/da)}{\partial\lambda} = F\left\{F\left[1 - \frac{A}{A_m}\right] - 1\right\}$$

(Clark, 1992), demonstrates this size dependency. Consider, for example, the case of full canopy closure ($F = 1$). This sensitivity is then $-A/A_m$, which ranges from near zero for small plants to -1 at maximum size. Thus, although canopy coverage reaches and maintains full occupancy in the absence of DI mortality in Figure 19.3b, the sensitivity of canopy-coverage change to any DI mortality that might occur is steadily increasing with stand age. This increasing sensitivity results from decreasing growth rates of individuals.

Figure 19.6 demonstrates the effect of a constant DI mortality rate on canopy coverage. Canopy coverage is driven toward full coverage by plant growth, while simultaneously being driven away from this value by DI mortality. If density compensation did not occur (i.e., where DC = 0 for all a), DI mortality would steadily displace density from the thinning Eq. (4), and canopy coverage would decrease. But canopy coverage increases from low values when plants are small and therefore growing rapidly (Fig. 19.6b). The negative effect of DI mortality on canopy coverage is insufficient to offset the tendency of growth to drive canopy coverage to full occupancy when the plants are small. This tendency prevails so long as DC is an important process. As density continues to decrease and growth rate declines as plants approach maximum size, however, declining DC (Fig. 19.6d) and the increasing sensitivity of canopy coverage to DI mortality causes canopy coverage to decline. Canopy coverage begins to decline late in stand development, despite the fact that total mortality rate is actually decreasing (total mortality rate is the sum of DD and DI mortality rates in Figure 19.7c (Clark, 1992).

Implications for Ecosystem Processes

These relationships among plant size, density, and age can be used to explore implications of indiviudal plant growth for such ecosystem-level processes as net ecosystem production (NEP), below- and aboveground resources, the relationship between maximum NEP and maximum individual plant-weight increment, and the self-thinning rule. Plants increase in size and the cohort thins according to Eq. (7). Together, plant weight and density imply a standing crop, $Y = Wx$, for which the rate of change, dY/da, is NEP. I assume that growth rate, dW/da, depends on the availability of a limiting nutrient, having soil concentration $R(a)$, that is taken up at a rate that depends on this availability, tissue-nutrient concentration $1/\eta$, proportionate root allocation w_r, and uptake parameters. By incorporating these processes of plant growth within this population framework, we will investigate ecosystem implications of individual growth. In extending these population-level consequences to ecosystems, we make two additional assumptions:

i) Plants retain geometric similarity as they increase in size. This assumption is contained within geometric interpretations of the 'self-thinning rule' (Yoda et al. 1963, Zeide 1987, Norberg 1988), being expressed as

$$W = \beta A^\alpha \qquad (9)$$

where W is the expected mass of an individual plant, and $1 < \alpha < 2$. Note that this is not the self-thinning rule itself, i.e., we have not made the assumption that a relationship exists between plant weight and density (see below). Given Eqs. (2) and (9), individual plant growth is given as

$$\frac{dW}{da} = \frac{\partial W}{\partial A}\frac{dA}{da}$$
$$= \alpha\beta A^\alpha r\left[1 - \frac{A}{A_m}\right].$$

This expression is isomorphic with Aikman's (this volume) growth equation (Appendix).

ii) For simplicity, long-lived plant tissues are assumed to maintain constant tissue nutrient concentrations (e.g., Vitousek et al. 1988).

Net ecosystem production

Individual plant growth has two contrasting effects on NEP. Increasing sizes of individuals represent a positive contribution to NEP. The previous section outlined how plant growth also drives the self-thinning process, a negative contribution to NEP. Which of these effects should prevail, why, when in stand development, and by how much? Using Eq. (9) together with the assumption that DD mortality compensates for DI mortality (i.e., λ approaches zero), we can solve for NEP as

$$
\mathrm{NEP} = \frac{dY(W, x)}{da} = x \frac{dW}{da} + W \frac{dx}{da}
$$
$$
= rY[\alpha - F]\left[1 - \frac{A}{A_m}\right], \tag{10}
$$

where dx/da is given by Eq. (4).

This relationship indicates that increases due to growth outweigh losses due to mortality so long as the 'self-thinning coefficient' α is greater than canopy coverage F. Given that α is greater than unity, and F is generally unity or less, NEP is positive. α is nearly always greater than unity (Yoda et al. 1963, Westoby 1984, Norberg 1988), because plants increase in height; growth is not confined to two dimensions. Thus, mortality losses cannot match growth gains, because biomass accumulation can continue at increasing heights above the ground surface, while mortality within the main canopy responds primarily to crowding in two dimensions. If growth occurred strictly in two dimensions, α might be expected to be closer to unity, and NEP would be closer to zero throughout stand development.

These contrasting effects of individual plant growth on NEP cause maximum NEP to occur at intermediate stand ages, because standing crop increases from zero as a result of plant growth, but eventually decreases due to declining density (Fig. 19.7a). For stands that maintain full canopy coverage, this maximum occurs when individual biomass is given by

$$
\frac{W}{W_m} \approx \left[1 - \frac{1}{\alpha}\right]^\alpha \tag{11}
$$

(Clark 1990), i.e., when plants are about 20% of maximum size, for $\alpha = 1.5$ (Fig. 19.7). Higher values of α result in slight delays in maximum NEP, e.g., this value is 25% for $\alpha = 2$ (see Clark 1990). At canopy coverage values much below or much above complete coverage, density-dependent processes cause this value to be greater, i.e. maximum NEP is delayed (Clark, 1992). Low DD mortality rate when the canopy is open allows biomass to accumulate without the mortality losses that occur in crowded stands. In extremely crowded stands maximum NEP is also delayed, because DD mortality removes excess density gradually. Dense stands in which crown differentiation is slow to develop can contain plants having reduced growth

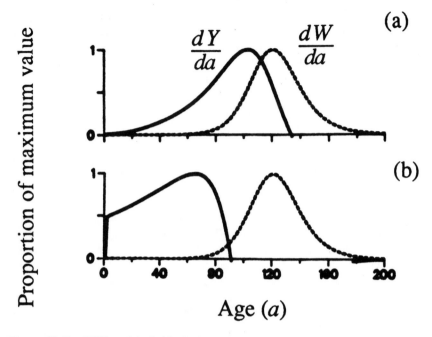

Figure 19.7. NEP and individual plant weight increment as a function of stand age. The examples differ in the relative importance of DI mortality. In (a) DC is sufficient to maintain full canopy coverage during most of stand development. In (b) DI mortality is sufficiently high that canopy occupancy is not maintained at full coverage. As a result, maximum NEP occurs earlier in stand development.

rates with less DD mortality than would occur with rapid crown differentiation (Smith 1962). In contrast, if an open canopy results from high DI mortality rate, maximum NEP will occur sooner in stand development (Fig. 19.7). The extent to which NEP is influenced by DI mortality, however, depends on DC. The greatest effects occur when DC is low and DI mortality is high. Although both of these conditions occur together in young and in old stands, either process alone can produce decreases in NEP.

Individual plant weight increment is maximized later in stand development, when

$$\frac{W}{W_m} \approx \left[\frac{\alpha}{\alpha + 1} \right]^\alpha \tag{12}$$

i.e., at about 45% of maximum plant weight W_m, for $\alpha = 1.5$ (Clark 1990, Fig. 19.7). This maximum occurs later, because individual plant weight does

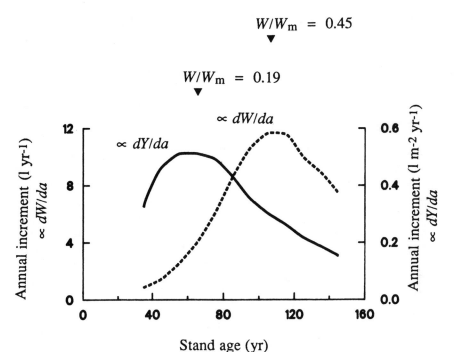

$$W/W_m = 0.45$$

$$W/W_m = 0.19$$

Figure 19.8. NEP and individual plant weight increment in the stand shown in Fig. 19.2. Peak values closely match predictions of Eqs. (11) and (12) for $\alpha = 1.5$. These volume data (in liters) are proportional to weight data.

not directly depend on population mortality as strongly as does standing crop. These predictions are supported by data taken from real stands (Fig. 19.8).

Nutrients

The soil-nutrient/plant interaction represents a good example of a relationship that operates at individual, population, and ecosystem levels. Nutrients influence individual plant growth, which has effects on population mortality rates, which in turn influence production and standing crop. If concentrations of nutrients in long-lived tissues remain relatively constant (constant nutrient-use efficiency η) (Miller 1984, Vitousek et al. 1988), then the contribution of plants to nutrient availability is proportional to accumulation of biomass of long-lived tissues. A highly idealized rate equation for changing availability (as a concentration) of the nutrient $R(a)$ can be written as:

$$\frac{dR(a)}{da} = S - \frac{rY[\alpha - F]}{\eta}\left[1 - \frac{A}{A_m}\right], \qquad (13)$$

where nutrient-use efficiency η is the reciprocal of tissue nutrient concentration. The two terms in this equation include an external source/sink term S and the effect of the cohort on the resource pool. For simplicity, I have not allowed for the lag required to mineralize nutrients returned to the forest floor in litterfall.

The second term, which describes the effects of the cohort on a soil nutrient, is negative ($F < \alpha$), but it becomes less negative with stand age, as A tends to A_m. This decreasing nutrient demand with stand age results in a tendency for the cohort to deplete the nutrient pool early in stand development. Late in stand development, however, uptake is so low relative to returns in mortality that the cohort tends to replenish the nutrient pool. The extent to which changes in the resource pool actually occur depends on changing nutrient supply that attends accumulation of mineralizable nutrient in the forest floor and changes in litter quality with stand age, so these results are not intended as predictions of soil-resource levels; they simply isolate the growth effect on those resources.

The other side of this plant-resource interaction is the effects of a limiting resource on the plants at individual, population, and ecosystem levels of organization. At the individual level, a growth-rate dependency on nutrient availability can be written as,

$$r(R) = \eta w_r \mu_m \left[\frac{R}{k_R + R}\right], \qquad (14)$$

where w_r is the proportion of total biomass allocated to root mass, μ_m is the resource-saturated rate of resource uptake per unit of root mass, and k_R is a half-saturation constant for nutrient uptake. The growth response to a constant resource availability becomes

$$\frac{d\text{RGR}}{dR} = \frac{\partial \text{RGR}}{\partial W}\frac{\partial W}{\partial R} + \frac{\partial \text{RGR}}{\partial R},$$

where RGR is the negative growth rate, $(dW/da)/W$. The first term represents the effect of the resource of RGR via its effect on plant size at a given age. This effect is positive and tending to zero with age for plants having deterministic growth. The second term is solved as

$$\frac{\partial \text{RGR}}{\partial R} = \frac{\alpha \eta w_r \mu_m k_R \left[1 - \dfrac{A}{A_m}\right]}{[k_R + R]^2}, \qquad (15)$$

which decreases with age for the same reason. Species having the ability to produce more biomass per unit of acquired resource (i.e., those with high η) place lesser demands on the soil pool of that nutrient per increment of new growth. Species best able to exploit resources will be those most responsive to fertilization, i.e., those with higher μ_m and low k_R. As plants approach maximum size, however, the advantage of increased resource availability declines, as the increase in growth per unit of resource diminishes.

Given that increased growth also increases mortality rate, however, why should the net response of NEP be positive? The easiest way to demonstrate this response begins with the relation

$$\text{NEP} = \frac{\partial Y}{\partial x}\frac{dx}{da} + \frac{\partial Y}{\partial W}\frac{dW}{da}.$$

Given low DI mortality, the added biomass represented by growth in the second term will always exceed the losses represented by mortality in the first term by $[\alpha - F]$. Thus, NEP is greater than or equal to zero [e.g., Eq. (10)]. Now growth rate contributes to both terms, but its negative effect on mortality rate is scaled by F, while its positive effect on growth is scaled by α. Thus, the net effect of fertilization on NEP is positive whenever $\alpha > 1$. As previously discussed, α is always > 1, because biomass accumulation occurs at increasing heights above the forest floor. Thus, the increasing mortality rate that results from fertilization is not sufficient to offset the positive effects on growth, because height growth allows continued exploitation of the vertical dimension.

The effects of fertility on NEP also depends on the nutrient-use efficiencies (η) of individual plants. The effect of η on NEP is positive, being strongest when resources are abundant. This effect declines, however, as plants approach maximum size. Thus, growth efficiency by individual plants and site fertility have their most important effects within stands of young age.

There is an important difference between nutrient responses at the individual level vs. the stand level that results from the role of crowding, F. v has less effect at the stand level (i.e., Y) than at the individual level (i.e., W), and NEP is less responsive to soil fertility as stands become more crowded (i.e., F increases). These differences result from the fact that mortality losses increases with F. NEP gains due to increased growth are offset to a greater degree by increased mortality, so that more fertile sites support fewer individuals of larger size (e.g., Harper 1977). The increased size per individual outweighs the decreased density, so high fertility and v both increase total standing crop, despite the fact that it is contained within fewer individuals. Nonetheless, while dominant individual plants may show relatively

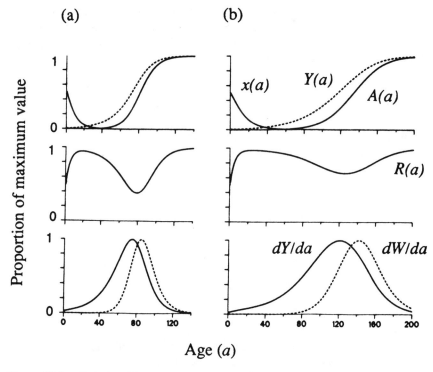

Figure 19.9. Relationships between NEP (dY/da), individual plant growth $A(a)$, cohort thinning $x(a)$, and availability of limiting soil nutrient $R(a)$ for two stands that differ in the degree of nutrient limitation. Nutrient limitation is weak in (a), and thus growth is rapid and resources are depleted to support this growth. Stronger limitation in (b) results in slower growth and attendant weaker demand on the limiting resource. All variables are expressed as fractions of maximum values.

large responses to changes in nutrient status, these effects are subdued at the stand level as a result of increased mortality.

The timing of maximum NEP, together with nutrient limitation, has several additional implications for resource availability. By decreasing NEP, resource limitation also decreases the degree to which the nutrient pool is drawn down during stand development (Fig. 19.9). Maximum NEP occurs earlier in stand development when soil fertility is higher. The maximum demand for belowground resources occurs near maximum NEP, because this is the time when the difference between rates of nutrient uptake for growth vs. return in litter and mortality is greatest. Stand-level processes are initially more sensitive to factors such as soil fertility than are the growth rates of the plants that make up the stand, because of the difference in stand ages when individual growth increment vs. NEP are maximized (Fig. 19.7). By the time individual plants are approaching maximum growth increment, the optimum growth of the stand is past (Fig. 19.8) and demand is lower relative to nutrient returns in dying tissues.

The self-thinning rule

Density compensation plays an important role in the "self-thinning rule." The power relationship between stand density and plant biomass,

$$W \propto x^{b_w},$$

or between stand density and standing crop per unit area,

$$Y = Wx \propto x^{b_y},$$

where

$$b_W = \frac{\partial W}{W} \bigg/ \frac{\partial x}{x},$$

and

$$b_Y = \frac{\partial Y}{Y} \bigg/ \frac{\partial x}{x}.$$

For the special case of a crowded stand ($F = 1$) with low DI mortality rate ($\lambda = 0$), parameters b_W and b_y are equal to $-\alpha$ and $1 - \alpha$, respectively. This is Yoda et al.'s (1963) "self-thinning rule." Decreasing canopy coverage steepens these coefficients to $b_W = -\alpha/F$ and $b_Y = 1 - \alpha/F$. The importance of DC results from the fact that DI mortality rate has the direct effect of decreasing the absolute values of these exponents and the indirect effect of increasing them through its effect on crowding F (Clark, 1992). When DC is sufficiently high to overcome the effects of DI mortality, the slope of the log x − log Y plot (i.e., b_Y) becomes steeper (more negative) than $1 - \alpha$. The steeper slope results from the fact that growth continues while DD mortality rate is low (Fig. 19.10a). This occurs when plants are growing rapidly. Where DC cannot offset the effects of DI mortality, however, the slope becomes less negative as losses exceed gains due to growth, and crowding declines. This occurs in stands where DI mortality is predictably high (Fig. 19.10b) and in all stands late in stand development (Fig. 19.10a).

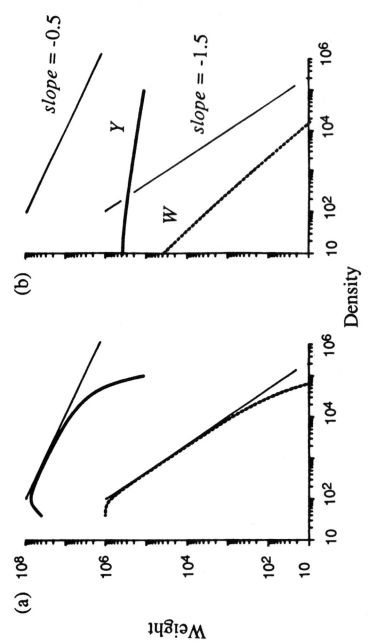

Figure 19.10. Self-thinning plots for two stands that experience different DI mortality schedules. (a) Density compensation is sufficiently high that canopy coverage is maintained throughout most of stand development. As canopy coverage begins to decline, the slope of the curves become less negative. (b) Density compensation is not sufficient to offset a high DI mortality rate, and the canopy remains open during all of stand development.

Discussion

In this paper I have reviewed some of the theory that links behavior of plant populations and ecosystems to growth of plants, and I have summarized several implications of individual plant growth for population- and ecosystem-level processes. The range of topics addressed are related through their dependencies on the growth of individual plants. The objective of producing an analyzable model requires a sacrifice of realism. All models and experimental studies make this sacrifice to varying degrees, and models ranging from simple abstractions like Eq. (7) to the individual-based models of Huston et al. (1989) should represent a complementary set of tools for understanding plant-population dynamics. The more important assumption used here is that of constant plant form, implied by Eq. (9). This assumption will always be violated to varying degrees (Ford 1975, Harper 1977, Ford and Diggle 1981), but the interpretations presented here hold, provided $1 \le \alpha \le 2$, which is generally true (Norberg 1988).

Density Compensation and Episodic Recruitment

The initial assumption that recruitment is episodic and confined to periods of short duration (Fig. 19.1) is supported by the analysis of cohort dynamics. For a gap-requiring species this episodic recruitment is a consequence of density compensation (DC), i.e., the response of DD mortality rate to DI mortality rate. DC maintains canopy closure at intermediate stand ages, when high density and rapid growth assure that any gaps that occur are rapidly closed by growth of individuals within the same canopy layer (Fig. 19.6b). Stand canopy coverage is rather insensitive to mortality that occurs at this time. Growth rates decline as plants approach maximum size, DC declines, and the canopy becomes increasingly sensitive to DI mortality. As a result, DD mortality rate cannot decrease sufficiently to offset the effects of DI mortality rate, and canopy gaps appear. With the formation of gaps, new recruitment can occur in the understory.

As a result of this episodic recruitment, demographic parameters are inherently stochastic, and it is logical to simplify population dynamics as a collection of rather even-aged cohorts. This approach is consistent with the "shifting mosaic" concept of forest populations (Watt 1947, Bormann and Likens 1979, Shugart 1984). In this paper I have focused on the dynamics within such a cohort. The dynamics that prevail between recruitment events, however, represent one of two important aspects of population dynamics. The density and age structure across the landscape mosaic depends on the distributions of these recruitment episodes in space and time (Fig. 19.1). That structure has been treated elsewhere (Clark 1991b), but it is important to mention that the landscape patterns of the processes discussed here are distributed in space and time, depending on the space/time pattern of recruitment episodes.

Phases in Stand Development

Taken together, the processes discussed here address many aspects of forest succession. Here, I place the time development of this composite process within the context of previous schemes of Bormann and Likens (1979), Oliver (1981), Sprugel (1985), and Oliver and Larson (1990) (Fig. 19.11). In doing so, I will not review these earlier treatments. Instead, I focus on insights provided by the theory that derives population- and ecosystem-level processes from the growth of individuals.

Reorganization or stand initiation

Canopy coverage increases until complete canopy occupancy is achieved (Fig. 19.11b). During this time, DD mortality rates are substantially lower than r (Fig. 19.11c), the maximum per-area rate of crown area increase. If the canopy is open (as in Fig. 19.11b), density compensation DC is low (Fig. 19.12d), despite rapid growth of individual plants. Thus, DD mortality rate is "insensitive" to DI mortality. This low canopy coverage means that openings will not be rapidly filled by lateral growth of neighbors in the same canopy layer. At the same time, DI mortality may be high as a result of juvenile sensitivity to environmental variance (Fig. 19.11c). In general, this period is characterized by a transition in mortality source, as DI mortality is declining and DD mortality is increasing.

The cohort acts as a potentially strong sink for nutrients at this time, because accumulation in new tissue occurs at a rate that is substantially greater than that at which it is returned in dying tissues (Vitousek and Reiners 1975). The rate of biomass accumulation is more sensitive to a given nutrient availability and to resource uptake and use by the cohort than it will be at later stages. The effects of these variables on NEP decline as plants increase in size and the canopy closes. At the same time, however, the demand for nutrients continues to increase with NEP (Fig. 19.11e).

Aggradation or the stem exclusion.

As the canopy closes, DD mortality approaches rate r, and density compensation achieves maximum values. Equations (3) and (4) establish the link between individual plant growth and cohort thinning, which Aikman and Watkinson (1980) suggested must exist, and they indicate why such thinning rates are observed to be so constant in crowded stands (Harper 1977). The sigmoid growth of individual plants implies a near-constant per-capita mortality rate, provided that the canopy remains closed (Fig. 19.2b). This rate is equal to the maximum per-area rate of crown-area increase [Eq. (2)]. Only

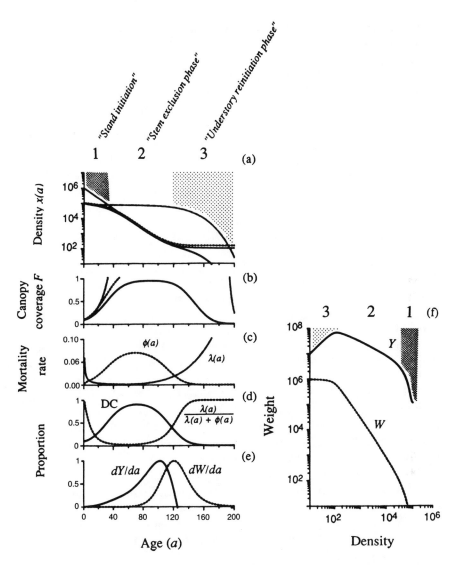

Figure 19.11. The phases of stand development compared with model predictions. The three phases are (1) Stand initiation, (2) Aggradation or Stem exclusion, and (3) Understory reinitiation. The curves in (a) are described in Fig. 19.4. The three curves in (b) are canopy coverage values corresponding to curves B, C, and D in Fig. 19.4.

when this crowded condition exists and when plants are young to intermediate age will rates be rather constant.

Any DI mortality that occurs at this stage tends to be compensated by decreased DD mortality; rapid lateral growth maintains gap area at a minimum (Fig. 19.11b). Theory predicts that the factors responsible for the rate of canopy closure [Eq. (8)] i.e., canopy coverage and a size effect on growth rate $(1 - A/A_m)$, are the same as those identified from empirical studies (Oliver and Larson 1990). Although DI mortality falls to low levels (Fig. 19.11c), if an episodic reduction in density does occur, DD mortality rate will decrease, permitting recovery of canopy coverage. Thus, although total mortality rate is highest at this stage, canopy gaps do not occur. Canopy coverage increases to maximum levels, but canopy coverage is becoming increasingly sensitive to DI mortality throughout this stage.

It is during this stage that NEP is maximized (Fig. 19.11e). Clark (1990) used models presented here to show that differences in the shape of the curve describing biomass accumulation can be explained by the shape of growth curve $A(a)$, depending on when the inflection occurs. Growth of most plants is initially rather exponential (e.g., Ingestad 1971), which is responsible for constant RGR. With the accumulation of non-meristematic and non-photosynthetic tissues in trees, however, RGR begins to decline, thus accounting for the inflection in $A(a)$; the growth pattern is therefore sigmoid. The earlier this inflection occurs, the earlier NEP is maximized. Clearly, this pattern varies among stands, so it is not unreasonable to expect the pattern of NEP, and thus standing crop, to vary (e.g., Peet 1981, Shugart 1984, Sprugel 1985). Higher nutrient availability at this stage would increase individual plant size, and it increases thinning rate (Harper 1977, White 1981, Clark 1990). Because nutrient turnover is proportional to NEP, maximum demands for both light and nutrients occur during this stage.

It is during this phase that the self-thinning rule is most applicable, primarily because canopy coverage remains high and rather constant and because it displays a low sensitivity to DI mortality (Fig. 19.11f). The rule makes sense at this stage because of the assumptions regarding canopy coverage and DI mortality originally used to explain it (Yoda et al. 1963) are least violated.

Understory reinitiation

This phase has been poorly understood in the past, I believe, because the density compensation effect has not been fully appreciated. The cohort "loses its grip" (Sprugel 1985) on the site at this time because of decreasing density compensation (Fig. 19.11d) and the increasing sensitivity of crown area to DI mortality. Declining density and slowing growth rates are together responsible for this transition. The maximum DD mortality rate, which approaches r in the younger stands, declines together with individual plant growth rates late in stand development (Fig. 19.11c). Increasing gap area results neither from the fact that mortality rate is increasing nor because a single dying individual represents a large area. It is important to distinguish between the effect of increasing mortality that results from senescence, vs. that which results from decreasing density compensation. Although increasing DI mortality rate at this age (Fig. 19.11c) contributes to increasing gap area (Fig. 19.11b), DI mortality could remain the same or even decrease, and gap area would still increase (e.g., Fig. 19.6). The importance of plant size results from its effect on declining growth rate. Gap area increases because the rate of individual plant growth is declining; plants could be huge and gaps would not occur provided that growth was still exponential. "Understory reinitiation" results from the fact that, as DD mortality fails to compensate for DI mortality rate, gaps left by dying individuals are not longer filled by lateral growth of neighbors (Oliver and Larson 1990).

Following maximum NEP, the cohort becomes more of a source than a sink for nutrients. While it is true that the cohort is still accumulating nutrients, the difference between uptake and return becomes small relative to previous times (e.g., Vitousek and Reiners 1975). This transition occurs because nutrients are returned in dying tissues nearly as rapidly as they are being accumulated in new growth. This transition begins before individual plants achieve their individual maximum biomass increments (Fig. 19.11e). Throughout this phase, sensitivities of individual plant growth and of NEP to nutrients is decreasing. Thus, allocation to nutrient acquisition structures becomes of decreasing benefit at the individual level, and the payoff from fertilization in terms of yield at the stand level decreases (Clark 1990).

The decreasing slope of the density-plant weight relationship at this time (Harper 1977, White 1981, Westoby 1984, Zeide 1987, Norberg 1988) (Fig. 19.11f) results from decreasing DC (Fig. 19.11b). Although it is generally claimed that these self-thinning relationships are always taken from "crowded" stands, I suspect that the inevitable decrease in canopy coverage that must occur with age plays a role. It is likely that crowns are less tightly packed in a stand of large, mature trees than they are at intermediate stand ages, when stands are dense and plants are growing rapidly. This effect would be difficult to identify in the field, unless different-aged stands could be observed simultaneously, or if canopy light penetration were compared. For these reasons, I believe that density compensation is a more parsimonious explanation for the decreasing slope of the self-thinning coefficient b_y in old stands than is Norberg's (1988) argument that soil depth eventually limits root penetration.

Old-growth phase.

The episode nature of recruitment opportunities that results from declining density compensation and increasing DI mortality late in stand development accounts for the loss of synchroneity, as a relatively even-aged cohort is transformed into a "shifting mosaic" (Bormann and Likens 1979, Shugart 1984, Oliver and Larson 1990). Because NEP is maximized at intermediate stand ages, the transition from an even-aged aggrading forest to a mosaic structure implies declining NEP. As the first cohort to become established on a large disturbed area begins to break up, population- and ecosystem-level processes depend on the distributions of recruitment events in space and time (Clark, 1991b).

Models of Plant-Population Dynamics

This simplistic theory is complementary to the simulation models that have become popular in recent years. While lacking the complexity of these empirical approaches, it provides several advantages, and it addresses some of the criticisms aimed at analytical theory (Huston et al. 1989). The principal advantage of this simplified approach is analyzability, which reveals the changing contributions of different processes with stand development. Many of the important relationships among individual plant growth, net primary production, DD and DI mortality, density compensation, and self-thinning can be investigated with four parameters. The simplicity of the expressions and of the analysis makes it possible to establish fundamental links across these scales of complexity that would be impossible with the gap models which contain dozen of equations and hundreds of parameters, or even many of the simpler numerical models of asymmetric competition, canopy differentiation, and size structure (e.g., Aikman and Watkinson 1980, Ford and Diggle 1981, McMurtie 1981).

Some of the shortcomings attributed to analytical models (Huston et al. 1989) may result more from approach than from any inherent inadequacy of such simplification. The poor predictive capacity of some plant-population models can result from a failure to accommodate several pervasive features of plant population. First, most plant populations are recruited in an episodic fashion, with local stands being composed of a small number of cohorts (Harper 1977, Warner and Chesson 1985, Comins and Noble 1985). Nearly all trees, for example, require some type of disturbance for recruitment (Grubb 1977, Spurr and Barnes 1980). This topic is discussed more fully in the previous section. Second, a "carrying capacity" in the usual sense does not exist, because density depends largely on individual size, and size is constantly changing. Even in a homogenous setting, density is best described as a distribution because of the mosaic nature of plant populations (Clark 1991b). Heterogeneity is produced by the population itself, as a consequence of density compensation. Third, density-dependent mortality that results from crowding is alleviated by density-independent mortality. Fourth, neither density nor biomass per unit area alone are sufficient to understand population dynamics or ecosystem processes, because size is a surrogate for competitive ability (Harper 1977, Ford and Diggle 1981, Weiner 1985, 1986). A given density can imply very different population interactions and eco-system processes, depending on the sizes of the individuals within the population.

Many of the qualitative differences between predictions of gap models and analytical theory may be a consequence of the ways in which they treat these processes. Because of their structure, gap models accommodate all of the considerations mentioned above. Single-gap simulations of small plots demonstrate episodic recruitment for many species (e.g., Fig. 5.4 in Shugart 1984). Only when the output of many gaps are averaged does recruitment appear continuous. The established cohorts thin exponentially and standing crop increases, usually in a sigmoid fashion. As trees approach maximum size, NEP declines, and the probability of gap formation increases. I can think of no way in which an analytical model that assumes continuous re-cruitment, some "carrying capacity" in the traditional sense, a composite mortality rate that does not account for interactions among DD and DI sources, and that fails to accommodate the simultaneous effects of growth and density could produce reasonable predictions at these different levels of complexity; yet analytical theory presented here predicts dynamics that have been re-peatedly observed during forest succession (see Stages in Stand Develop-ment, above), and extensions of it to a shifting mosaic landscape result in logical distributions of cohort densities and age classes, and predictions of optimal life histories that are closely matched by those of species that ac-tually occur in very different disturbance regimes (Clark, 1991a, 1991b).

An aspect this model shares with other theory criticized by Huston and Smith (1987) and Huston et al. (1989) is the use of state variables that apply to all individuals to summarize aspects of a population. It is certainly true that such a simplification fails to capture certain characteristics of a popu-lation (e.g. asymmetric competition, canopy differentiation, size variability within an age class, spatial pattern). If the time-development of a process results primarily from attributes that are ignored in this manner (e.g., in-dividual size differences), then this simplification is not useful. If those con-sequences result instead from the average effect of individual attributes, there is little justification for complicating the analysis with intractable models. Theory summarized here suggests that many aspects of plant-population dy-namics at a local scale fall in the latter category. Important and repeatable patterns observed in plant populations can be summarized by the way in which the "average" canopy plant grows.

This approach is complementary to simpler analytical theory and to stand simulations. Most analytical models (e.g., a Lotka-Volterra equation, or Tilman 1982, 1985) assume that individuals are identical. Some models contain size or age structure, but that structure is everywhere the same (Schaffer and Leigh 1976). Births and deaths depend on time indirectly through their dependencies on state variables such as population size. Nonetheless, such simple models have produced many valuable insights when applied to problems that lend themselves to this type of simplification. At the other extreme are complex gap models that monitor each individual within a small cohort. The landscape is assumed to support many such cohorts that are independent of one another. These elaborate models are among the few tools that address processes that are inherently complex and/or that develop too slowly to be observed by ecologists.

Theory presented here represents an intermediate approach. Individual differences within a small cohort are ignored, but the landscape consists of many such cohorts, each with its own unique history. The metapopulation is described by distributions of cohort densities, patch ages, and cohort ages (Hastings and Wolin 1989, Clark, 1991b). These distributions acknowledge the stochastic nature of recruitment in space and time, but they treat deterministically dynamics that occur between recruitment events. I believe this level of complexity to be well-suited to the analysis of plant-population processes, because it captures the basic dynamics, while retaining the virtue of analyzability.

All of these approaches contain gross simplifications, and each addresses somewhat different problems. Much analytical theory is difficult to apply to real plants, because it not only simplifies — it assumes a homogeneous structure and time-invariate demographic parameters that may apply only in restricted cases, such as clonal herbaceous plants. Thus, analyses of such models seem open to several interpretations. For this reason, I share Huston and Smith's (1987) and Huston et al.'s (1989) dissatisfaction with some analytical theory. Conversely, I do not see that simplification through use of state variables that apply to all individuals is necessarily the problem, and I would place more emphasis on the development of theory that better describes the process, as simply as a particular application will permit. Individual-based (e.g., "gap") models will continue to represent an important tool that will be most powerful when used together with simpler theory that abstracts components of plant populations in a way that permits analysis.

Appendix

In this appendix I demonstrate the relation between my growth model and that provided by Aikman (this volume). Aikman's modified Richard's equation can be written as:

$$\frac{dW}{da} = aWf[1 - (bW)^c], \qquad (I.1)$$

where f is the proportionate reduction in growth that becomes increasingly severe with crowding, being near unity in an open stand and approaching zero with increasing crowding. The degrees to which species respond to crowding by reduced growth and mortality varies widely (Harper 1977). Weiner (this volume) provides examples for several species. The case where f remains near unity represents species that display minimal plastic growth response with increasing density. For this case, Eq. (I.1) can be written as:

$$\frac{dW}{da} = \alpha rW[1 - (W/W_m)^{1/\alpha}],$$

where $a = \alpha r$, $b = 1/W_m$, and $c = 1/\alpha$. This expression is equivalent to the expression used in this chapter under the assumption $W \propto A^\alpha$.

Literature Cited

Aber, J. D., D. B. Botkin, and J. M. Melillo. 1979. Predicting the effects of different harvesting regimes on productivity and yield in northern hardwoods. *Can. J. For. Res.* **9**:10–14.

Aber, J. D., J. M. Melillo, and C. A. Federer. 1982. Predicting the effects of rotation length, harvest intensity, and fertilization on fiber yield form northern hardwood forests in New England. *For. Sci.* **28**:31–45.

Aikman, D. P. and A. R. Watkinson. 1980. A model for growth and self-thinning in even-aged monocultures of plants. *Ann. Bot.* **45**:419–427.

Assmann, E. 1970. *The Principles of Forest Yield Study*. Pergamon, New York.

Bormann, F. H. and G. E. Likens. 1979. *Pattern and Process in a Forested Ecosystem*. Springer-Verlag, New York.

Botkin, D. F., J. F. Janak, and J. R. Wallis. 1972. Some ecological consequences of a commputer model of forest growth. *Journal of Ecology* **60**:849–872.

Botkin, D. B. 1981. Causality and succession. In *Forest Succession: Concepts and*

Management. D. C. West, H. H. Shugart, and D. B. Botkin (eds.), pp. 36–55. Springer-Verlag, New York.

Charles-Edwards, D. A., D. Doley, and G. M. Rimmington. 1986. *Modelling Plant Growth and Development.* Academic Press, New York.

Clark, J. S. 1990. Integration of ecological levels: individual plant growth, population mortality, and ecosystem processes. *J. Ecol.* **78:**275–299.

Clark, J. S. 1991a. Disturbance and tree life history on the shifting mosaic landscape. *Ecology* **72:**1102–1118.

Clark, J. S. 1991b. Disturbance and population structure on the shifting mosaic landscape. *Ecology* **72:**1119–1137.

Clark, J. S. 1992. Density-independent mortality, density compensation, gap formation, and self-thinning in plant populations. *Theoretical Population Biology* in press.

Comins, H. N. and I. R. Noble. 1985. Dispersal, variability, and transient niches: species coexistence in a uniformly variable environment. *Am. Nat.* **126:**706–723.

Ford, E. D. 1975. Competition and stand structure in some even-aged plant monocultures. *J. Ecol.* **63:**311–333.

Ford, E. D. and P. J. Diggle. 1981. Competition for light in a plant monoculture modelled as a spatial stochastic process. *Ann. Bot.* **48:**481–500.

Grubb, P. J. 1977. The maintenance of species richness in plant communities. The importance of the regeneration niche. *Biol. Rev.* **52:**107–145.

Hara, T. 1988. Dynamics of size structure in plant populations. *Trends Ecol. Evol.* **3:**129–133.

Harper, J. L. 1977. *Population Biology of Plants.* Academic Press, New York.

Hastings, A., and C. L. Wolin. 1989. Within-patch dynamics in a metapopulation. Ecology **70:**1261–1266.

Husch, B., C. I. Miller, and T. W. Beers. 1972. *Forest Mensuration.* Wiley, New York.

Huston, M., D. DeAngelis, and W. Post. 1989. From individuals to ecosystems: a new approach to ecological theory. *BioSci.* **38:**682–691.

Huston, M. and T. Smith. 1987. Plant succession: life history and competition. *Am. Nat.* **130:**168–198.

Loehle, C. 1988. Tree life history strategies: the role of defenses. *Canadian Journal of Forest Research* **18:**209–222.

McMurtie, R. 1981. Suppression and dominance of trees with overlapping crowns. *J. Theor. Biol.* **89:**151–174.

Miller, H. G. 1984. Dynamics of nutrient cycling in plantation ecosystems. In *Nutrition of Plantation Forests.* G. D. Bowen and E. K. S. Nambiar (eds.), pp. 53–78. Academic Press, New York.

Norberg, Å. 1988. Theory of growth geometry of plants and self-thinning of plant populations: geometric similarity, elastic similarity, and different growth modes of plant parts. *Am. Nat.* **131:**220–256.

Oliver, C. D. 1981. Forest development in North America following major disturbance. *For. Ecol. Manage.* **3:**153–168.

Oliver, C. D. and B. C. Larson. 1990. *Forest Stand Dynamics.* McGraw-Hill, New York.

Pastor, J. and W. M. Post. 1986. Influence of climate, soil moisture, and succession on forest carbon and nitrogen cycles. *Biogeochem* **2:**3–27.

Peet, R. K. 1981. Changes in biomass and production during secondary forest succession. In *Forest Succession: Concepts and Management,* D. C. West, H. H. Shugart, and D. B. Botkin (eds.), pp. 324–338. Springer-Verlag, New York.

Peet, R. K. and N. L. Christensen. 1980. Succession: a population process. *Vegetatio* **43:**131–140.

Reiners, W. A. 1986. Complementary models for ecosystems. *Am. Nat.* **127:**59–73.

Schaffer, W. M. and E. G. Leigh. 1976. The prospective role of mathematical theory in plant ecology. *Sys. Bot* **1:**209–232.

Shugart, H. H. A. 1984. *Theory of Forest Dynamics: the Ecological Implications of Forest Succession Models.* Springer-Verlag, New York.

Smith, D. M. 1962. *The Practice of Silviculture.* Wiley, New York, New York.

Smith, T. and M. Huston. 1989. A theory of the spatial and temporal dynamics of plant communities. *Vegetatio,* in press.

Sprugel, D. G. 1984. Density, biomass, productivity, and nutrient-cycling changes during stand development in wave-regenerated balsam fir forests. *Ecol. Monog.* **54:**165–186.

Sprugel, D. G. 1985. Natural disturbance and ecosystem energetics. In *The Ecology of Natural Disturbance and Patch Dynamics,* S. T. A. Pickett and P. S. White (eds.), pp. 335–352. Academic Press, New York.

Spurr, S. H. and B. V. Barnes. 1980. *Forest Ecology.* Wiley, New York.

Tait, D. E. 1988. The dynamics of stand development: a general stand model applied to Douglas-fir. *Can. J. For. Res.* **18:**696–702.

Tilman, D. 1982. *Resource Competition and Community Structure.* Princeton University Press, Princeton, N.J.

Tilman, D. 1985. The resource-ration hypothesis of plant succession. *The American Naturalist* **125:**827–852.

Tilman, D. 1988. *Plant Strategies and the Dynamics and Structure of Plant Communities.* Princeton University Press, Princeton, N.J.

Valentine, H. T. 1988. A carbon-balance model of stand growth: a derivation employing pipe-model theory and the self-thinning rule. *Ann. Bot.* **62:**389–396.

Vitousek, P. M., T. Fahey, D. W. Johnson, and M. J. Swift. 1988. Element interactions in forest ecosystems: succession, allometry and input-output budgets. *Biogeochem.* **5:**7–34.

Vitousek, P. M. and W. A. Reiners. 1975. Ecosystem succession and nutrient retention: a hypothesis. *BioSci.* **25:**376–381.

Vogt, D. A., D. J. Vogt, E. E. Moore, B. A. Fatuge, M. R. Redlin, and R. L. Edmonds. 1987. Conifer and angiosperm fine-root biomass in relation to stand age and site productivity in Douglas-fir forests. *J. Ecol.* **75:**857–870.

Waring, R. H. and W. H. Schlesinger. 1985. *Forest Ecosystems: Concepts and Management*. Academic Press, Orlando, FL.

Warner, R. R. and P. L. Chesson. 1985. Coexistence mediated by recruitment fluctuations: a field guide to the storage effect. *Am. Nat.* **125:**769–787.

Watt, A. S. 1947. Pattern and process in the plant community. *Journal of Ecology* **35:**1–22.

Weiner, J. 1985. Size hierarchies in experimental populations of annual plants. *Ecology* **66:**743–752.

Weiner, J. 1986. How competition for light and nutrients affects size and variability in *Ipomoea tricolor* populations. *Ecology* **67:**1425–1427.

Weller, D. E. 1987. A reevaluation of the $-3/2$ power rule of plant self-thinning. *Ecol. Monog.* **57:**23–43.

Westoby, M. 1984. The self-thinning rule. Advances in Ecological Research **14:**167–225.

White, J. 1981. The allometric interpretation of the self-thinning rule. *J. Theor. Biol.* **89:**475–500.

Yoda, K., T. Kira, H. Ogawa, and K. Hozumi. 1963. Self-thinning in overcrowded pure stands under cultivated and natural conditions (intraspecific competition among higher plants). *J. Biol. Osaka City Univ.* **14:**107–129.

Zeide, B. 1987. Analysis of the 3/2 power law of self-thinning. *For. Sci.* **33:**517–537.

20

A Comparison of Models to Simulate the Competitive Interactions between Plants in Even-Aged Monocultures

Laurence R. Benjamin and R. Andrew Sutherland

ABSTRACT. **Several types of models have been developed to describe the competitive interactions between plants, and they can be classified into five types based on the assumptions they make about the utilization of resources by individual plants. There is no type of model in which the underlying set of assumptions is entirely realistic. In this paper, we fit models representative of the different types to a standard data set from an extensive row spacing experiment using carrots. The comparative agreement between observed and fitted values when different models were fitted was dependent on the data. An overlapping domain model gave the best description of root weights derived from plants growing in the inner rows of plots, but a nonoverlapping domain model gave better fits to root weights derived from plants growing in edge rows. Possible explanations of this observation are discussed.**

The persistence of several contrasting spacing models in the literature might thus be explained because the fit of models is sensitive to the data to which they are fitted.

Introduction

The many models that have been developed to describe the growth of individual plants in crops have been classified by Benjamin and Hardwick (1986) according to the assumptions made about how resources are shared. They identified five types of model; "nonoverlapping domain," "overlapping domain," "unbounded areas of influence," "diffuse population effect," and "tiers of vegetation." Most models concentrate on describing the influence of spatial arrangement of neighbors on the weight of a reference plant at a given time, and in all types the quantity of resources per unit area of ground is assumed to be constant.

In the nonoverlapping domain type of model, the territorial extent over which a plant draws upon resources is determined by the distance to its immediate neighbors. Often the limits of this territory are assumed to be restricted to a Voronoi polygon, formed by the perpendicular bisectors of the lines joining the center of the plant to its immediate neighbors. Within each polygon, the single plant has sole use of resources. This type of model readily describes the effects of spatial arrangement of the nearest neighbors. However, the number and location of the nonimmediate neighbors has no effect on plant growth.

In the overlapping domain type of model, a circle, or other shape, centered on each plant bounds the territorial extent over which resources are drawn. Where the circles of two plants overlap, the resources in the overlapping areas are shared between them. In this type of model the effect of plant size on the competition with its neighbors can be easily accommodated by assuming that the radius of the circle is proportional to some attribute such as height, stem diameter or, more usually, weight. Also, the presence of plants other than the immediate neighbors has an effect on the growth of each plant, but very distant plants have no effect at all. However, the angular dispersion of neighbors has only a small effect on the growth of the reference plant. This type of model includes those in which the shape of the area of influence around the plant is elliptical rather than circular (for example, Wixley 1983). When the radius of the area of influence becomes large compared with the between-plant spacing, changes in the spacing of the immediate neighbors around the reference plant would be predicted to have a negligible effect on plant growth. Alternatively, when the radius of the area of influence is very small, the areas of influence between adjacent individuals do not overlap until the very last stages of growth, and very little plant-to-plant interaction is simulated.

The unbounded area of influence type of model assumes that there is no distinct boundary to the area over which each plant draws upon resources, but the ability of the plant to capture resources gradually decreases with distance from the plant. The weakening ability to utilize distant resources can be described by empirical equations. This type of model has many of the advantages and disadvantages of the overlapping domain type, but is completely unable to describe the effect of angular dispersion of plants around the reference plant, unless some specific modification is introduced.

The diffuse population effect type of model assumes that there is no territorial component to the capture of resources by each plant. The priority for the use of resources available in a given area by each plant is determined by the state of the population, such as its density or yield per unit area, and by the ratio between the individual and the population mean for a given characteristic such as weight. This type of model unrealistically assumes that the precise location of plants is irrelevant to the competition between them. This might not be a serious impediment. For example, in *Dactylis glomerata* growth of individuals is most influenced by the density of the previously emerged neighbors, and spatial distribution of neighbors exerts less influence (Ross and Harper, 1972). These models can use simple equations to simulate the more important processes that generate plant-to-plant variation, because they do not require the location of each individual to be mapped.

The tiers of vegetation type of model include mechanistic models which simulate the photosynthetic and respiratory activity of individuals in relation to their morphology and that of their neighbors. These models concentrate nearly entirely on the metabolism of the shoot, as there is not sufficient data for a mechanistically based model for the roots. This type of model aids understanding of the growth of individuals in terms of physiological processes, but is too complex to allow simulation of subtle effects of plant interactions, such as spatial distribution of neighbors.

Only the first four types of models have the potential to describe the effects on spatial arrangement of neighbors on the growth of an individual in a mathematically tractable form. However, none of these four types is without obvious defect in the assumptions made.

Previous attempts to evaluate the relative merits of different competition models have determined the degree of correlation between increment in tree growth, and some measure of competition index (Opie 1968, Adlard 1973, Daniels 1976, Noone and Bell 1980). In these earlier works the fit to the data was not clearly superior for any type of model.

The earlier studies on trees often relied upon the variation in spacing that occurred naturally in woodlands. Also, comparisons between plantations grown at different densities might be confounded with differences in site fertility. Herbaceous plants are useful for competition studies because of the relative ease with which their spatial pattern of sowing can be varied, but their individual growth is more difficult to measure than in trees. A number of studies have examined the relative fits of different forms of the same type of model. For example, Silander and Pacala (1985) examined different forms of unbounded areas of influence type of model using different densities of *Arabidopsis thaliana*. Matlack and Harper (1986) have discussed the relative merits of polygon shape and area as predictors of plant growth for *Silene dioica* in non-overlapping domain models. For overlapping-domain models, assuming an elliptical rather than circular domain gave better fits to some sets of *Nicotiana tabacum* data (Wixley 1983). However, we are not aware of any other comparison of different types of spacing model using a common set of data derived from herbaceous plants grown at a wide range of spatial patterns.

To determine the importance and nature of spatial pattern on plant competition, models, based on the different sets of assumptions of the first four types of model, are fitted to a standard set of experimental data from an extensive row-spacing experiment on carrot (*Daucus carota*, L.).

The Data

The data used in this paper are mean weights and densities of carrots cultivar Ideal in individual rows from experiment 2 of a series of experiments described by Benjamin and Sutherland submitted.

Briefly, the plants were grown in beds 1.52 m wide from wheeling center to center. Each bed contained eight rows, 120 mm apart, with a 340 mm gap at either side between the outer row of each bed and the center of the wheeling. Each plot consisted of a length of bed, with the number of seeds sown in each row adjusted to give a target plant density of 100, 200, 400 or 800 plants m^{-2}, and a ratio of the number of seeds m^{-1} of row in each of the two outer rows to the number of seeds m^{-1} of row in each of the six inner rows of 1.0:1, 1.5:1 or 3.0:1. The experiment was laid out as a randomized block design with two replicate blocks of each of these 12 treatments at both a high and low nitrogen application site. There were no differences between the two sites, so the experiment can be regarded as having four replicate plots of each treatment. The plants were intended to be uniformly spaced along each row, but a random displacement was caused by the seed drill method used. For the purpose of fitting the four types of models it was assumed that the within-row spacing was uniform, and was equal to the harvested plot length divided by the number of plants at harvest. One of the plots had a incorrectly recorded value in one row, so this plot was omitted from the analysis, leaving 47 plots in total.

The main interest was to discover the extent to which models are capable of expressing the competition between rows within a plot rather than the weight of the plants. Consequently, the mean competitive effect per plant in each row predicted by the different models was expressed as a ratio to the sum of mean competitive effects of all rows in the plot. This ratio was then compared with the ratio of mean plant weight in a row to the sum of mean plant weights in a plot. For each plot, the sum of the ratios was unity, and hence only differences in plant weight within plots were simulated by the models, and not differences between plots.

Models

For all models, the following criterion of goodness of fit was used.

$$\sum_{i=1}^{i=47} \sum_{j=1}^{j=8} (O_{ij} - f_{ij})^2 \tag{1}$$

where O_{ij} for the jth row in the ith plot is given by the expression

$$O_{ij} = \frac{w_{ij}}{\sum\limits_{j=1}^{j=8} w_{ij}}, \tag{2}$$

and where f_{ij} is given by the expression

$$f_{ij} = \frac{\hat{w}_{ij}}{\sum\limits_{j=1}^{j=8} \hat{w}_{ij}}. \tag{3}$$

w_{ij} is the observed mean root weight in each row of each plot, and $\hat{w}_{ij}$ is the mean of the estimated root weight in each row of each plot. The value of f_{ij} was determined for four types of spacing model.

The "non-overlapping domain" model assumed that the plant weight is determined by the area of the Voronoi polygon. In the present data, this area, A_{ij}, surrounding each plant in row j of plot i, was determined from the between and mean within-row spacing. The calculated area of the Voronoi polygon was modified to allow for an unused compacted strip of soil in the wheelings. Allowing this strip to be 0.3 m wide, approximately the width of a tractor wheel, gave a better fit to the data. The following relationship was fitted

$$f_{ij} = A_{ij} \bigg/ \sum\limits_{j=1}^{j=8} A_{ij} \tag{4}$$

The "overlapping domain" model assumed that there was a zone of influence around each plant from which it draws upon resources. This was taken to be a circle of radius, r, given by

$$r = D.w^{1/3} \tag{5}$$

where w is the mean root weight at harvest and D is a parameter to be estimated. However, if an area is overlapped by the circles of m plants, then each plant receives $1/m$ of the resources in it. To fit this type of model, an array of points was constructed for a representative sample area of plot. For each plant in the sample area, the value of the points within its circle of influence was summed. The value of each point was $1/m$, where m was the number of circles that overlapped it. The estimated weight, $\hat{w}$, was assumed to be proportional to the summated value. The parameter D was estimated by minimizing the sum of squares in Eq. (1).

The "unbounded areas of influence" model assumed that the intensity of competition, C_{ij}, suffered by a representative plant in the jth row of the ith plot, was given by

$$C_{ij} = \sum_{k=1}^{k=n} \frac{1}{d_{ijk}^2},\qquad (6)$$

where there are n other plants in the plot, with the kth plant a distance d_{ijk} away. In practice, only plants within a distance r of the representative plant were included, where r was chosen so that including more plants had little effect on the value of C_{ij}. For the plants in row j of plot i, their estimated weight $\hat{w}_{ij}$, was assumed to be inversely proportional to C_{ij}. Replacing the numerator in Eq. (6) with the observed weight, w_k, gave a poorer fit to the data, and is not considered further in this paper.

The diffuse population effect type of model assumes that the different spatial patterns present in the data have no effect on plant growth, but initial differences in individual plant weights are important in determining the competition for resources. Up to about 50 days after sowing, there were no differences in mean plant weight between the different rows (Benjamin and Sutherland, submitted), so this type of model was not appropriate for these data, and not considered further.

Two modified non-overlapping domain models were also fitted to the data. These models have been developed previously by the authors (Sutherland and Benjamin 1987, Benjamin and Sutherland, submitted) and have features of the overlapping and non-overlapping domain models. In the earlier of these models, the expected weight, $\hat{w}$, was given by

$$\hat{w} = \int\int_A L(c.r^2 + 1)^{-2}.dA\qquad (7)$$

where L is the level of resources at distance r from the plant, $(c.r^2 + 1)^{-2}$ is the efficiency of their use, c is a parameter and A is the Voronoi polygon surrounding each plant. The parmeter c was determined by minimizing the residual sum of squares in Eq. (1), the value of L having no effect on (1).

In the later model, $\hat{w}$ was given by

$$\hat{w} = \int\int_A L\left(c.\left(\frac{x^2}{R} + R.y^2\right) + 1\right)^{-2}.dA,\qquad (8)$$

where x and y are the coordinates of the position of the resources relative to the plant in the direction across and along the rows, respectively. In the former model, the term for the efficiency of use of resources had circular contours around each plant, but in the later version this term has been modified to give elliptical contours. The shape of the ellipse is dependent on R, the ratio of between- to within-row spacing.

An analysis of variance (anova) model was also fitted to the data to give an estimate of the residual variation due to experimental error. The ratio, f_{ij}, for the jth row of the ith plot, as defined in eqn (3), is given by

$$f_{ij} = P_i + T_{f.d.s.j} \tag{9}$$

where P is a term for plot, and T is a term for the different combinations of imposed fertility level (subscript f), target density (subscript d), seeding rate (subscript s), and row position (subscript j). The resulting residual sum of squares had 161 degrees of freedom.

Results

The residual sum of squares, after fitting the different spacing models, varied from 0.1869 to 0.5976, but all values were much greater than the value of only 0.1298 obtained from fitting the anova model (Table 20.1). A formal lack of fit analysis revealed that the lack of fit for Eqs. (5) and (8) was not statistically worse than the analysis of variance model [Eq. (9)] ($P > 0.05$); for the other equations, the lack of fit was significantly worse than for Eq. (9) ($P < 0.001$) (Table 20.2). The greater sum of squares associated with Eqs. (5) and (8) was largely due to greater sum of squares associated with the outer rows. For example, for Eq. (5), the sum of squares associated with the outer rows was 4.2 times greater than that associated with outer rows for Eq. (9), whereas the sum of squares associated with the inner rows for Eq. (5) was actually less than that associated with the inner rows for Eq. (9) (Table 20.1). For Eqs. (4), (6) and (7), the sum of squares associated with both the inner and outer rows were greatly inflated compared with the sum of squares for Eq. (9).

The overlapping domain model gave a low residual sum of squares when fitted to these data, but on the outermost rows (1 and 8) the fitted ratios were systematically less than the observed, especially at high densities (Fig. 20.1). For the inner rows (2 to 7 inclusive) the agreement between observed and fitted for this model was generally excellent (Fig. 20.1). The non-overlapping domain model gave the greatest residual sum of squares (Table 20.1), with a wide scatter of points around the diagonal lines in all rows (Fig. 20.2). In general, the fitted ratios were too low in the inner rows and too high in the outer rows. All points would lie on the diagonal lines if the fit was perfect.

Table 20.1. Summary of different models fitted to the data

Equation	Model Type	Estimated Parameter Values	Residual Degrees of Freedom	Residual Sum of Squares for the Outer Rows	Residual Sum of Squares for the Inner Rows	Total Residual Sum of Squares
4	Non-overlapping Domain	—	329	0.2220	0.3756	0.5976
5	Overlapping Domain	$D = 0.04 \text{ g}^{-1/3} \text{ m}$	328	0.1039	0.0930	0.1969
6	Unbounded areas of influence	$r = 0.26$ m	328	0.1420	0.2242	0.3662
7	Modified non-overlapping domain	$c = 2300 \text{ m}^{-2}$	328	0.1740	0.1700	0.3440
8	Modified non-overlapping domain	$c = 700 \text{ m}^{-2}$	328	0.0616	0.1253	0.1869
9	Analysis of Variance		161	0.0246	0.1052	0.1298

Table 20.2. The relative performance of the models

Equation	Sum of Squares due to Lack of Fit	Degrees of Freedom due to Lack of Fit	Mean Square	Variance Ratio
4	0.4678	168	0.002783	3.452***
5	0.0671	167	0.0004018	0.498[ns]
6	0.2364	167	0.001416	1.756***
7	0.2142	167	0.001283	1.591***
8	0.0571	167	0.0003419	0.424[ns]

Note: Variance ratio of mean squares due to lack of fit and residual mean square from Eq. (9) in Table 20.1.
ns = no significant difference ($P > 0.05$)
*** = significant difference ($P < 0.001$)

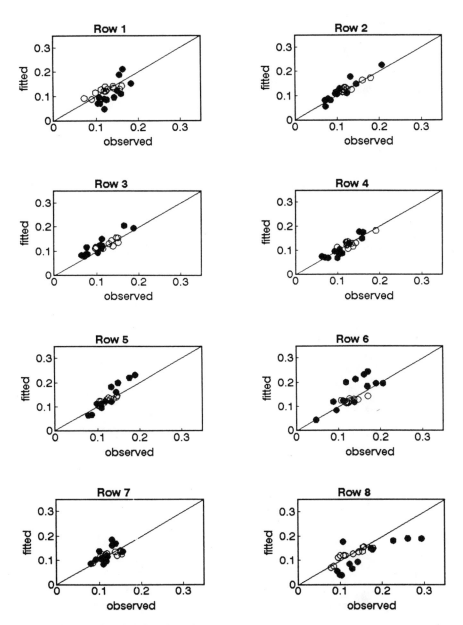

Figure 20.1. Overlapping domain model. Relationship between observed and fitted mean root weights per row expressed as a proportion of the observed or fitted mean weights summed over a plot. Diagonals are the lines of perfect agreement, in which observed = fitted. 100 plants/m² ○; 800 plants/m² ●. (200 and 400 plants/m² excluded for clarity).

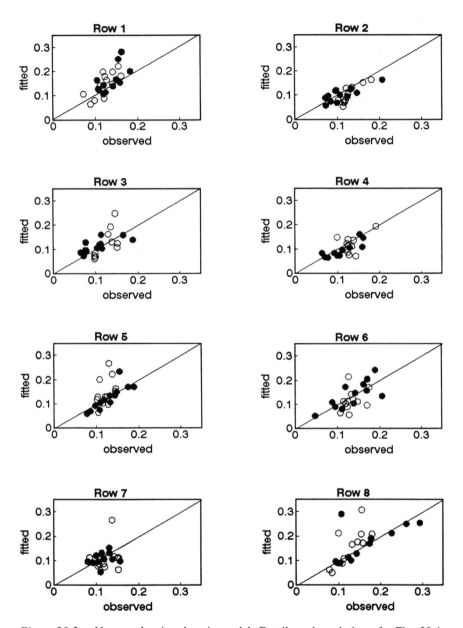

Figure 20.2. Non-overlapping domain model. Details and symbols as for Fig. 20.1.

The first modification to the nonoverlapping domain model by Sutherland and Benjamin (1987) gave a better fit to the data than the unmodified model (Table 20.1), but there was an underprediction of ratios in the outer rows (Fig. 20.3). The later modification by Benjamin and Sutherland (submitted) corrected this systematic deviation, but on row 1 the correction tended to be too great, and on row 8 the correction was not great enough, especially at high density (Fig. 20.4). Rows 1 and 8 are spatially equivalent, and the presence of a difference between them indicates an extraneous source of variation in these data.

The unbounded areas of influence model gave a poor fit to the data. The scatter of points around the diagonal lines was large for the inner rows (Fig. 20.5), and in the outer rows the fitted weights were under-predicted at high target densities and over-predicted at low target densities.

On row 8 in the high density plots there is one point with a much higher fitted value than would be expected from the other points for that density (Figs. 1 through 5). Elimination of all the points for this plot reduced the residual sum of squares by about 7 percent for all models, but did not influence the ranking of the sums of squares (data not shown).

The overlapping domain model [Eq. (5)] and the non-overlapping domain model modified by Benjamin and Sutherland (submitted) [Eq. (8)], gave similar residual sums of squares, but are based on contrasting sets of assumptions. Plotting the fitted values for these two equations against one another (Fig. 20.6) reveals that on the outer rows, Eq. (8) predicts larger weights than Eq. (5). Also, on all row types Eq. (5) tends to predict larger weights than Eq. (8) when the weights are large, but the opposite occurs when the weights are small.

Discussion

Earlier comparisons of competition using data from trees has shown that no one model is consistently superior to others (Opie 1968, Adlard 1973, Daniels 1976, Noone and Bell 1980). Our study has identified a possible cause of these inconsistent results. In our data, an overlapping domain model gave the best fit to data for plants in the inner rows, that is, for plants surrounded by others, but this model gave a poorer fit than the modified nonoverlapping domain models for plants in the outer rows.

The overlapping domain type of model appears to be based on the most accurate assumptions, but the shape of the domain can be influenced by the spatial arrangement of neighbors (Wixley, 1983). Models which had assumed that the efficiency of resource assimilation had circular contours whose center is at the location of the plant (overlapping domain and non-overlapping domain model modified according to Sutherland and Benjamin 1987) gave poorer fits to data from the outer rows than the non-overlapping domain

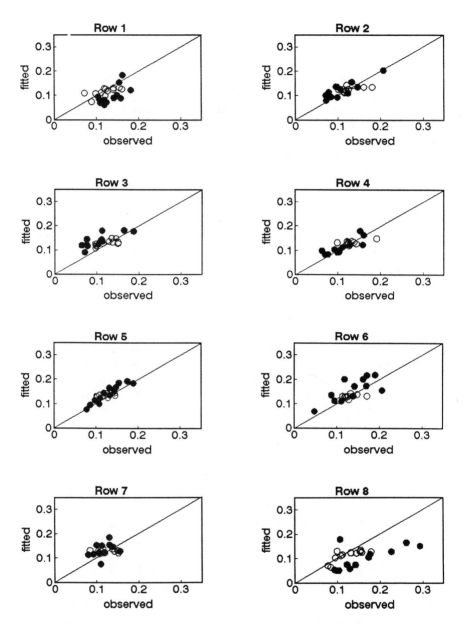

Figure 20.3. Non-overlapping domain model, modified by Sutherland and Benjamin 1987. Details and symbols as for Fig. 20.1.

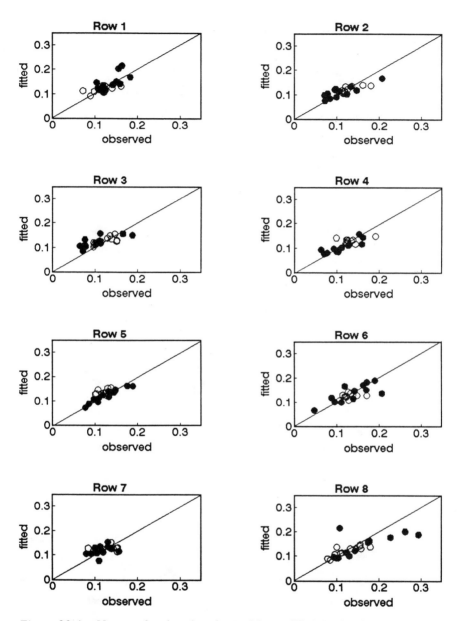

Figure 20.4. Non-overlapping domain model, modified by Benjamin and Sutherland (submitted). Details and symbols as for Fig. 20.1.

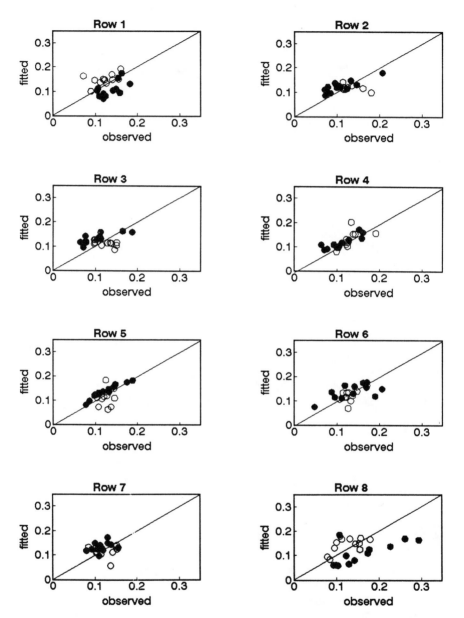

Figure 20.5. Unbounded areas of influence model. Details and symbols as for Fig. 20.1.

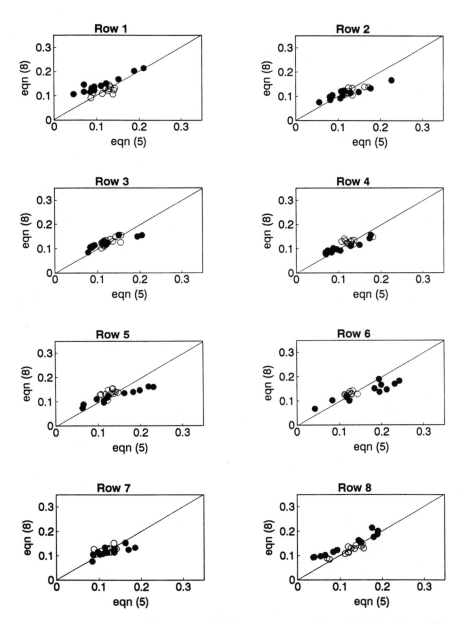

Figure 20.6. Relationship between fitted mean root weights for Eqs. (5) and (8) per row expressed as a proportion of the fitted mean weights summed over a plot. Symbols as for Fig. 20.1.

model modified according to Benjamin and Sutherland (submitted), which allowed the contours of efficiency of resource use to vary with the spatial pattern of neighbors.

Roots will proliferate in zones of greater mineral nutrient supply (Drew et al. 1973) and leaves are orientated to minimize self-shading (Fukai and Loomis 1976). These morphological responses to spatial variation in resource abundance give an explanation for the better fit of those models which allow shape of the contours of efficiency of resource use to depend on the presence of neighbors. However, the poor fit of the unmodified non-overlapping domain model is surprising given the observed plasticity of morphology in response to spatial variation in resource supply. This difficulty can be resolved if roots and/or shoots of adjacent plants are assumed to intermingle, rather than be confined by their Voronoi polygon. This would allow plants in the inner rows to exploit some of the abundance of resources beyond the edge of the crop, giving an explanation for the tendency of the unmodified non-overlapping model to under-predict weights in the inner rows.

The poor fit of the unbounded areas of influence model has three possible explanations. First, the form of the relationship between competitive effect and distance between neighbors may be inaccurate; second, if this relationship is wrong then competitive effects might be erroneously included or excluded from distant neighbors; third, the competition function takes no account of the spatial pattern of neighbors.

The much better fit to the data of the anova model, compared with the spacing models, suggests that there is great scope for the improvement in spacing models, and that alternative types of model should be sought. The better performance of the anova model might be because this form of model can accommodate extraneous sources of variation, independent of the imposed spatial arrangement treatments. Some extraneous sources of variation were present in the data because there were differences between spatially equivalent outer rows in the agreement between fitted and observed values, possibly due to effects arising during sowing.

This source of variation probably explains only a small proportion of the difference in fit between the anova and spacing models. For a given within-row spacing, the mean weight of roots in either rows 1 or 8 is greater than that in any of the inner rows, and differences between outer rows are much smaller than the difference between inner and outer rows (Benjamin and Sutherland, submitted).

This paper has illustrated some of the weaknesses in the types of model that have been developed to describe the response of plants to the spatial pattern of neighbors. The future development of more accurate models will depend on the thorough examination of the performance of different types of model to data sets with wide ranges of spatial patterns. In addition, the implicit assumption in the literature, that similar models can be applied to species varying greatly in their morphology, needs to be examined in great detail.

Acknowledgements

We thank colleagues at HRI for their comments in preparation of this paper. This work was funded by the UK Ministry of Agriculture, Fisheries and Food under Project FO2 & 3B.

Literature Cited

Adlard, P. G. 1973. *Development of an Empirical Competition Model for Individual Trees Within a Stand.* Proceedings of IUFRO meeting—mensuration of growth and yield. Nancy, France 1973.

Benjamin, L. R. and R. C. Hardwick. 1986. Sources of variation and measures of variability in even-aged stands of plants. *Ann. Bot.* **58:**757–778.

Benjamin, L. R. and R. A. Sutherland. Control of mean root weight in carrots (*Daucus carota,* L) by varying within- and between-row spacing. *J. Agric. Sci. Camb.* (submitted).

Daniels, R. F. 1976. Simple competition indices and their correlation with annual lobolly pine growth. *For. Sci.* **22:**454–456.

Drew, M. C., L. R. Saker, and T. W. Ashley. 1973. Nutrient supply and the growth of the seminal root system in barley. I. The effect of nitrate concentration in the growth of axes and laterals. *J. Exp. Bot.* **24:**1189–1202.

Fukai, S. and R. S. Loomis. 1976. Leaf display and light environments in row-planted cotton communities. *Agric. Meteorol.* **17:**353–379.

Matlack, G. R. and J. L. Harper. 1986. Spatial distribution and the performance of individual plants in a natural population of Silene dioica. *Oecologia* **70:**121–127.

Noone, C. S. and J. F. Bell. 1980. *An Evaluation of Eight Inter-tree Competition Indices.* Research Note No. 66. Oregan State University Forest Research Laboratory, Corvallis.

Opie, J. E. 1968. Predictability of individual tree growth using various definitions of competing basal area. *For. Sci.* **14:**314–323.

Ross, M. A. and J. L. Harper. 1972. Occupation of biological space during seedling establishment. *J. Ecol.* **60:**77–88.

Silander, J. A. and S. W. Pacala. 1985. Neighborhood predictors of plant performance. *Oecologia* **66:**256–263.

Sutherland, R. A. and L. R. Benjamin. 1987. A new model relating crop yield and plant arrangement. *Ann. Bot.* **59:**399–411.

Wixley, R. A. J. 1983. An elliptical zone of influence model for uneven-aged row crops. *Ann. Bot.* **51:**77–84.

21

Modeling of Growth and Competition in Plant Monocultures

David P. Aikman

ABSTRACT. Aikman and Watkinson (1980) proposed a theoretical model for the dynamic effect of competition on the time-courses of growth and survival of plants within an even-aged population. This early example of an individual based model achieved a link between effects at the level of the individual plant and those observable at the population level: e.g., computer simulations could generate the minus 3/2 power law limitation between the logarithms of mean mass of surviving plants and their number per unit area observed in some communities of plants. The specific relations in the original model to not give good fits to some sets of data and, inappropriately, give a positive stimulus to the simulated growth of larger plants from an addition of extra small plants. The structure of the model, however, is shown to provide an effective basis for a more empirical application to a multi-aged set of data. Data from even-aged carrot plants, grown at low density, are used to compare the fit from 7 plant growth relations, a Richards model giving the best fit. The data set is then expanded to include higher densities, and 8 options for describing competition between evenly sized plants were compared: a competition function related to the efficiency of use of Leaf Area Index proved effective. Expanding the data set to include multi-aged populations, fits are obtained for uneven competition relations.

Introduction

The competitive interactions between plants during the growth of a population can be modeled from an individual basis. Aikman and Watkinson (1980) proposed a novel theoretical model in which the assumptions were that there was a dynamic growth process whose rate was dependent on the size of the plant, diminishing as a plant matured towards an upper size limit, that there was variability between the individual plants in a population, and that competition (particularly for the smaller plants in a dense population), reduced the growth rates, possibly resulting in mortality. It was an early example of an individual based biological model. The original model was successful in achieving some objectives, but the specific functions in it do not necessarily give a satisfactory fit to experimental data. In this paper, an applied model illustrates the manner in which the structure of the original model may be used to select a set of relations that give a better fit to a set of experimental data. The original Aikman-Watkinson model made the *a priori* assumption of a Logistic growth relation for the isolated plant, modified in that the initial growth rate was taken as proportional to the area or zone of influence of the plant, i.e., to the two thirds power of the volume or mass, since the intrinsic productivity of the plant was assumed to be proportional to a zone area of canopy or root. The power of the negative term of the logistic growth equation was not modified. The model assumed that there was variability between plants. For the isolated plant,

$$\frac{dm_i}{dt} = a_i s_i - bm_i^2, \tag{1}$$

with $m_i = m_{i,0}$ at $t = 0$, where m_i is the mass of the ith plant, s_i is an estimate of the zone area of the plant, and a_i and $m_{i,0}$ are drawn from independent Normal Distributions, but, for simplicity, b was not varied between plants.

Assuming that plants retain the same shape as they grow, then

$$s_i = km_i^{2/3} \tag{2}$$

where k is the zone area for a plant of unit mass.

In a population, the original model assumed that competition reduced the efficiency, $f(s_i)$, with which the ith plant exploited its associated zone and Eq. (1) was modified to give

$$\frac{dm_i}{dt} = a_i s_i f(s_i) - b_i m_i^2 \tag{3}$$

The function $f(s_i)$ was taken as 1 for a plot at low occupancy but reducing as the sum of the zone areas of the n plants, $\Sigma_n s_j$, tended to the total ground area available, S. It was assumed that plants of below average size, $\bar{s}$, would have a greater reduction in the efficiency factor than those of above average size. Giving mathematical expression to these assumptions, the original model used is

$$f(s_i) = \left\{ 1 + \left[\frac{\sum\limits_n s_j}{S} \right]^{\varphi_1} \left[\frac{\bar{s}}{s_i} \right]^{\varphi_2} \right\}^{-1} \tag{4}$$

The parameters φ_1 and φ_2, both positive, control, respectively, the severity of the area constraint upon competition and the relative advantage or disadvantage of plants differing in size from the average.

For mathematical simplicity, the individual plant was assumed to die if dm_i/dt becomes negative, due to competition reducing its growth capability below that necessary for maintaining its biomass. In reality, of course, there would be a lag while reserves were depleted (Mahmoud and Grime 1974, Britton 1982).

With a high value of φ_2, competition will tend to amplify differences in size between individuals and, as a high density community nears the constraint on area, will cause the smaller plants to be progressively eliminated with little impairment to the growth of the largest plants towards their upper size limit. Conversely, with a low value of φ_2, there is a tendency for competition to reduce differences between plants and, as a high density community nears the constraint on production, for a uniformly-sized community to cease growth. Through competition, the model provides what may be termed a distribution-modifying function (Westoby, 1982).

Values for $m_{i,0}$ and a_i were drawn from their Normal Distributions for each of the n plants interacting in a simulation and their subsequent growth and competition was computed. The equations are simple and the computer can readily handle the ensemble, but such an approach to modeling would probably not have been chosen prior to the accessibility of the computer. The objective was to provide theoretical background for the dynamics of plant competition. The model proved successful in linking the processes assumed at the individual plant level with effects observable at the population level. Depending on the value chosen for φ_2, the model was able to develop a bimodal size distribution and mortality, such as was observed with marigold (*Tagetes patula* L. var. "Naughty Marietta") by Ford (1975), or to reduce dispersion as was found for a grass (*Festuca paradoxa* Desv.) by Rabinowitz (1979). Predictably, given its allometric assumptions, it exhibited the $-3/2$ constraint between the logarithms of plant mass and number of surviving plants *per* unit area noted for a number of different plants (White and Harper, 1970). It provided a fit to the growth of a stand of loblolly pines (Harms, 1981).

For simplicity, Eq. (4) assumed that the individual plant is influenced by the general community, rather than its specific immediate neighbors. This has the advantage of speed in computation, and would assume a degree of plasticity in the growth of plants in the population in which the growth form of the neighboring plants is influenced by their neighbors, etc. However it might be more accurate to use the more localized effects of neighbors. For example, Wixley and Shaw (1981), using data from field tobacco grown in rows with vacancies, fitted the yield per individual plant to estimates of the locally available areas. With greater competition, such localized effects could result in space-related dynamics of self-thinning (Hughes, 1988).

A more serious criticism of Eq. (4) is that an increase in the number of very small plants would reduce the value of $\bar{s}$, and thus, in a simulation, unrealistically increase the value $f(s_i)$ for a larger than average plant, unless the values of the competition parameters are changed at the same time. Thus the model would not be sufficiently robust to extend to cover multi-aged populations. Indeed the potential accuracy for populations with disparate distributions of plant size could be limited even for even-aged situations.

Benjamin (1988) found that this model did not give such good fits to some sets of multi-aged crop data as those obtained by a different modification of the Logistic equation:

$$\frac{dm_i}{dt} = am_i(1 - bm_i)\left(1 - \frac{y}{Y}\right)\{1 + \ln(m_i/\bar{m})Kn\} \tag{5}$$

where $y = \Sigma_n m_j$, n being the number of plants per unit area, Y is the limit on the yield per unit area, and $\bar{m} = y/n$, the average plant mass.

This model differs from the Aikman-Watkinson model in that i) the intrinsic growth term of the logistic is not changed to a 2/3 power, ii) the competition effect is applied to the whole logistic differential equation rather than the intrinsic growth term only, and iii) a different formulation is used for the occupancy or yield constraint and its relative effect on plants of differing size.

With the competition factor applied to the positive term only, as in the Aikman-Watkinson model, mortality follows when the smaller plants are unable to derive sufficient benefit to maintain their biomass from their zones of influence due to the small value of the efficiency factor. Application of a competition expression to the net growth relation, as in the Benjamin formulation, would lead to such mortality only if the competition expression itself becomes negative, but no self-thinning occurred in the trials analysed by Benjamin. In a theoretical interpretation of the Logistic equation, the second term with its negative sign may be considered as reflecting a catabolic process which partially offsets an anabolic growth rate of the first term, and hence the original model applied the competition relation to the first term only. Alternatively, the reduction in relative growth rate of the isolated plant may be considered as reflecting limitations on the growth process within the individual plant through some internal or intrinsic efficiency factor which decreases with increasing mass. It is possible that the external limitation acts in a similar manner, as a factor on the net growth rate. Hence, in the application of an empirical model to data, it is reasonable to examine either option.

Although it gave improved fits, the Benjamin model has similarities to the Aikman-Watkinson model, and suffers from the same criticism as above, namely, an inappropriate stimulation to growth of large plants from an increase in the number of plants of small size. It also relies on a general competition effect rather than on neighborhood effects. The modeling should be improved by removing the inappropriate behavior of the competition expression and, perhaps, by using detailed neighborhood effects.

Theory

Conceptually, the structure of the models may be considered as consisting of three components: a relation describing the growth of isolated plants, a relation giving an effect of competition between equal sized plants, and a relation modifying its effects over plants of differing size. It will be seen that this framework permits a useful approach in the application that follows.

Modeling the Growth of an Isolated Plant

A number of functions have been found to fit data on the growth rates of plants: five examples are given below (Hunt, 1982). In each a differential equation, relating mass m and time t, is expressed in terms of some function of m.

   i)   Monomolecular or Exponential Saturation:

$$\frac{dm}{dt} = a(1 - bm)$$

   ii)   Logistic or Lotka-Volterra:

$$\frac{dm}{dt} = am(1 - bm)$$

   iii)   Richards' growth relation:

$$\frac{dm}{dt} = am\{1 - (bm)^c\}$$

   iv)   Gompertz growth relation:

$$\frac{dm}{dt} = am\{1 - b\ln(m)\}$$

v) Weibull growth relation:

$$\frac{dm}{dt} = ac(1 - bm)\{-\ln(1 - bm)\}^{(1-1/c)}$$

The common features are that the models each define some specific time course between a starting value, m_0, and an asymptotic final value, m_∞, at which the net growth rate is zero. For models ii) and iii) the growth rate at small m is approximately proportional to the value of m resulting in an early exponential phase. Most of these relations were originally applied to the growth of animal organs or populations. Their development or their application to plant growth is usually empirical (e.g., Richards 1959).

The original model assumed a 2/3 power for the first term, and so the list is extended with the following:

vi) modified Richards growth equation:

$$\frac{dm}{dt} = am^d\{1 - (bm)^c\}$$

vii) modified Gompertz growth equation:

$$\frac{dm}{dt} = am^c\{1 - b\ln(m)\}$$

Modeling Competition Among Even-Sized Plants

Competition results in a complex set of effects in the root and shoot zones. The models of Diggle (1976) and Gates (1978) use nearest neighbor effects but are not dynamic. Nearest neighbor effects cannot be the whole story, as can be seen from edge effects; for example, in crops grown in rows, the end plants typically show the highest growth rates. According to the nearest neighbor effect, the adjacent plants should be the smallest since they have the largest neighbors, but in plots of trees there is a positive benefit from the edge to the second tree and at least as far as the third (Zavitkovski 1981, Meldahl *et al.* 1984). In plots of cereal crops, there is a positive edge effect to the second row of plants as well as the first (Hadjichristodoulou 1983, Wright *et al.* 1986). Thus one should consider using a function with an effect over a longer range than just nearest neighbor, and the factor $f(s_i)$ could be modified. As an example, one could use an empirical function such as

$$f(s_i) = \left[1 + \left\{ \sum_{j \neq i} s_j/r_{ij}^2 \right\}^{\varphi_1} \right]^{-1}, \tag{6}$$

where r_{ij} is the distance between plant i and plant j.

Weiner (1982) used such an expression, as did Britton (1982), who combined it with a model of plant growth by Thornley (1977). This function does not result in an unrealistic effect on the large plants from the addition of very small ones. Competition is controlled by the local density rather than the general community. However it only gives the severity of space pressure as a parameter to be fitted, and does not allow for differential effects on plants of differing size.

There is no obvious simple theoretical equation to represent such effects. One option is an empirical competition relation. Since the growth equations are based upon mass, one could modify Eq. (6) for mathematical convenience and to speed computation:

$$f(m_i) = (1 + x_i^{\varphi_1})^{-1} \quad \text{where} \quad x_i = \sum_{j \neq i} \varphi_2(m_j/r_{ij}^3) \tag{7}$$

φ_1 and φ_2 are empirical parameters to be estimated by fitting to data.

Functions such as *a*) $(1 - x^{\varphi_1})$, *b*) $(1 - x)^{\varphi_1}$, *c*) $(1 + x^{\varphi_1})^{-1}$, *d*) $(1 + x)^{-\varphi_1}$, *e*) $\exp(-\varphi_1 x)$, and *f*) $\{1 + \ln(1 + x^{\varphi_1})\}^{-1}$ approximate to 1 when x is small and decrease towards zero as $x \rightarrow 1$, in the case of *a*) and *b*), or as $x \rightarrow \infty$ for the others. Such empirical functions could be explored for the best fit for the competition effect applied to a particular set of data.

Alternatively, aiming for a function with some theoretical basis, a uniform canopy of leaf area index L intercepts a fraction $\{1 - \exp(kL)\}$ of the incident light where k is the leaf extinction coefficient (Monsi and Saeki, 1953). Thus the relative efficiency of light interception for photosynthesis and transpiration by leaf area may be estimated using $\{1 - \exp(-kL)\}/kL$. Although this would give an overestimate of the efficiency of light interception prior to the development of a closed canopy, self-shading would be included, even if implicitly, in the modeling of the growth of the isolated plant. Hence, as additions to the list of competition functions, one may consider *g*) $f(m_i) = \{1 - \exp(-x_i^{\varphi_1})\}/x_i^{\varphi_1}$ and *h*) $f(m_i) = [\{1 - \exp(-x_i^{2/3})\}/x_i^{2/3}]^{\varphi_1}$, the latter explicitly allowing for an allometry between area and mass.

Modeling Competition Among Unevenly Sized Plants

Variation in factors such as inter-plant spacing, seed mass, time of emergence, and initial plant growth rate will lead to unevenly sized plants and a modification to the competition relation will be required. To allow for size differential, one may consider using an expression like

$$f(s_i) = \left[1 + 2 \sum_{j \neq i} \{(s_j/r_{ij}^2)^{\varphi_1} s_j^{\varphi_2}/(s_j^{\varphi_2} + s_i^{\varphi_2})\} \right]^{-1} \tag{8}$$

where the parameter φ_2 allows some control over the asymmetrical effect of competition between plants of differing size. The factor 2 ensures that the expression in Eq. (8) is consistent with Eq. (6) for equally sized competing plants.

A suggested modification to Eq. (7) is:

$$x_i = \varphi_2 \sum_{j \neq i} \{(m_j/r_{ij}^3)2m_j^{\varphi_3}/(m_j^{\varphi_3} + m_i^{\varphi_3})\} \tag{9}$$

Neighborhood relations may involve greater detail in programming, can be slower in fitting, and involve more experimental recording. A diffuse competition relation may have some advantages, and the following is one suggestion:

$$x_i = \varphi_2 \sum_{j} m_j 2 \sum_{+} m_j^{\varphi_3} \Big/ \left(\sum_{+} m_j^{\varphi_3} + \sum_{-} m_j^{\varphi_3} \right) \tag{10}$$

In this equation, a yield constraint is based on a comparison of the summation of plant mass over unit area, $\Sigma_j m_j$, with an upper constraint associated with the reciprocal of φ_2, and modified for a plant of mass m_i by a relative size term in which Σ_- is the sum over the smaller plants, $m_j < m_i$, and Σ_+ is the sum over the larger plants, $m_j > m_i$. For classified data, the class to which the plant belongs could be split equally between the upper and lower sums.

Example of Application

The computing time required for optimizing the fit of a model and evaluating its parameters increases disproportionately with the number of parameters. Fortunately, if the structure proposed for the model reflects that present in experimental data, then a particularly efficient strategy is to use subsets of the data to make selections of the mathematical relationships and to obtain initial estimates of the associated subsets of the parameters. The strategy will be demonstrated by its application to a set of data from a multi-aged monoculture of carrot (Benjamin 1988, experiment 2). The plants were grown in 3 replicate blocks at each of 4 densities of plants from 3 sowing dates grown in each of 4 combinations: i) even-aged, or paired with ii) the later or iii) the earlier of the other two sowings, or iv) combined with both other sowings. The mean dry masses of six plants from each of seven harvests were recorded to give $3 \times 4 \times 3 \times 4 \times 7 = 1008$ observed results (less one missing value), or 336 means over blocks. Some details are given in Table 21.1.

To provide an estimate of the precision within the data set, a factorial

Table 21.1. Detail of the interplant spacing (square lattice), sowing dates (relative to 18 June), their percent emergence, and harvest dates for experimental observations of dry masses of carrot plants (Benjamin 1988, experiment 2).

Plant spacings (4):	5.0, 7.5, 10.0, 15.0 cm
Sowing dates (3):	0, 14, 21 days
Emergence %:	70, 59, 96
Harvest dates (7):	27, 41, 50, 62, 70, 83, 97 days

Anovar on the natural logarithms of the masses from the blocks, spacings, combinations, and harvests was performed for each of the sowings. Pooling the third order sums of squares, other than that for the blocks × spacing × combinations, with the fourth and fifth orders gave a root mean square (RMS) of 0.35 for the error term for the fit of the grand mean and 131 other lower order terms. For data averaged over the 3 blocks, 84 statistical terms would be required, and the comparable RMS would be 0.20. Modeling should aim to get close to such a RMS, as this is an estimate of the precision in the data set, but should, hopefully, require many fewer parameters to fit the trends in the data.

Fitting the Growth of an Isolated Plant

Initially, data from the even-aged plots of lowest density were used to fit each of the suggested growth models, optimizing the estimates of parameters by minimizing the residual sum of squares (RSS) of the logarithms of masses, and the models compared using the associated RMS values after allowing for the number of parameters fitted. There was little difference between the plants grown with even-aged neighbors and those grown, at the same density, with neighbors from later sowings, even for the longest growth times. Thus there does not appear to be much effect from competition at 15 cm spacing. As expected from the initial linear trend of the points in Figure 21.1, the Logistic, Richards, and Gompertz equations had lower RMS values than the Monomolecular or Weibull functions. There was an improvement when plants in the first sowing were given a separate value for initial mass, m_0, and the growth constraint, b. The Richards function was the best, fitting well to all but the first harvest of the first sowing, as shown by the lines in Figure 21.1. There was little improvement to the RMS value when the modified form vi) was used.

Fitting Competition Among Even-Sized Plants

The Richards plant growth relation, with the initial estimates of the 6 associated parameters, was combined with the competition models for the next stage of the investigation. The data set consisted of the means over the

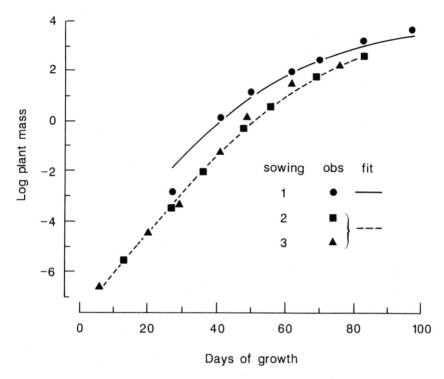

Figure 21.1. The fit of Richards Growth Relations to observed growth of plant mass (mg) from even-aged plots of carrot at low density for 3 sowings. A separate fit is required for sowing 1, excluding its first harvest, but data from sowings 2 and 3 are combined to fit a common predicted curve.

blocks from the even-aged plots for all densities, sowings, and harvests, but omitted the means from the first harvest of the first sowing, i.e. a total of $4 \times 3 \times 7 - 4 = 80$ values. For an evenly spaced and even-aged monoculture, there is no detailed neighborhood information, and, hence, $\varphi_2 m_i / r^3$ can be used for x_i in the competition relations.

The two parameters of the even-sized competition relations were estimated by minimizing the RSS. To check whether the competition term should modify the whole plant growth differential equation or just the first (positive) term, the eight competition relations were optimized when applied to the positive growth term and to net growth rate. For the carrot data, four relations gave a lower RMS when applied to the positive growth and four to the net growth. As a possible compromise, separate factors of type c) were applied to the two terms of the plant growth equation and the 4 parameters fitted, but there was only a small improvement to the RMS.

The best fits were obtained with expressions d) applied to the net growth, and h) applied to the positive term or the net growth. Each of these three relationships for even competition was then refitted, this time including the Richards plant growth parameters in the set to be optimized. Expression d) gave the best fit with an RMS of 0.199, but h), on net growth, was almost as good (RMS = 0.200), and its theoretical basis could make h) the preferred choice.

The reduced emergence of plants in sowings 1 and 2 meant that there were missing plants. The experimenter had used appropriate plants from the guard areas to make up the complete set. However, competition upon the plants present will be altered by the absence of some neighbors. To make some allowance for this effect, the competition terms were multiplied by the emergence proportions for the corresponding sowings. This economizes on computing compared with calculating the weighted average for all possible combinations of neighboring site occupancy. If the relevant details had been recorded for each plant, the growth of each could have been calculated separately, although the preceding method would be of value for the initial selection from the competition relations. The fit of d) was not as good when this allowance for nonemergence was omitted.

The fits obtained by the Richards function modified by the LAI-related function h), with allowance for non-emergence, are presented in Fig. 21.2a, b, and c. There is little effect from competition to the early growth, and this phase is omitted from the figure for clarity.

The effect of competition from plant mass is expressed relative to the cube of distance. Relaxing this constraint by fitting a third parameter as the power of r achieved no reduction to RMS with d), and the fitted value for the extra parameter scarcely differed from 3. This result supports the allometry assumed between m and r in Eq. (7). Fitting a separate value of φ_2 to the first sowing, as compared with a common one to the two later sowings, did not give any significant improvement to the RMS.

Fitting Competition Among Unevenly Sized Plants

For the mixed-age plots of this experiment, unlike the even-aged plots, there is sufficient information for fitting a neighborhood competition model. For such a model, an assessment is required of the number of neighbors or distance to be included in the calculated effect. The layout of the experiment was rectangular, and there were combinations of two sowings and of all three sowings. In the latter, a plant had two nearest neighbors from each of the other two sowings, and, as the four next nearest, one from each of the other two sowings and two from the same sowing: plants in combination with one other sowing had four nearest neighbors from the other sowing, but four from their same sowing at the next nearest distance. From the data for these combined sowings at the high densities, plants from the first sow-

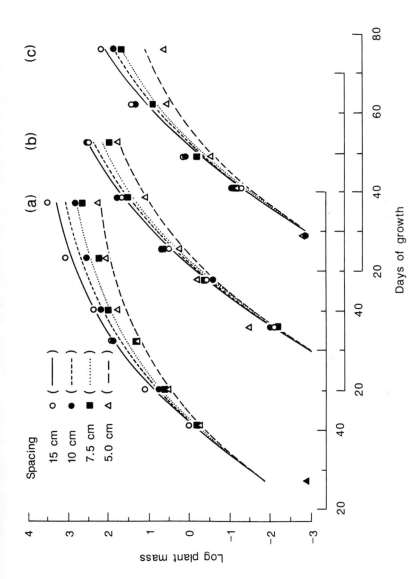

Figure 21.2. The fit of Richards Growth Relations modified by a competition function to observed growth of plant mass (mg) from even-aged plots of carrot from 3 sowings and 4 different spacings with fitted curves (a) for sowing 1, (b) for sowing 2, and (c) for sowing 3.

ing, when combined with the other two sowings, apparently derived benefit from the reduced number of other large plants located diagonally [see data represented by triangles in Fig. 3(a–d)]. Hence, in the neighborhood models, the effects from the 4 diagonally nearest plants were included along with those from the 4 nearest neighbors. The value of φ_2, estimated previously, needs adjusting to allow for the 4 neighbors at distance r and the 4 at $\sqrt{2}r$. With this adjustment, the estimates of the eight terms of the combined functions for plant growth plus equal competition can be kept fixed, and the data from all densities, sowings, combinations, and harvests, except those of harvest 1 from sowing 1 (a total of 336 values less 16), can be used to get an initial estimate for the uneven competition parameter of the asymmetric competition model given in Eq. (9); the RMS was 0.356. Refitting the three competition parameters improved the RMS to 0.322, and improving the fit over the complete set of 9 parameters reduced the RMS to 0.320.

Fitting the three parameters of the diffuse competition model (10) gave a RMS of 0.335, improved to 0.306 with adjustment on all nine parameters. The fitted lines are shown in Fig. 3 together with most of the data, although some early harvests are omitted for clarity.

Discussion

The residual mean square error for the fit of the Richards growth relations with the 6 parameters to the low density data from even-aged plants, averaged over plots, was only marginally greater than that which would be expected from fitting an Analysis of Variance with 20 parameters to the data. The procedure of fitting the growth models showed that plants from the first sowing tended towards a higher mass limit than do those from the two later sowings, and that they received a boost in their initial mass gain. Indeed, the increase to their specific growth rate persisted beyond the time of the first harvest, apparently associated with a previous increase in mineral uptake. Benjamin (1988) noted that the sterilization treatment of the soil could have transiently raised its fertility and, hence, affected the growth of plants in the first sowing. Thus the growth modeling gives a good fit with a much smaller number of parameters and gives some indication of processes underlying the effects observed.

The addition of expressions for competition generated good fits to the data for even-aged plants. These added only two extra parameters, although the size of the data set was increased by the results from the 3 higher densities over the 3 sowings and the 7 harvests. The two relations derived from considering leaf area efficiency gave fits comparable to the range of those obtained from the more empirical relations. The limit to the yield, as reflected by φ_2, from the combinations that included sowing 1 did not appear to differ from those which excluded it. The extra fertility resource, which stimulated

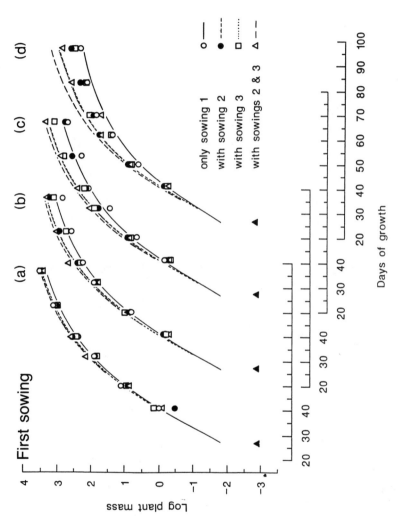

Figure 21.3. The fit of Richards Growth Relations modified by a competition function with asymmetric size dependence for mixed-age combinations to the observed plant mass (mg) from 3 sowings of carrot at 4 spacings: (a), (e), and (i) at 15 cm; (b), (f), and (j) at 10 cm; (c), (g), and (k) at 7.5 cm; and (d), (h) and (l) at 5 cm.

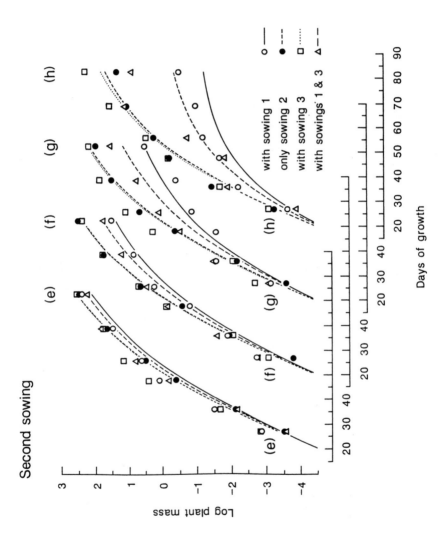

Figure 21.3 (Continued)

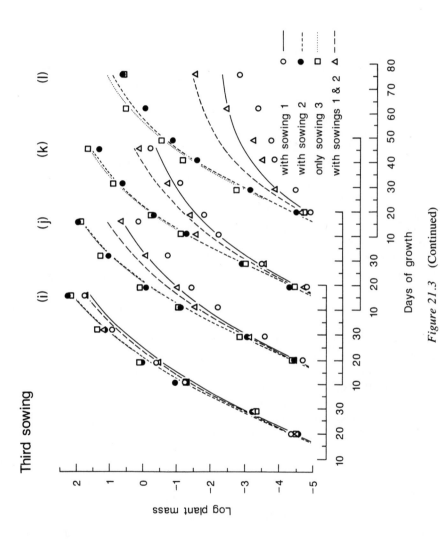

Figure 21.3 (Continued)

the growth rate for the isolated plants of the first sowing, is not apparently reflected in a higher bound to the plot yield under competition.

The extension to competition between multi-aged plants did not produce such good fits when judged against those obtainable by a complete Analysis of Variance. However, the model only used one additional parameter for the uneven competition, making a total of 9, whereas, even ignoring blocks and the first harvest from sowing 1, the Analysis of Variance would require 80 statistical terms. Moreover, the fits obtained compared favorably with those achieved by others who have modeled the same data set. The fit obtained to this set of data from the diffuse competition model was better than that obtained from the more detailed neighborhood model.

The effect of competition in this set of data is asymmetric. The growth of plants from a later sowing is strongly affected by plants from an earlier sowing, more than from the same sowing. On the other hand, the growth of plants from the first sowing in even-aged plots at 15 cm spacing (open circles in Fig. 21.3a) is very similar to that from the first sowing grown in combination with either the second or third sowing at 10 cm spacing (closed circle or open square in Fig. 21.3b): apparently, the major effect is from the similarly large plants located diagonally at $10\sqrt{2}$ cm rather than the nearest neighbor smaller plants. The same relation is true for the comparisons of the first sowing plants at 10 and 7.5 cm even-aged spacing, (Figs. 21.3b and 21.3c) with those in combinations at 7.5 and 5 cm, respectively (Figs. 21.3c and 21.3d). However, the competition effect is not completely one-sided, as a similar set of comparisons shows that there is some effect of sowing three plants on those of the second sowing.

Further improvements could probably be made in fitting a model to the particular data set. There are some complexities of the experiment that have not been covered in the modeling. Only a first approximation is made for the effect of missing competitors rather than evaluating a weighted average of the possible occupancy combinations in detail: regrettably, the actual neighborhood details were not recorded. At each harvest, the removal of rows of plants left only a single row of plants guarding the next harvest row: no allowance is made for the influence of these less constrained guards on the next sample, or for a possible offsetting benefit from the appearance of the space beyond the new guards, and no allowance has been made for difference in growth of guard plants used to substitute for absent plants. Although experimenters have traditionally sought to minimize or avoid effects from edges and gaps, such situations contain more useful information for competition modeling than do well-guarded uniform plots.

Differences between blocks were not explored. The dates of emergence of the individual plants were not recorded, nor their individual masses at harvest. Hence, it is not possible, with these data, to move from an *i*-state distribution model to an *i*-state configuration model, although, for the particular management application with carrots, one would wish to predict distributions of plant mass at harvest, as the economic crop value depends on the total yield occurring within the most valuable price band of fresh size.

The plants in the experiment did not exhibit self-thinning; the only losses were at germination. The biennial carrot builds up reserves of biomass in the storage root in its first year and, hence, would be protected to some extent from competitive stress. Fitting a model to experiments in which self-thinning did occur would require changes in the fitting criterion. The data would consist of time sequences of the numbers of surviving plants and their mass distributions, together with the numbers of plants dying in successive intervals and their masses at death. Since the model is designed to predict the masses of individual plants, and since the time of plant death has no precise determination in practice, a reasonable approach would be to generate the simulations with the proviso that, once any plant shows a negative growth rate and is deemed to have died, its mass is fixed constant. Thus the model would generate a frequency distribution of predicted masses at each time (or of predicted mass increments) which could be compared with corresponding observations. A fit could be obtained non-parametrically by minimizing the discrepancy between the two cumulative distributions, as in a Kolmogorov-Smirnov 1-sample statistic. Of course, if the plants could be compared individually with their own predicted masses, then minimizing the sum of squares might be preferable.

In the wider perspective, it is demonstrated that the structure of the dynamic competition model can be applied in practice, and that it is effective in providing relatively good fits to experimental data. Although the modeling proceeded on an empirical basis, the data supported some mechanistic interpretations. It was interesting, for example, that a satisfactory fit for interplant competition was obtained from a shading expression, inducing speculation that some of the reduction in relative growth rate of the isolated plant could be associated with self-shading. It should be possible to perform simulations of plant growth and competition on firmer mechanistic bases with an adequate level of detail. These might support the modeling of multi-species communities, picking out intermediate variables whose values could be checked in sets of observations: light interception from direct sunlight and diffuse sky for photosynthesis and transpiration from the soil moisture, acquisition of mineral nutrients, growth and interactions, and so forth. Assuming some validation, sets of predictions could be generated for certain applications by computer simulation for ranges of parameter values, empirical functions suggested, and the dependence of their coefficients on model parameters obtained—the aim being to produce a mechanistic basis for models that could be used statistically, but which would rely on the fitting of a small number of parameters.

Acknowledgements

Mathematical modeling in Biology is most likely to be effective when a modeler with awareness of the biology interacts with a biologist with an appreciation of the mathematics. I wish in particular to thank Dr. Andrew Watkinson of the University of East Anglia, for his collaboration in developing the original model, and Dr. Laurence Benjamin of Horticulture Research International for challenging modelers with his useful set of data.

Literature Cited

Aikman, D. P. and A. R. Watkinson. 1980. A model for growth and self-thinning in even-aged monocultures of plants. *Ann. Bot.* **45**:419–427.

Benjamin, L. R. 1988. A single equation to quantify the hierarchy in plant size induced by competition within monocultures. *Ann. Bot.* **62**:199–214.

Britton, N. F. 1982. A model for suppression and dominance in trees. *J. Theor. Biol.* **97**:691–698.

Diggle, P. J. 1976. A spatial stochastic model of inter-plant competition. *J. Appl. Probab.* **13**:662–671.

Ford, E. D. 1975. Competition and stand structure in some even-aged plant monocultures. J. Ecol. **63**:311–333.

Gates, D. J. 1978. Bimodality in even-aged plant monocultures. *J. Theor. Biol.* **71**:525–540.

Hadjichristodoulou, A. 1983. Edge effects on yield components and other traits in mechanized durum wheat and barley trials. *J. Agric. Sci.* **101**:383–388.

Harms, W. R. 1981. *A Competition Model for Tree and Stand Growth Models.* pp. 179–183, General Technical Report, Southern Forest Experiment Station, USDA Forest Service No. SO-34.

Hughes, G. 1988. Spatial dynamics of self-thinning. *Nature* **336**:521.

Hunt, R. 1982. *Plant Growth Curves: The Functional Approach to Plant Growth Analysis.* Edward Arnold, London.

Meldahl, R. S., E. V. Nordheim, and A. R. Ek. 1984. Alternative methods for determining edge effects on small plot samples. *Can. J. For. Res.* **14**:874–878.

Mahmoud, A. and J. P. Grime. 1974. A comparison of vegetative growth rates in shaded seedlings. *New Phytol.* **73**:1215–1219.

Monsi, M. and T. Saeki. 1953. Uber den Lichtfaktor in den Pflanzengesellschaften und seine Bedeutung für die Stoffproduktion. *Jap. J. Bot.* **14**:605–614.

Rabinowitz, D. 1979. Bimodal distributions of seedling weight in relation to density of *Festuca paradoxa* Desv. *Nature* **277**:297–298.

Richards, F. J. 1959. A flexible growth function for empirical use. *J. Exp. Bot.* **10**:290–300.

Thornley, J. H. M. 1977. Growth, maintenance and respiration: a re-interpretation. *Ann. Bot.* **41:**1191.

Weiner, J. 1982. A neighborhood model of annual-plant interference. *Ecology* **63:**1237–1241.

Westoby, M. 1982. Frequency distributions of plant size during competitive growth of stands: the operation of distribution-modifying functions. *Ann. Bot.* **50:**733–735.

White, J. and J. L. Harper. 1970. Correlated changes in plant size and number in plant populations. *J. Ecol.* **58:**467–485.

Wixley, R. A. J. and M. J. P. Shaw. 1981. A model relating yield to individual plant area for predicting growth compensation in Central African field tobacco. *Ann. Appl. Biol.* **98:**339–346.

Wright, D., S. S. Ali, and L. G. Hughes. 1986. Edge effects in cereal yield trials. *J. Natl. Inst. Agric. Bot.* **17:**179–186.

Zavitkovski, J. 1981. Small plots with unplanted borders can distort data in biomass production studies. *Can. J. Forest Res.* **11:**9–12.

22

Individual Behavior and Pollination Ecology: Implications for the Spread of Sexually Transmitted Plant Diseases

Leslie A. Real, Elizabeth A. Marschall, and Bernadette M. Roche

ABSTRACT. The study of pollinator-plant interactions has tradition-ally taken an individual-based approach. The patterns of movement of individual pollinators, patterns of floral choice, and the evolution of floral traits in individual plants have been major foci for field and laboratory investigations. While considerable research has addressed the *evolutionary* implications of individual behavior in pollinator and plant traits, little attention has been directed at elucidating the *eco-logical* implications of pollinator behavior on plant populations. For example, we still do not possess a clear picture of the relationships between pollinator visitation frequencies, seed production, and re-cruitment of new individuals into the plant population. In many cases, pollinator behavior can act as a strong selective force on floral traits with no appreciable influence on overall plant population dynamics. In some cases, the effects of pollinator behavior on plant population dynamics can be quite direct, as in the transmission of pollinator-borne diseases.

As a model system, we have examined the interaction between the anther smut fungus, *Ustilago violacea*, and its host plant, *Silene alba*. Plants infected with the anther smut are rendered sterile and produce flowers with anther sacs filled with spores. Insect pollinators which land on infected flowers act as vectors of the fungus. We have dem-onstrated experimentally that pollinators exhibit behavioral prefer-ences with respect to disease status of the flowers. In particular, bum-ble bees preferentially visit healthy rather than diseased *Silene* plants. We have used an individual-based modeling approach to evaluate some of the intricacies of this complex plant-vector-pathogen interaction. We ultimately hope to understand how patterns of vector movement and transfer of spores may account for the spatial distribution and tem-poral spread of another smut disease within the plant population.

Thus far, we have constructed simple computer simulation models of individual vector movement among a patch of healthy and diseased plants with a spatial structure identical to a particular natural population. Our analysis of spatial structure in this natural population, using Morisita's index of dispersion and the parameter k of the negative binomial, demonstrates that clumping is greater in the diseased subset of the population than in the total population of healthy and diseased plants. In our simulations, each plant has just one flower. During each time interval in the simulation there is a probability α that a bee will enter the patch. Each simulation begins with a disease-carrying bumble bee entering the patch of flowers, at which point the bee has equal probability of landing on any particular flower. Further movements of the bee differ between simulations of nondiscriminating vectors, which move randomly with respect to disease status of flowers encountered, and those of discriminating vectors, which visit diseased flowers at a lower probability than healthy flowers. Results indicate that a discriminating vector can spread disease as rapidly as the nondiscriminating vector given enough disease initially present in the population. However, where disease was absent prior to the introduction of the first smutted bee, the nondiscriminating vector spread the disease at a higher rate than the discriminating vector.

Introduction

Investigations into the evolution and ecology of pollinator-plant interactions often begin with an explicit study of the behavior of individual pollinators. It is the individual pollinator that decides which floral species to visit, how many flowers to visit, how often to switch among competing floral resources, how long to stay within a given patch or field, whether to collect pollen or nectar, etc. All of these behaviorally-based interactions can, in principle, profoundly influence patterns of evolutionary and ecological interaction.

From the evolutionary perspective, pollinators act as important, often strong, selective agents, shaping adaptations in both the plants being exploited and in the exploiting pollinator population. This view is not particularly new and is represented in Darwin's writings (1859, 1876), as well as many of his contemporaries (e.g. Müller 1883, Knuth 1906). The analysis of pollinator-plant interactions has traditionally focused on the adaptations observed in flowering plants for efficient pollen transfer (Baker 1963, Baker and Hurd 1968, Grant and Grant 1965). Stebbins (1970, 1981) goes so far as to suggest that most of the adaptative radiation in the angiosperms and divergence and diversification of floral morphology can be tied to the selective regime created by a single most effective pollinator or small set of closely related pollinators (also see Regel 1977). Ample evidence suggests that many floral traits are specific adaptations to enhance or facilitate pollinator visitation (Waser 1983); e.g., nectar guides (Jones and Buchmann 1974, Waser and

Price 1983a), floral morphology, color, and architecture (Kugler 1943, Willson and Price 1977, Thompson 1981, Gerber 1982, Augsperger 1980, Galen 1989), nectar reward distributions (Real and Rathcke 1991, Pleasants and Chaplin 1983) and flowering time (Linsley 1958, Waser 1979, Bertin 1982). The absence of pollinators can also exert strong selection pressure and has been implied in the evolution of autogamy in isolated populations of plants or in species that are inferior competitors in attracting insect visitors (Lefebvre 1970, Wyatt 1983).

Pollinators can also influence the pattern of movement of genes within populations, thereby acting as a contemporary force that may affect the genetic structure of plant populations (Handel 1983, Waser and Price 1983a,b). Often the movements of insect pollinators and associated patterns of gene dispersal can be coupled to aspects of pollinator foraging behavior and floral preferences (Turner et al. 1982, Ott et al. 1985, Schmitt 1980).

The role of pollinators (and specifically their behavior) in shaping evolutionary interactions is apparent, uniquitous, and well documented. The case for the role of pollinators in determining ecological interactions is certainly less well established, and what results we have in hand are rather conflicting.

The ecological implications of individual behaviors depend on whether one is looking from the plant population perspective or from the perspective of the pollinator population. The ecology of pollinator populations is very poorly understood. We have very scanty evidence to suggest that pollinator population sizes are in any way controlled by the availability of floral resources. Richards (1975) provides some evidence that bumble bee population sizes fluctuate with the densities of local floral resources. The major sources of mortality in bumble bees may, however, be flooding and parasitism.

The interaction between the colony's stored energy reserves, local floral resources, and the ability to withstand major sources of mortality needs further study. Pyke (personal communication), as part of a long-term study of bird-pollination systems in Australia, has concluded that floral resources have little if any effect on the densities of nectarivorous bird populations. Pollination studies have not ascertained the direct influence of floral resource availability on the reproductive biology of the attendant pollinators and the associated translation of observed reproductive patterns into population dynamics. Any investigation into pollinator population ecology will demand such an inquiry.

Much more attention has been focused on the role of pollinator behavior in determining the ecological dynamics of plant populations. Primarily, such studies have explicitly determined the relationship between pollination and reproductive success in individual plants. For example, Real and Rathcke (1988, 1991) assessed the effect of variation in per flower nectar production rate on rates of pollinator visitation to individual shrubs of the Mountain laurel, *Kalmia latifolia*. Individual shrubs that were high nectar producers received the highest rates of visitation. These same high quality individuals showed the highest reproductive success when measured as percentage fruit set. Similar relationships between nectar production rates, visition rates, and reproductive success have been observed in milkweeds (Pleasants and Chaplin 1983).

While it is clear from these findings that reproductive output in individual plants can be affected by pollinator behavior, it is not clear that differential individual patterns of reproduction will translate into different patterns of population growth or changes in equilibrium population size. In many plant species, populations are not limited by the number of offspring produced by adult individuals, but rather by the number of offspring that can find suitable sites for germination, i.e., "safe-sites" (Harper 1977, Grubb 1977). In species that are limited by safe-sites, increased reproductive success at the individual level may have no effect on the species' local ecological dynamics.

While differences in individual reproductive success (mediated through pollinator behavior) may have little effect on plant population dynamics, these differences will profoundly affect the evolutionary process and genetic structure of the plant population. Individual plants that experience higher reproductive success will more likely be represented in subsequent generations since they will have the advantage in colonizing safe-sites even when the overall population size is constant.

An increase in the pollinator visitation rate to a particular individual plant may not lead to an increase in reproductive output by that individual. Some plant species appear to be limited by maternal resources for the maturation of seeds and fruits rather than by number of pollinator visits (Stephenson and Bertin 1983). When pollination is not limiting seed set, the connection between aspects of individual pollinator behavior and long-term plant population dynamics becomes even more tenuous. However, even in plant species that exhibit substantial fruit abortion, the evolutionary implications of differential pollinator behavior can be pronounced. Plants attracting larger numbers of pollinators, and at a greater rate, will be able to select from a greater diversity of potential male pollen donors. This increase in gametic diversity should increase male competition and thereby produce higher quality offspring (Mulcahy 1983). Also, the simple increase in the fraction of male gametes added to the male gemete pool will provide advantageous for those plants receiving higher visitation rates. Many issues in the evolution of plant mating systems depend on the fraction of an individual plant's pollen that is added to the total male gamete pool (Holsinger 1988), and this fraction is to some degree under the direct influence of pollinator behavior.

Consequently, and individual-based analysis of pollinator-behavior has proven very useful in resolving evolutionary issues. Individual-based approaches have not proven as useful in understanding ecological dynamics of either the pollinator or the plant species. To some degree this discrepancy may merely reflect the proclivities of investigators, and in the future, when we pay more attention to ecological interactions in pollination systems, this discrepancy will disappear. Alternatively, demonstrating links between individual behavior and ecological dynamics may be intrinsically more difficult for the following reasons. First, an evolutionary approach revels in examinations of individual differences. Evolutionary change can not occur without these differences. Studies of ecological dynamics, on the other hand, often average across individual differences. The analysis of ecological dynamics often does not change if we ignore individuals. Many of the articles in this book are related to uncovering those conditions when we cannot average over different individuals, and the conclusion seems to be that often we can average without loss of information or predictive power (Caswell, this volume, Metz, this volume). Second, ecological dynamics may often be controlled by external factors not directly under the influence of the behavior of individual pollinators; for example, when recruitment of individual plants is limited by the availability of germination sites. The individual analysis of pollinator behavior will only be important in understanding plant population dynamics where recruitment into the plant population is pollination limited, and where ecological interactions are primarily under biotic control.

In many cases, the mechanisms of biotic control in plant populations involve aspects of individual behavior in insect pests or pollinators. For example, some plant populations may be regulated by disease agents (Burdon 1987), and often these diseases are transmitted by insects moving from infected to healthy populations or individuals. Such insect-born plant diseases are of enormous importance in agriculture, and an analysis of the manner by which individual behavior of the insect vector affects the transmission dynamics within the plant population is both intellectually challenging and economically important. As a model system for exploring these kinds of issues, we have examined the interaction between the anther smut fungus, *Ustilago violacea*, and its host plant, *Silene alba* (Baker 1947). Both male and female plants infected with the anther smut are rendered sterile and produce flowers with anther sacs filled with spores. Thus when insect pollinators land on infected flowers they act as vectors of the anther smut fungus. We have undertaken an individual-based modeling approach to elucidate the intricacies of this plant-vector-pathogen interaction.

Floral Biology of the *Silene-Ustilago* System

Flowers of plants infected with *Ustilago* differ from healthy flowers in a number of ways, the most obvious of which is the presence of purple spores rather than yellow pollen in the center of the flower. In a greenhouse study, Alexander and Maltby (1991) showed that fungal infection stimulates flower production and reduces flower size. Another plant trait that is likely to influence pollinator behavior is the amount of floral nectar. Roche (1991) conducted experiments to determine the effect of the fungus on nectar production. She measured bagged nectar production by *Silene alba* in both natural and manipulated populations. The two populations differed in that the natural population was subject to natural sources of disease transmission, while in the manipulated population individual *Silene* plants were randomly inoculated with *Ustilago* spores. In the natural population, disease transmission was due to vector movement while in the experimental population, disease was assigned randomly while plants were seedlings. The absence of the role of vector behavior in the experimental population allows for some interesting comparisons between the two populations.

In the natural population, diseased plants produced significantly more nectar than healthy plants ($F_{1,210} = 43.84$, $p < .001$), but there was no significant effect of sex on nectar production. In the manipulated population, there was no significant effect of disease status ($F_{1,212} = 1.24$, $p = .27$) or sex ($F_{1,212} = .36$, $p = .55$) on nectar production. There is a possible explanation for the difference in the results from the two populations reflecting the role of vector behavior in the transmission of the disease: plants which produce more nectar will receive more insect visits, and thus have a higher probability of contracting the disease. The behavior of nocturnal insect visitors, monitored through the movement of fluorescent dyes, supports this conjecture.

During summer 1988, Roche monitored the entire suite of nocturnal visitors to *Silene* in a natural population by dusting a set of diseased and healthy flowers, each with a distinctly colored fluorescent dye. Before daylight, ultraviolet lamps were used to survey the entire population for flowers with dye. The disease status of the dye donor and the disease status of the recipient plant were noted. The frequency of diseased plants in the population was 0.182. Results of the dye movement patterns suggest non-random insect movements: (1) insects moved from diseased to diseased individuals more frequently than expected by chance ($\chi^2 = 5.33$, $df = 1$, $p < 0.05$) and (2) diseased plants received a disproportionate number of visits relative to their frequency in the population ($\chi^2 = 7.25$, $df = 1$, $p < 0.01$). This supports the notion that some insects were majoring on diseased plants, and that diseased plants are disproportionately attractive.

Nocturnal visitors to *Silene,* such as moths, visit these plants primarily as sources of nectar. The increased attractiveness of diseased plants to nocturnal visitors may reflect their increased profitability. Diurnal visitors, such as honeybees and bumble bees, on the other hand, visit the plants for both nectar and pollen. Diseased plants may be less profitable for these insects since they are no longer a source of pollen. Roche (personal observation) found that bumble bees preferentially visited healthy rather than diseased *Silene* plants (see also Alexander 1990). There was a great deal of individual variation among bumble bees: 50% visited only healthy plants, while the other 50% visited both healthy and diseased plants. If a bumble bee visits both healthy and diseased plants, it visits them according to their frequency in the habitat. We are in the process of analyzing the honeybee data to see if they show similar preferences for healthy over diseased individuals.

Modeling Disease Transmission

Our ultimate goal is to understand the manner by which patterns of insect movement and transfer of spores may account for the spatial distribution and temporal spread of the anther-smut disease within the plant population. Our initial efforts focus on constructing very simple computer simulations of individual vector movement among healthy and diseased plants with a spatial structure identical to the natural population already described. In this paper, we emphasize patterns of bumble bee movement, and will expand our analysis at some future time to include the dual influences of diurnal and nocturnal insect visitors.

Patterns of bumble bee movement among plants are non-random, with healthy male plants receiving most of the insect visits. This observed pattern of rare visits to diseased flowers within long sequences of visits to healthy male flowers may have interesting effects on the pattern and rate of spread of the disease. To spread the disease, an insect must first visit a diseased plant and then visit a healthy plant. The rate of disease spread will depend on the probability of both of these events occurring. Clearly, both vector behavior and relative proportions of diseased and healthy plants initially in the population will affect these probabilities.

To test whether the observed visitation pattern by bumble bees indeed affects disease spread, we simulated a population of *Silene alba* visited by a randomly moving (nondiscriminating) bumble bee vector, and compared patterns and rates of disease spread in this population to one in which bumble bees concentrated their visits on healthy male flowers with only rare visits to diseased flowers. We repeated this at several initial levels of disease. The simulation was designed to allow easy incorporation of observed insect behaviors; the parameters described below are all measurable and can be readily ascertained from field data.

Analysis of Spatial Distribution in Natural Populations

Our simulations are similar in form to percolation models (Zallen 1983, Durrett 1988). Individual plants occur in two possible states—healthy or

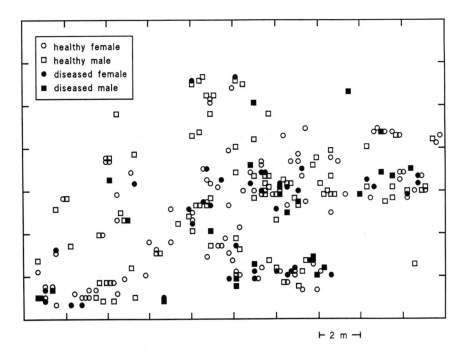

Figure 22.1. Spatial distribution of individual plants of *Silene alba* from a field near Mt. Lake Biological Station, Giles County, VA (Alexander, unpublished data). Sex and disease status of individuals are indicated on the map (healthy males □, diseased males ■, healthy females ○, diseased females ●).

diseased. The transition from healthy to diseased status is irreversible, and the probability of transition is a function of the disease status of a plant's neighbors.

Neighborhood processes are extremely sensitive to the underlying spatial distribution of interacting individuals. To account for this sensitivity, flowers in our simulation were arranged to mimic the spatial pattern of a particular patch of 246 *Silene alba* in Giles County, Virginia (Site 1, Alexander and Antonovics 1988) (Fig. 22.1).

We analyzed the spatial structure of this population using both Morista's index of dispersion I_δ (Morisita 1962, Krebs 1989) and the parameter k of the negative binomial distribution (Southwood 1978). Morisita's index is a measure of the variance to mean ratio of the number of plants in a quadrat. Values greater than one indicate clumping, whereas values aproaching zero are characteristic of uniform distributions of individuals. The parameter k of the negative binomial distribution is an inverse measure of clumping; as clumping decreases, k increases. A random (Poisson) distribution of individuals has an infinitely large value of k.

We analyzed spatial structure separately for (1) the total population (healthy plus diseased individuals) and (2) the subset of diseased individuals within the population. Because both I_δ and k are dependent on quadrat size, we calculated the values of each index for each of 7 quadrat sizes.

Our analysis suggests that as quadrat size increased, clumping among individual plants decreased (Figure 22.2). Although there was some measurable amount of clumping at all quadrat sizes, Morisita's index approached 1 (Poisson) for the largest quadrats. This pattern is similar to the spatial distribution of tropical dry-forest trees (Hubbell 1979). Distributions of this sort are characteristic of populations with density declining away from high density centers. Both the diseased subset and the total population showed the same general pattern of decreased clumping with increased quadrat size.

Clumping was greater in the diseased subset of the population than in the total population at all but one quadrat size (Figure 22.2). Low values of k in the negative binomial distribution are characteristic of "point-cluster processes" (Diggle 1983). Such processes result when there are multiple centers randomly distributed in a spatial field, but with local contagion at each center. This is the exact process that we would expect to see if pollinators act as the major vector for the disease, and where pollinators enter fields at random, contaminate plants, and then infected plants act as sources of contagion within the population.

Our analysis supports our belief that pollinators influence the process of disease spread in natural populations of *Silene*. To explore this role further, we present simulations of individual foraging behavior and its effect on the pattern and process of disease spread.

Individual-based Simulations of Disease Spread

In the following simulations, each plant has just one flower. Female and male *Silene* are dispersed exactly as in the natural population (Figure 22.1). In simulations in which some plants were diseased initially, the diseased individuals were chosen randomly from the population.

Vector Movement

During each time interval in the simulation, there is probability α that a bumble bee will enter the patch of flowers. Each simulation begins with a disease-carrying bumble bee entering the patch of flowers. At this point, the bee has equal probability of landing on any particular flower. Further movements of the bee differ between simulations of random movement (nondiscriminating vector) and those of non-random movement (discriminating vector).

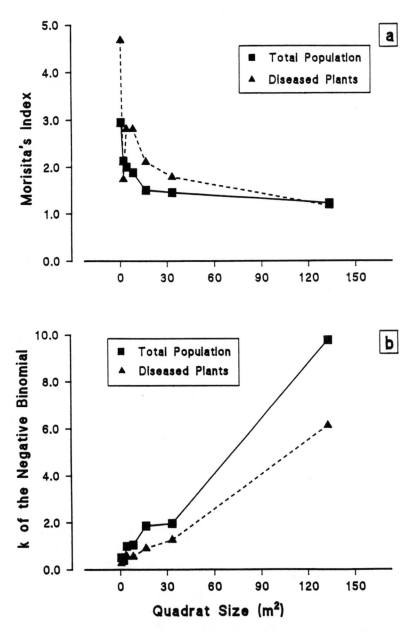

Figure 22.2. Analysis of clumping in the total population of *Silene* (■———■) and the diseased subset of individuals within the population (▲---▲) measured by (a) Morisita's Index and (b) the clumping parameter *k* of the negative binomial distribution. Quadrats were measured in square meters.

Random Vector Movement

In simulations of random vector movement, the bumble bee leaves a plant in a random direction. We have divided the 2π radians circle into 12 equal angles, each equally likely to be the bee's flight direction. The bee encounters all flowers in this $\pi/6$ radian sector in order from nearest to farthest from the original flower. As the bumble bee flies away from the flower, it has some constant probability β of landing on each male flower it encounters in this sector. If it encounters a female flower, it lands with only some small probability γ, the probability of mistaking a female for a male flower. Once it has landed it repeats this process, choosing a new random flight direction, until its flight takes it beyond the limits of the patch of flowers.

Non-random Vector Movement

In these simulations, bumble bees move just as in simulations of random movement, except that they avoid diseased plants. When a diseased plant is encountered, rather than visiting it with probability β, a bee visits it with a much smaller probability $\delta \cdot \beta$, where δ is the probability of mistaking a diseased flower for a healthy male flower. We assume rare visits to diseased and female flowers are "mistakes" by bumble bees specializing on healthy male flowers, although the actual parameter estimation requires no such assumption. These parameters can be estimated by measuring the relative frequencies of visits to healthy male, healthy female, and diseased plants in relation to their proportions in the population.

In all simulations, a unit of time was the increment necessary to encounter or visit one flower. Bees were allowed to enter the patch of flowers during the first 100 increments; simulations ended when the last of these bees exited the patch.

Disease Transmission

Each simulation begins with a smutted bumble bee (i.e., one that is carrying fungal spores) infecting the first flower it visits. Bees entering the patch subsequent to this do not carry the disease; they become smutted only by visiting an infected flower. In these simulations, an infected flower is immediately infectious. This differs from the observed rate in which an initially healthy plant expresses the disease, on average, 38 days after inoculation (Alexander and Antonovics 1988). Once smutted, a bee's probability of transmitting disease can decay exponentially (with parameter η) with successive flower visits; however, in the following simulations, smutted bees transmitted disease with probability 1 to all flowers visited before leaving the population.

Table 22.1. Parameter values used in simulations. Discriminating and nondiscriminating vectors differ in the value of δ. Random visitation corresponds to δ = 1.0; non-random visitation has δ = 0.1.

Parameter	Value
α	0.1
β	0.85
γ	0.1
η	0.0
ρ	0.0, 0.1, 0.2, 0.5
δ	0.1, 1.0

Results

This model is a combination of many stochastic processes, including initial disease location, location of first plant getting disease, flight direction of the bee, etc. Because of this, we expect the results of our simulations to be highly variable, even under a specific set of parameter values. The initial locations of the diseased plants, in both natural and simulated populations, is of particular importance in determining the rate of disease spread. If a diseased plant is surrounded by healthy plants, it is more likely to have its spores carried to another plant than is a diseased plant isolated near the edge of the population. To examine this variability, we repeated the simulation under different initial conditions and different sequences of random number draws. We simulated diseased spread in populations having initial disease proportions (ρ) of 0, .1, .2, and .5. All other parameters were held constant (Table 22.1). For each ρ, we used two populations of plants, each with a unique initial spatial distribution of disease. Within each of these two populations, we simulated vector movement and disease spread three times by starting the simulation with three different random number seeds. A random number seed always returns the same sequence of random numbers using a QuickBASIC (version 4.0) psuedorandom number generator. By using three different seeds, we get three different sequences of random numbers determining when bees enter a patch, which flowers are visited, what directions bees fly, etc. At each level of ρ within each of the two initial populations, we calculated the difference in rate of disease spread between random and discriminating vectors at each random number seed. This resulted in six simulations at each ρ for each behavior type. Because the results at ρ = .2 were quite variable, we simulated vector movement three more times on an additional initial populationn for a total of 9 simulations. At ρ = 0, there is only one possible initial population (i.e., one with no initial disease but subject to incoming diseased vectors). Consequently, we repeated this simulation with six random number seeds to get six simulations for each behavior type. Rate of disease spread was measured as the difference between final and initial proportions diseased in each simulation.

Rate of spread was highly variable in any combination of initial disease proportion and vector behavior type (Figure 22.3). At $\rho = 0$, the randomly moving vector spread the disease at a higher rate than the discriminating vector (Figure 22.3a, t-test, $p < .05$). With even moderate amounts of disease present initially, though, the two vector behaviors did not differ in their effect on rate of spread (t-test, $p > .05$ for $\rho = 0$, .1, .2, and .5).

Rate of spread of disease is determined by two probabilities; that of encountering diseased flowers and that of moving from a diseased to a healthy flower. A random vector encounters disease in proportion to its occurrence in the population. A discriminating vector has a lower probability of encountering disease. Once smutted, a randomly moving vector moves to a healthy plant with a probability equal to the proportion of healthy plants in the population. A discriminating vector, however, has a high probability of moving to a healthy flower once it has visited a diseased flower. At low levels of disease in the population, a discriminating vector rarely visits a diseased flower, and, consequently, spreads disease very slowly. At higher levels of disease, the avoidance of diseased flowers is compensated by the higher likelihood of spreading it to healthy flowers once it has been encountered. As a result, the discriminating vector can spread disease as rapidly as the nondiscriminating vector, given enough disease initially present in the population.

Conclusion

Models of disease spread often are based on random association or encounter between infected and susceptible individuals within the host population. A variety of behavioral preferences and mechanisms exhibited by the transmitting vector often will lead to a substantial violation of this assumption. In our *Silene* example, the preference behavior of pollinators can alter the rate of spread of the disease, and possibly influence the spatial distribution of the disease within the natural population. We are presently adding more realism to the model, including lag times in infection, multiple flowers per plant, and more elaborate choice behavior by bees. The current model illustrates the potential of the individual-based modeling approach to this type of problem. The behavioral parameters needed to construct individually-based models of disease transmission and spread are readily asertained from field observations and experiments. Disease spread may generally show considerable sensitivity to the behavior of individual vectors and the ecology of disease may prove an ideal arena for assessing the merits of the individually-based modeling approach.

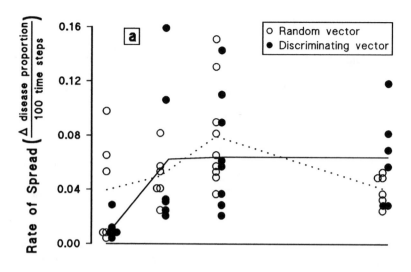

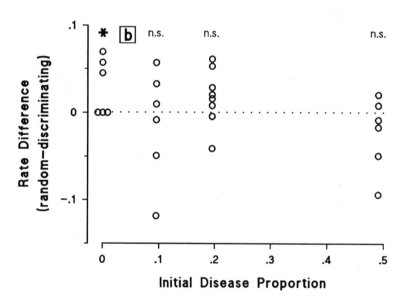

Figure 22.3.   (a) Rate of disease spread as a function of initial disease proportion in simulated populations of *Silene alba* having either randomly moving or discriminating disease vectors. Lines connect mean values for each vector behavior type (randomly moving vector O · · · O, discriminating vector ●———●). (b) Difference in rate of disease spread between random and discriminating vectors where vectors were paired by random number seeds (see text for details). Significant differences in rate of spread under these two types of pollinator behavior only occur at low initial frequencies of the disease.

Acknowledgments

We would like to thank Helen Alexander for allowing us to use her unpublished data. This research was supported by NSF grant BNS 9096209 to LAR, and the Electric Power Research Institute through a fellowship in Population Dynamics to EAM.

Literature Cited

Alexander, H. M. 1990. Dynamics of plant-pathogen interactions in natural plant communities. In *Pests, Pathogens, and Plant Communities*. J. J. Burdon and S. R. Leatherer (eds.), pp. 31–45. Blackwell Scientific Publishers, Oxford, England.

Alexander, H. M. and J. Antonovics. 1988. Disease spread and population dynamics of anther-smut infection of *Silene alba* caused by the fungus *Ustilago violacea*. *J. Ecol.* **76**:91–104.

Alexander, H. M. and A. Maltby. 1991. Anther-smut infection of *Silene alba* caused by *Ustilago violacea:* factors determining fungal reproduction. *Oecologia,* in press.

Augsperger, C. K. 1980. Mass-flowering of a tropical shrub (*Hybanthus prunifolius*): influence on pollinator attraction and movement. *Evolution* **34**:475–488.

Baker, H. G. 1947. Infection of a species of Melandrium by Ustilago violacea (Pers.) Fuckel and the transmission of the resultant disease. *Ann. of Bot.* **11**:333–348.

Baker, H. G. 1963. Evolutionary mechanisms in pollination biology. *Science* **139**:877–883.

Baker, H. G. and P. D. Hurd. 1968. Intrafloral ecology. *Ann. Rev. Entomol.* **13**:385–414.

Bertin, R. I. 1982. The ruby-throated hummingbird and its major food plants: ranges, flowering phenology, and migration. *Can. J. Zool.* **60**:210–219.

Burdon, J. J. 1987. *Diseases and Plant Population Biology*. Cambridge University Press, Cambridge, England.

Darwin, C. H. 1859. *On the Origin of Species by Natural Selection*. Murray, London.

Darwin, C. H. 1876. *On the Effects of Cross and Self Fertilization in the Vegetable Kingdom*. Murray, London.

Diggle, P. J. 1983. *Statistical Analysis of Spatial Point Patterns*. Academic Press, London.

Durrett, R. 1988. *Lecture Notes on Particle Systems and Percolation*. Wadsworth & Brooks/Cole, Pacific Grove, CA.

Galen, C. 1989. Measuring pollinator-mediated selection on morphometric floral traits: Bumblebees and the alpine sky pilot. *Polemonium viscosum. Evolution* **43**:882–890.

Gerber, M. 1982. Architecture, size, and reproduction in plants: a pollination study of *Mertensia ciliata*. (James) G. Don. MS Thesis, Oregon State University, Corvalis, OR.

Grant, V. and K. A. Grant. 1965. *Flower Pollination in the Phlox Family*. Columbia University Press, New York.

Grubb, P. 1977. The maintenance of species richness in plant communities: the importance of the regeneration niche. *Biol. Rev.* **52**:107–145.

Handel, S. N. 1983. Pollination ecology plant population structure, and gene flow. In *Pollination Biology*, L. Real (ed.), pp. 163–212. Academic Press, New York.

Harper, J. L. 1977. *The Population Biology of Plants*. Academic Press, London.

Holsinger, K. E. 1988. Inbreeding depression doesn't matter: the genetic basis of mating-system evolution. *Evolution* **42**:1235–1244.

Hubbell, S. P. 1979. Tree dispersion, abundance, and diversity in a tropical dry forest. *Science* **203**:1299–1309.

Jones, C. E. and S. L. Buchmann. 1974. Ultraviolet floral patterns as functional orientation cues in hymenopterous pollination systems. *Anim. Behav.* **22**:481–485.

Knuth, P. 1906. *Handbook of Flower Pollination*. Oxford University Press (Clarendon), Oxford, England.

Krebs, C. J. 1989. *Ecological Methodology*. Harper and Row, New York.

Kugler, H. 1943. Hummeln als Blütenbesucher. *Ergeb. Biol.* **19**:143–323.

Lefebvre, C. 1970. Self fertility in maritime and zinc mine populations of *Armeria maritima* (Mill.) Willd. *Evolution* 24:571–577.

Linsley, E. G. 1958. The ecology of solitary bees. *Hilgardia* **27**:543–599.

Morisita, M. 1962. I_8-index, a measure of dispersion of individuals. *Res. Popul. Ecol.* **4**:1–7.

Mulcahy, D. 1983. Models of pollen tube competition in *Geranium maculatum*. In *Pollination Biology*, L. Real (ed.), pp. 152–162. Academic Press, New York.

Müller, H. 1883. *The Fertilization of Flowers*. Macmillan, London.

Ott, J. R., L. A. Real, and E. M. Silverfine. 1985. The effect of nectar variance on bumblebee patterns of movement and potential gene dispersal. *Oikos* **45**:333–340.

Pleasants, J. M. and S. J. Chaplin. 1983. Nectar production rates of *Asclepias quadrafolia*: causes and consequences of individual variation. *Oecologia* **59**:232–238.

Real, L. A. and B. J. Rathcke. 1988. Patterns of individual variability in floral resources. *Ecology* **69**:728–735.

Real, L. A. and B. J. Rathcke. 1991. Individual variation in nectar production and its effects on fitness in *Kalmia latifolia*. *Ecology*, in press.

Regel, P. J. 1977. Ecology and evolution of flowering plant dominance. *Science* **196**:622–629.

Richards, K. W. 1975. Parasitid mites associated with bumble bees in Alberta, Canada (Acavina, Parasitidae, Hymenoptera, Apidae) PhD thesis, University of Kansas, Lawrence.

Roche, B. M. 1991. The cost of attractiveness: the dynamics of a sexually transmitted plant disease. *Oecologia* in press.

Schmitt, J. 1980. Pollinator foraging behavior and gene dispersal in *Senecio* (Compositae). *Evolution* 34:934–943.

Southwood, T. R. E. 1978. *Ecological Methods, 2nd Ed.* Chapman and Hall, London.

Stebbins, G. L. 1970. Adaptive radiation of reproductive characteristics in angiosperms. I. Pollination mechanisms. *Ann. Rev. Ecol. Syst.* 1:307–326.

Stebbins, G. L. 1981. Why are there so many species of flowering plants? *BioScience* 31:573–576.

Stephenson, A. G. and R. I. Bertin. 1983. Male competition, female choice, and sexual selection in plants. In *Pollination Biology*, L. Real (ed.), pp. 110–151. Academic Press, New York.

Thompson, J. D. 1981. Spatial and temporal components of resource assessment by flower-feeding insects. *J. Anim. Ecol.* 50:49–59.

Turner, M. E., J. C. Stephens, and W. W. Anderson. 1982. Homozygosity and patch structure in plant populations as a result of nearest-neighbor pollination. *Proc. Nat. Acad. Sci. USA* 79:203–207.

Waser, N. M. 1979. Pollinator availability as a determinant of flowering time in ocotillo (*Fouquieria splendens*). *Oecologia* 39:107–121.

Waser, N. M. 1983. The adaptive nature of floral traits: ideas and evidence. In *Pollination Biology*, L. Real (ed.), pp. 241–285. Academic Press, New York.

Waser, N. M. and M. V. Price. 1983a. Pollinator behavior and natural selection for flower color in *Delphinium nelsonii*. *Nature* 302:422–424.

Waser, N. M. and M. V. Price. 1983b. Optimal and actual outcrossing in plants, the nature of plant-pollinator interaction. In *Handbook of Pollination Biology*. C. E. Jones and R. J. Little (eds.), pp. 341–359. Van Nostrand-Reinhold, New York.

Willson, M. F. and P. W. Price. 1977. The evolution of the inflorescence size in *Asclepias* (Asclepiadaceae). *Evolution* 31:495–511.

Wyatt, R. 1983. Pollinator-plant interactions and the evolution of breeding systems. In *Pollination Biology*, L. Real (ed.), pp. 51–96. Academic Press. New York.

Zallen, R. 1983. *The Physics of Amorphous Solids*. Wiley, New York.

Part V
Summary of Working Group Discussions

23

Individual-Based Modeling: Summary of a Workshop

Louis J. Gross, Kenneth A. Rose,
Edward J. Rykiel, Jr., Webb Van Winkle, and
Earl E. Werner

Introduction

A major part of the Workshop on Populations, Communities, Ecosystems: An Individual Perspective, consisted of roundtable discussion groups which focused on specific aspects of individual-based models (IBMs). Key objectives were to identify areas of consensus regarding the formulation and application of IBMs, and to define potentially important future directions for research. The questions assigned to discussion groups were: (1) What are the most useful mathematical approaches to individual-based modeling? (2) Are some programming languages or machines better than others for use in individual-based models? (3) How successful has the approach been at levels other than the population (e.g., multi-species)? (4) What have been the management applications of individual-based approaches to date and what are some new areas of application on the horizon?

In this paper, we briefly summarize the many points debated during these group discussions. Much of the material included in these discussions has been incorporated in the papers in this volume, and we make no attempt to be comprehensive in these brief remarks. Rather, our purpose is to emphasize the major areas of consensus reached and the most hopeful future directions. Our remarks have been facilitated by the work of the group discussion leaders and rapporteurs. Our objective in presenting this summary of the discussions is to stimulate others to address some of the issues that emerged during the workshop.

What Are the Most Useful Mathematical Approaches to Individual-Based Modeling?

Approaches to IBMs were categorized as analytical and simulation. Analytical approaches refer to modeling based on the use of mathematical tech-

Figure 1. Six criteria influencing the usefulness of analytical and simulation approaches to individual-based modeling. As each of the criteria becomes more important, one is forced towards the simulation approach.

	Approach		
Criteria	Analytical		Simulation
spatial environment	homogenous	→	heterogenous
demographic stochasticity	unimportant	→	important
rare events	unimportant	→	important
biological and/or environmental discontinuities	unimportant	→	important
numer of individuals	large	→	small
biological complexity	simple	→	complex

niques to solve the model equations. Partial differential equation models (e.g., Sinko and Streifer 1967) were commonly cited as typifying the analytical approach. Analytical models have the advantages of relative simplicity, generality, and, when solvable, exact solution. When exact solutions cannot be derived by mathematical techniques, numerical approximation methods are employed to obtain solutions. For the most part, analytical approaches have involved the solution of equations that do not represent individuals in the population as discrete entities. Rather, analytical approaches tend to focus on tracking how metrics representing the aggregate of individuals in the population (e.g., size distribution) change through time and/or space.

The simulation approach refers to the solution of the model equations by directly representing individuals as discrete entities; the aggregate of the many discrete individuals at any one point in time and/or space is the solution to the model equations. Simulation approach IBMs typically use Monte Carlo techniques to stochastically assign attributes, behaviors, and responses to individuals. Forest community models based on individual trees (Shugart 1984) and fish population models based on tracking the attributes of many individual fish (e.g., DeAngelis et al. 1991) were cited as specific illustrations of the simulation approach. Analytical approaches generally correspond to *i*-space distribution models, whereas the simulation approach corresponds to *i*-space configuration models (Metz and Diekmann 1986, Metz and De Roos, this volume, Caswell, this volume).

Selection of the appropriate approach (analytical or simulation) depends on the importance and balance among the six criteria listed in Fig. 1 [see DeAngelis and Rose (this volume) for additional discussion]. Each of the six criteria lie on a continuum ranging from relatively unimportant to very important (going from right to left in Fig. 1). Four of the criteria refer to the relative importance of: local interactions among individuals, stochasticity, rare events, and discontinuities. The remaining two criteria are the degree of biological (and therefore mathematical) complexity necessary to address the problem, and the numbers of individuals in the population of interest. We assume *a priori* that the simplest modeling approach (that achieves the research or assessment objectives) is the best approach. Thus, one starts with the presumption of using an analytical approach (i.e., a model with relatively simple governing equations). As the importance of each of these critera increases, one is forced to the right in Fig. 1, towards the simulation approach.

A sharp dichotomy of opinion arose over the relative merits of the analytical approach versus the simulation approach. At one extreme of opinion, several people viewed the simulation approach as an inefficient and, in some cases, incorrect solution to some analytical model. At the other extreme was the opinion that the only correct IBM was one that followed individuals as discrete entities. This dichotomy of opinion is, by now, the tiresome case of the "mathematical types" (theoretical biologists) versus the "biological types" (empirical biologists). The power of analytical approaches is directly related to the degree to which the system can be simplified (i.e., its dimensionality reduced). The power of simulation approaches is directly related to the degree to which the "important" components and processes of the system can be realistically represented. Biologists must consider the mathematicians view of logic, low dimensionality, and simplicity; mathematicians must recognize the biologist's tendency to believe that everything is important. If the less mathematically-inclined biologists can be persuaded that sometimes the system can be simplified and treated with mathematical rigor, and if the mathematical biologists might be persuaded that mathematical rigor is not always synonymous with ecological rigor, it might be possible to resolve this conflict.

There was general agreement that mathematical theory should be used as a common language for describing IBMs. IBMs are being developed in many diverse disciplines of science. Each discipline has its own nomenclature and jargon that hinders easy and rapid inter-disciplinary communication of model structures and results. IBMs are in their infancy; yet, we already have difficulty comparing different IBMs in terms of their structure and generality of prediction results. The number and types of IBMs will likely increase dramatically in the near future, further complicating intermodel comparisons. Keeping adequately informed of the approaches taken by others, and how they relate to each other, is critical for rapid and efficient progress in the general field of IBMs. Using mathematics as a common language for describing IBMs and generalizing model results will greatly increase and ease communication among IBM developers and users. Mathematical description of IBMs would impose a high level of rigor in the use of clear definitions and terminology. Describing IBMs using the common language of mathematics will permit rapid determination of the general structure of a specific model and its relationship to other models, and allow model-specific results to be generalized and placed into a broader context.

Strong emphasis was placed on increasing the use of mathematics for providing insights into model behavior. Developers and users of IBMs, especially simulation IBMs, should borrow and adapt techniques from among the many mathematical techniques that are available. While the application of mathematics (and mathematical rigor) to analytical approaches is obvious and widespread, the application of mathematical theory to simulation IBMs has been lacking. Many simulation IBMs give the appearance of being a "brute-force" approach with little regard to mathematical rigor. Developers of simulation IBMs often give the impression of having immediately turned to the simulation approach rather than the analytic one, with little consideration to the mathematics underlying the simulation approach. Simulation IBMs appear conceptually simple, and thus developers wrongly assume that one can ignore the mathematics of the model. Analytical solutions are not possible for IBMs of even moderate complexity. However, it seems that once investigators decide to take the simulation approach to IBMs, the mathematical tools associated with analytical approaches are dismissed as being inappropriate. While many of the mathematical tools cannot be applied to simulation IBMs *sensu stricto,* the concepts underlying the tools can, and should, be applied in an approximate sense. In particular, one method to ensure a simulation approach has been correctly formulated and coded is to take a very special case for which an analytic approach suffices, and compare the results. This method of model verification has been little used to date in *i*-state configuration models.

Simulation IBMs generally follow many individuals through time and/or space, thereby generating large amounts of information (model "data") that need to be adequately summarized and visualized. How should the large amounts of model data be aggregated and summarized into easily interpretable metrics that capture the important features of the results? Software and hardware currently available, or likely to be available in the near future, are adequate to perform the required number crunching and graphical display of model results. Deciding on what outputs to use and how to summarize the extensive model data generated, however, is an open-ended issue and deserves more attention. Again, IBM developers need to borrow approaches, concepts, and techniques from other fields, such as pattern recognition and spatial statistics.

Are Some Programming Languages or Machines Better Than Others for Use in Individual-Based Models?

There was little interest at first in the question of selecting a programming language or machine. The attitude was that constructing an IBM was a biological problem and technology should not drive how the model is constructed. Essentially, the initial opinion of the group was that one should find a language and machine that works for you and use it. Upon further discussion, selection of the software (programming language) and hardware (machine) were acknowledged to be important because different languages and availability of computing power can affect the conceptualization of the model and encourage or inhibit exchange of models between investigators. While the questions of selecting a programming language and machine were eventually recognized as important, there were no obvious answers or resolution to these questions.

Software and hardware technologies for constructing IBMs are rapidly improving. Recent developments in software include parallel programming (Haefner, this volume), object-oriented languages (Palmer, this volume), and artificial intelligence. Hardware improvements include increased computing speed, parallel processing (Haefner, this volume), and special purpose computer boards (e.g., transputers, array processors). Advances in computer-aided software engineering now provide programming environments (rather than just programming languages) to facilitate model development, implementation, and application. Typically, these environments have a graphical user interface or windowing system that includes integrated facilities for editing and debugging code. Technological advancements can be expected to enhance and accelerate the develoment of IBMs, just as it has mathematics, physics, and engineering.

How Successful Has the Approach Been at Levels Other than the Population?

As is evident from the papers in this volume, the vast majority of work to date on IBMs has been at the population level. This is a natural result of the relative complexity of adding inter-specific behavioral interactions to the already diverse set of behaviors possible between individuals within a single population. Yet there are some advantages and disadvantages which apply across scales. For example the large amount of data required to parameterize and test IBMs is of concern at all scales, though particularly difficult to obtain above the population level. These data requirements are a significant bottleneck in the development and use of these models, yet they can also be a stimulant to the collection of appropriate ecological data. The process of model construction can lead to more efficient data collection, for the models may be very helpful in determining which data are more worth collecting. They may also be very effective in motivating people to collect additional types of data on individuals, something which has been very difficult to do in some areas. The models might also lead to useful correction methods, by suggesting methods to adjust data that are biased in some way.

A very important feature of IBMs is their ability to make connections between phenomena at different levels of biological organization, e.g., elucidating the consequences of individual growth to population and community levels. Such organization levels do not scale up in obvious ways in ecology. For example, much of competition theory in ecology is at the population level, yet most of the predictions of this theory are tested with surrogates (e.g., growth rate) at the individual level. IBMs offer a conceptually appealing approach to link the individual level to the population and community levels.

Individual-based models are one of the most effective ways to investigate the mechanisms of species interactions, and to quantitatively associate these mechanisms with the phenomena in question. Because these models make predictions about a wide range of individual and population attributes, and potentially of community and ecosystem attributes, IBMs are readily testable and should be easy to falsify. It is often possible to test the models explicitly at two different stages. First the assumptions on the mechanisms of interaction can be tested directly against field and lab data collected on individuals. Secondly, model predictions at higher levels of aggregation (population, community, ecosystem) can be compared with observations. Despite the availability of these alternative modes of testing, as in much of ecological modeling the IBMs constructed to date have been tested relatively infrequently at either stage.

The construction of an IBM can also be of considerable heuristic value because mechanisms are generally explicit and components more readily defined in terms of observable biological properties. An additional advantage of IBMs is that they can be used more realistically to analyze questions associated with small population sizes and demographic stochasticity. As a corollary of this point, it can often be the case that (relatively few) atypical individuals can have disproportionate effects on a system, and an IBM is an effective tool to discover and examine these effects. Aggregated models are just not capable of examining such effects.

IBMs also offer excellent potential to examine questions of natural selection and evolution. Their value resides in the fact that the models are constructed at the level (individual) on which natural selection operates. It would be straightforward, with available information, to build genetics into simulation IBMs models and use them to explore the natural selection consequences of multispecies interactions. This has already been well established for analytic IBMs of age-structured populations (Charlesworth 1980). One advantage of the simulation IBMs that especially recommends their use in the context of applied work is that it is easier to communicate the system, mechanisms, etc. between the biologist and the modeler than with many of the partial differential equation systems currently used. The more detailed dynamics of the IBM may help resolve questions that have been difficult to resolve with traditional approaches; for example, the point was made that in the stock-recruitment problem only one data point per year contributes to the question at hand. However, the events determining recruitment could involve a few individuals, episodic phenomena, and seasonal dynamics, all situations more easily dealt with using a simulation IBM.

Several difficulties in the application of IBMs to the analysis of interspecific interactions were discussed. One prominent concern involved problems of how to deal with components that operate on very different time or spatial scales, including multiple trophic levels. For example, consider the problem of predator-prey interactions where the predator and prey are of very different sizes, with consequently very different densities and time scales of turnover rate in the system. The spatial dimension is also critical; interacting local populations may differ in one or more spatial dimension, and an exploration of this problem is likely to greatly change our perspective on a number of critical ecological interactions. The question of how to deal with the scale problem was a recurrent theme, and perceived as a critical question for further research.

A second concern was how questions of aggregation and scale interact in the modeling of multispecies systems. A related question concerned the abstraction problem as addressed by Schaffer (1981) and Bender et al. (1984). Which components of a system can we safely ignore given our question, and under what conditions can we ignore these components? When we can abstract components with little or no loss of information? Some preliminary guidelines were developed by Schaffer and Bender et al., but much more work needs to be done. Much of this concern, of course, is tied up with the relative time scale problem noted above.

A methodological concern that arose in the context of multispecies interactions was how to deal with simulations of interactions where the two (or more) species involved are very different in their densities. For example, it would not be unreasonable for a large piscivorous species to be an order to several orders of magnitude less abundant than its prey. In this case is it necessary to keep track of all of the prey individuals? What detail is needed on the various components? Again, the question asked will be the primary guide here, but we need clarification and guidance on general issues, such as those for lumping components in the models. Multispecies IBMs obviously have the potential to become very complex, and will require significant lumping within at least some populations. We need studies of the effects of lumping, e.g., age classes, in i-space distribution models.

Further methodological problems discussed included the order of computation, selection of a time step, and the representation of higher order interactions in simulation IBMs. Since computations must be done sequentially, it isn't clear which behavioral interactions should be handled first in a sequence. Additional questions concern how we attack the constellation of problems surrounding the indirect effect or higher order interaction question which plagues community ecologists today. Little agreement was reached on how IBMs would provide any aid in clarifying or organizing our approach to these difficult questions without considerable further explorations of various approaches.

The group was aware of very few models of multispecies interactions in the literature that employed an IBM approach. The forest gap model (Huston, this volume) and the neighborhood interaction model (Ford, this volume) are exceptions. However, the group generally agreed that there was much promise for these models in elucidating the nature of species interactions, and attacking prominent lacunae in ecological theory that are difficult to approach with traditional tools (e.g., some of the spatial dependencies of interactions). It was felt that IBMs will not supplant existing approaches, as the complexity inherent in this approach can be daunting and not always needed to answer the questions of interest. In the case of many ecological phenomena, however, these models could be used to make important advances at this point, and in fact might be an important aid in the production of simpler, more tractable models.

What Have Been the Management Applications of Individual-Based Approaches to Date and What Are Some New Areas of Application on the Horizon?

The appropriate form of models for management depends to a great extent upon who is actually doing the managing. If the models are oriented towards use by individual managers (something which is relatively rare, except in certain agricultural situations), the model form will be quite different than if the model is being used to provide insight to a researcher who is acting in an advisory capacity to a manager. In either case, however, a clear advantage of simulation IBMs is that they are intuitively easier for managers to understand. An individual is a "real" entity, while population properties are a much more amorphous concept to explain to managers.

Regarding specific applications of IBMs, it was clear that many if not most of the essential models useful in management are of some age- or stage-structured form, and that this was true across all taxonomic lines. Therefore, the focus of the discussion was on i-space configuration models. For these, it appears that at present we are still focused mainly on understanding process, rather than developing predictive capacity. However, there were several examples of i-state configuration models which were either in or close to direct application for management decisions. These include:

(1) A size-of-stocking model is being evaluated by the Wisconsin Department of Natural Resources to compare the cost of stocking as

a function of hatchling length (Madenjian and Carpenter 1991, Madenjian et al. 1991).

(2)   A variety of models in human demography are used to investigate alternative birth rate quotas and their effect, as well as problems associated with disease transmittal.

(3)   In conservation biology, models are used to manage very rare species for which there are often very detailed data on the few remaining individuals, often in zoos. These details include specific genetic information, as well as past matings.

(4)   In a variety of crop situations, though not in general use, the models could be applied directly to problems of appropriate planting densities and investigating edge effects.

A key advantage to *i*-space configuration models is that they inherently provide output in the form of probability distributions, and thus predictions can be interpreted in a risk analysis format. This advantage may be useful to illustrate to managers the nature of our inability to predict deterministic outcomes in real situations. It also allows us to address issues that don't involve just means. For example, having the full probability distribution allows us to determine how policy changes might affect the probability that the outcome is outside of some range that it is desirable to stay within. In conservation biology, this would apply to cases such as the probability that a particular local population persists for at least some time period, rather than just dealing in mean persistence times.

Interactions outside the realm of the species of interest often need to be considered. Resource managers often find themselves managing people, not the resource. So it may be useful to couple a resource model with a model for individual humans to account for human behavioral changes through time. As mentioned above, there has been relatively little application of the IBM approach to multi-species management. One example of this which brings up another future direction concerns intercropping. This is a case where an IBM could be coupled to a control problem, and is an area in which there has been relatively little analytical work as well. Control of spatially-structured populations is quite difficult to investigate analytically, but is at least potentially tenable from a simulation IBM perspective.

Managers of large-scale systems tend to have a low tolerance for high levels of uncertainty and for promises of useful model results "tomorrow." IBMs do not score very highly with regard to these two intolerances. The only published examples of the application of IBMs to large-scale systems involve vegetation dynamics, used to evaluate the regional and global consequences to vegetation structure and function of changing climate and land use. These are based upon the responses of individual plants to environmental variables, and illustrate the approach of "soft linkages." Here, a sequence of models, one of which could be an IBM for individual plants or animals, operating at different spatial and temporal scales, are linked by expert judgement.

Regarding new applications of IBMs on large scales, the consensus was that the present situation was not likely to change greatly. This is because the current limitations are not due primarily to technological or methodological issues such as computer hardware or instrumentation for making measurements on individuals, even simultaneously in large numbers. Nor are the limitations due to our scientific understanding of mechanisms at the level of individuals. Rather, the limitations are primarily due to the realities of the structure and function of large-scale systems in which detailed consideration of the role of individual organisms in a single simulation model has not been shown to be either necessary or even helpful in addressing questions relating to large-scale systems.

Conclusions

Two major themes that emerged from the Group discussions were: (1) the importance of mathematical rigor in all types of IBMs, and (2) the necessity of borrowing concepts and techniques from other disciplines for use with IBMs. A schism is developing between the analytical (or *i*-space distribution) approach and the simulation (*i*-space configuration) approach. A consequence of this schism is a lack of mathematical rigor in simulation IBMs. Analytical and simulation approaches are highly related (DeAngelis and Rose, this volume) and much could be gained by recognizing the relationships among the approaches, and applying mathematical concepts and rigorous analysis developed for the analytical approach to simulation IBMs.

Additional rigor is also needed in describing IBMs and analysis results. We need to go beyond just collecting specific models and results, and be able to generalize across models. Generalization requires effective communication. This is especially difficult with IBMs because they are being developed in a wide variety of different fields. Mathematics offers a common language for describing model structures, a framework for describing model results so that emergent patterns can be detected, methods for analyzing model behavior, and techniques for summarizing and displaying complex model output. Software and hardware technologies are rapidly improving, and developments such as object-oriented languages (Palmer, this volume) and parallel processing (Haefner, this volume) will undoubtably have significant impacts on the development and interpretation of IBMs.

A key advantage of simulation IBMs is the ability to account for rare individuals, as well as rare events or circumstances which can come together in a few individuals. For some situations, rare individuals of a certain size, physiology or genetic composition have an effect on the population far beyond their relative biomass or numbers. This advantage of IBMs crosses taxonomic lines. There was consensus that IBMs will be generally more useful for sedentary organisms rather than free ranging ones, due to how the plasticity of form is often coupled to local interactions and environmental conditions in the sedentary situation.

Whether the IBM approach is needed clearly depends on the question posed by the investigator, and the tradeoffs involved between complexity of the models and the intensive data requirements. There was much discussion of the potential complexity of these models and the conundrum that this can lead us into, although other views were articulated that argued for use of these models because they can also be inherently quite simple. Participants felt that it was also likely that in the course of developing these models we will be able to identify critical aspects of various interactions, and subsequently simplify the models to allow more effective or tractable analytical exploration. It was proposed that generally small subsets of interactions drive major patterns, and that these then can be determined and the models appropriately simplified. Clearly we need more studies, such as those by Pacala and Silander (1990), that collect data so that predictions may be made with both aggregated and individual based models to see if, and under what conditions, we need be concerned with following individuals, given the nature of the system and questions attacked.

Literature Cited

Bender, E. A., T. J. Case, and M. E. Gilpin. 1984. Perturbation experiments in community ecology: theory and practice. *Ecology* **65**:1–13.

Charlesworth, B. 1980. *Evolution in Age-structured Populations*. Cambridge University Press, Cambridge.

DeAngelis, D. L., L. Godbout and B. J. Shuter 1991. An individual-based approach to predicting density-dependent compensation in smallmouth bass populations. *Ecol. Model.*, in press. *Ecol. Modell.* **57**:91–115.

Madenjian, C. P. and S. R. Carpenter. 1991. Individual-based model for growth of young-of-the-year walleye: a piece of the recruitment puzzle. *Ecological Applications* **1**:268–279.

Madenjain, C. P., B. M. Johnson and S. R. Carpenter. 1991. Stocking strategies for fingerling walleyes: an individual-based approach. *Ecological Applications* **1**:280–288.

Metz, J. A. J., and O. Diekman (eds.). 1986. *The Dynamics of Physiologically Structured Populations*. Lecture Notes in Biomathematics 68. Springer-Verlag, Berlin.

Pacala, S. W. and J. A. Silander. 1990. Field tests of neighborhood population dynamic models of two annual weed species. *Ecol. Monogr.* **60**:113–134.

Schaffer, W. M. 1981. Ecological abstraction: the consequences of reduced dimensionality in ecological models. *Ecol. Monogr.* **51**:383–401.

Shugart, H. H. 1984. *Theory of Forest Dynamics: The Ecological Implications of Forest Succession Models.* Springer-Verlag, New York, NY.

Sinko, J. W. and W. Streifer. 1967. A new model for age-structure for a population. *Ecology* **48**:910–918.

Index